A SHORT HISTORY OF CHEMISTRY

A SHORT HISTORY OF CHEMISTRY

J. R. Partington

Late Professor of Chemistry
in the University of London

THIRD EDITION, REVISED AND ENLARGED

DOVER PUBLICATIONS, INC.
New York

This Dover edition, first published in 1989, is an unabridged and unaltered republication of the 1957 revised and enlarged third edition of the work first published by Macmillan & Co. Ltd., London, in 1937.

Manufactured in the United States of America
Dover Publications, Inc., 31 East 2nd Street, Mineola, N.Y. 11501

Library of Congress Cataloging-in-Publication Data

Partington, J. R. (James Riddick), 1886–1965.
A short history of chemistry / J. R. Partington.
p. cm.
Reprint. Originally published: 3rd ed., rev. and enl. London : Macmillan ; New York : St. Martin's Press, 1957.
Bibliography: p.
Includes index.
ISBN 0-486-65977-1
1. Chemistry—History. I. Title
QD11.P3 1989
540′.9—dc19 89-1216
CIP

" Fortunately science, like that nature to which it belongs, is neither limited by time nor by space. It belongs to the world, and is of no country and of no age. The more we know, the more we feel our ignorance; the more we feel how much remains unknown; and in philosophy, the sentiment of the Macedonian hero can never apply,—there are always new worlds to conquer." (Sir Humphry Davy, 30th November, 1825.)

EXTRACTS FROM THE PREFACE TO THE FIRST EDITION

THE intention of the present work is to give a concise survey of the history of Chemistry. The period before Boyle has been treated briefly, although greater prominence than usual is given to van Helmont on account of his undeniable importance. The development of Chemistry in the later part of the nineteenth and in the twentieth century is only briefly sketched and illustrated by the work of a few outstanding chemists.

Limitations of space have excluded any but brief biographies of chemists; in Tilden's *Famous Chemists* only about twenty names are dealt with in nearly 300 pages, and a similar limitation is found in Thorpe's *Essays in Historical Chemistry*, with eighteen topics in 571 pages. In subject matter also, an attempt has been made, by a careful division of topics, to find a better balance than has sometimes been attained. Most histories of Chemistry lay too much emphasis upon Organic Chemistry, and there are several special histories of this subject; more consideration has now been given to the important development of Physical Chemistry, no special history of which has yet appeared.

The greater part of the material has been drawn from original sources. A selection of references and a short bibliography of about a hundred items are given in the hope that they may assist the reader seeking further information. These references and bibliography are not intended to be complete, and could not have been so without adding largely to the size of the book.

The reading of original memoirs can never be replaced by recourse to text-books, and it is hoped that the student, and especially the teacher, will supplement the information in the text by reference to some of the more important sources indicated. The information in books on the history of Chemistry is still largely derived, either directly or through later books, from Kopp's classical *Geschichte der Chemie*, published in 1843-7. Later historical research has considerably altered the views held in Kopp's time. An attempt has been made to utilize the results of modern scholarship and of the researches of recent authors on the history of science. It has been said * that "accuracy is as fundamental in the historical field as in the scientific one, and . . . has the same meaning in both fields"; it may be added that it is just as difficult to attain it in both fields.

The great chemists, almost without exception, have been singularly rich in personal qualities of a kind which can win our admiration and respect. A study of their correspondence, or other more intimate documents, often removes a less favourable impression derived from superficial acquaintance. This applies, for example, to Berzelius and Liebig, both of whom had real nobility of character.

The point of view in the present work is intended always to be objective, impartial, and free from conventional platitudes, but this is an ideal very difficult to attain. Some of the publications of the last century contain bitter polemics, and the reader may sometimes wonder whether a desire to achieve the truth was the sole guiding motive in many heated discussions of theories which have long since passed away. Laurent could say in 1853: "I was an impostor, the worthy associate of a brigand, etc., etc., and all this for an atom of chlorine put in the place of an atom of hydrogen, for the simple correction of a chemical formula," and in saying this he is simply stating a fact.

In the preparation of the book, I have received great kindness and assistance from some friends. My teacher, the late Professor H. B. Dixon, and my fellow-worker in the Manchester laboratories, the late Dr. A. N. Meldrum, had stimulated an earlier interest in the history of chemistry by lectures, and discussions, respectively. Dr. D. McKie has given much time to a study of the complete proofs of the book. Mr. L. F. Gilbert, B.Sc., has made a critical examination of some of the later part of the book.

* G. Sarton, *The Study of the History of Science,* Cambridge, (Mass.), 1936 p. 11.

Mrs. K. Stratton, M.Sc., has rechecked a number of references in the proofs with the originals, and has collaborated in finding and arranging the material in the Summary and Supplement to Chapter XIII. The Index has been prepared by my wife, Mrs. M. Partington, M.Sc. All this generous and unselfish help is most sincerely and gratefully acknowledged.

The original photographs of Laurent and Gerhardt were very kindly lent by Professor Tiffeneau, that of Kekulé by Frau Professor Anschütz. The portrait of Moissan is from a photograph in the Roscoe collection in the Chemical Society, and the blocks of the portraits of Emil Fischer and T. W. Richards were lent by the Chemical Society, which also gave permission to reproduce the portraits of Stas, Hofmann, Lothar Meyer and Victor Meyer from their Memorial Lectures, the necessary arrangements being most willingly and helpfully made by the Assistant Secretary, Mr. S. E. Carr.

The staffs of the libraries of the Chemical Society, Queen Mary College, and Manchester University, and of the London Library, have given valuable assistance, and the Pharmaceutical Society kindly lent a volume from its Library.

As on previous occasions, I am very glad to express my thanks to Sir Richard Gregory, Bt., F.R.S., and to Mr. A. J. V. Gale, for assistance in proof reading, and advice on many points.

J. R. PARTINGTON.

WEMBLEY, MIDDLESEX.

PREFACE TO THE THIRD EDITION.

In this edition many small changes have been made in various parts of the book and some sections have been rewritten so as to take account of recent studies by myself and others. This applies particularly to the sections on Mayow and Lavoisier. Some parts of the text have been extended to include more information, and a new section on the modern theory of valency has been added.

J. R. PARTINGTON.

CAMBRIDGE.

CONTENTS

Note. Symbols representing atomic weights differing from those now used are underlined and the value of the atomic weight is always stated (e.g. $\underline{C}=6$; $\underline{O}=8$). In all cases where the ordinary symbols are used, the modern atomic weights are to be understood. The barred symbols used by Wurtz and Kekulé are represented by ordinary symbols, to which they are equivalent, but barred symbols used by Berzelius represent " double atoms ", i.e. in most cases twice the modern atomic weight (~~H~~ = 2, ~~N~~ = 28, ~~C~~ = 24, ~~O~~ = 32, etc.), and are represented with bars. (In some older English books these symbols are underlined instead of barred.) In a few cases symbols and formulae have been rewritten in modern form, but not often, since one of the objects of a course in the history of chemistry is to impart facility in understanding older literature, which in the actual practice of chemistry is consulted far more frequently than the beginner might suspect

A SHORT HISTORY OF CHEMISTRY

CHAPTER I

THE ORIGINS OF APPLIED CHEMISTRY

Early Applied Chemistry

THE earliest applications of chemical processes were concerned with the extraction and working of metals and the manufacture of pottery. These arts were carried out without any theoretical background, but often with considerable skill, indicating long practice and a sound appreciation of the properties of materials. A survey of the industrial activities of the ancient nations * shows that the technical arts of the Classical Period in Greece and Rome, formerly regarded as the spontaneous expression of a higher civilization, are really rather decadent forms of crafts practised many centuries before in the Bronze Age cultures of Egypt and Mesopotamia. The irruption of an iron-using race or races into Mediterranean sites occupied long before by peoples of greater culture, which occurred about 1000 B.C. and introduced the Iron Age, was in many ways a break in the continuity of craftsmanship, but many of the oldest arts still survived in almost their original form. The potter, for example, still used nearly the same materials and appliances as Neolithic man.

The following very brief sketch attempts a survey of the development of chemical arts in the earliest periods.†

Early Knowledge of Metals

In the early period of his life, man was not acquainted with the use of metals: his implements were made of stone, horn or bone. The first metal known was probably gold, which occurs in the

* Partington, *Origins and Development of Applied Chemistry*, London, 1935; *ibid.*, in *Essays . . . in honour of Charles Singer*, ed. E. A. Underwood, Oxford, 1953, i, 35-46; Warren, *J. Chem. Education*, 1934, xi, pp. 146, 297.

† A good account of the Classical Period is given by Hoefer, *Histoire de la Chimie*, Paris, 1866, vol. 1.

native metallic form in some river sands and would attract attention by its colour and lustre. The earliest gold was probably obtained as small nuggets by washing alluvial deposits (Fig. 1). Gold ornaments are found with remains of polished and worked stone implements dating back to a very early period, the so-called Neolithic Age. The next metal known was probably copper, and some think it was known even before gold in Egypt. Native copper was worked by the American aborigines, but the Egyptian copper was probably obtained by reducing the ore malachite (the basic carbonate) from Sinai in charcoal fires. Copper appears in the earliest remains in Egypt and in Mesopotamia in the form of cast objects dating back to about 3500 B.C.

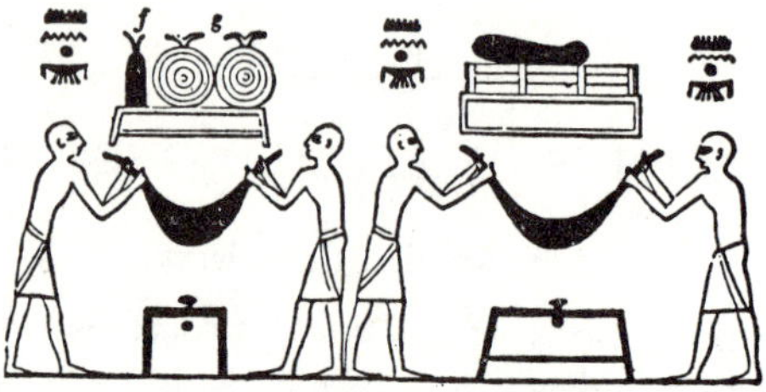

FIG. 1.—GOLD WASHING IN ANCIENT EGYPT.

The earliest known working in metals appears before 3400 B.C. (First Dynasty in Egypt) in Egypt and in Mesopotamia (modern Iraq), and rather later in the island of Crete, in the Mediterranean. Egypt and Mesopotamia are rivals in the claim for the origin of the working of metals, although both may have learnt it from another race or races.

FIG. 2.—SUMERIAN COPPER SCIMITARS FROM TELLO. ABOUT 3000 B.C.

The ancient inhabitants of Mesopotamia, the Sumerians, who probably migrated there from further east, had an advanced culture at least as early as the First Dynasty in Egypt, and were expert in the working of gold, silver and copper. Very fine specimens of early Sumerian metal work have been found on the site of Ur of the Chaldees, where there are the remains of a great temple. Specimens of good tin bronze were found in remains at Ur of 3000 B.C., this alloy later giving way to copper.* The copper scimitars shown in Fig 2, the finely engraved silver vase shown in Fig. 3, and the copper and gold objects in Fig. 4 are good examples of early Sumerian work. An early culture very like, if not the same as, the Sumerian existed in the Indus Valley at Mohenjo-daro and Harappā.

FIG. 3.—SILVER VASE OF ENTEMENA, RULER OF LAGASH, 2850 B.C., WITH ITS COPPER STAND. (The Louvre, Paris.)

The ancient Egyptians probably obtained their copper from ores in the peninsula of Sinai, which were very easily reduced to the metal, and these ores

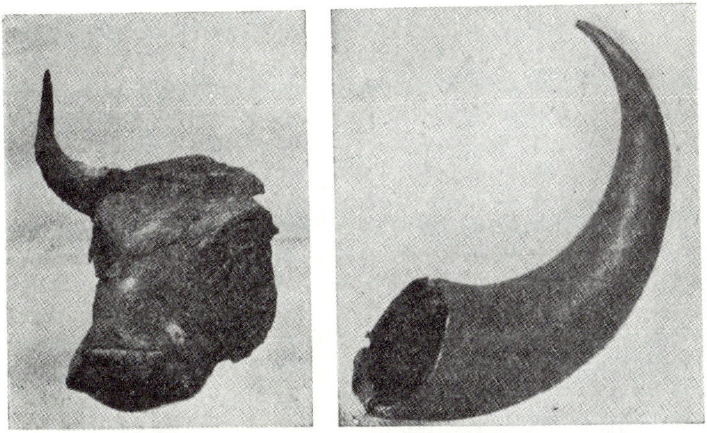

FIG. 4.—COPPER BULL'S HEAD AND GOLD BULL'S HORN. TELL AL 'UBAID, EARLY SUMERIAN, ABOUT 3000 B.C. (British Museum.)

are known to have been worked at a very early period. Copper was in use in the Predynastic Period, i.e. before 3400 B.C.

* Partington, *Scientia*, 1936, p. 197.

Fig. 5 shows an early Egyptian vessel of copper (about 3000 B.C.) and Fig. 6 some early metal objects of a slightly later period, in-

FIG. 5.—EARLY DYNASTIC EGYPTIAN COPPER VESSEL, ABYDOS.
(From Evans's *Palace of Minos at Knossos.*)

cluding a lump of iron. Lead is represented by the archaic statuette shown in Fig. 7. Small quantities of iron have also been found on early Sumerian sites.

FIG. 6.—COPPER MIRROR, COPPER TOOLS, AND (LOWER RIGHT-HAND CORNER) LUMP OF IRON FOUND AT ABYDOS (2700-2500 B.C.).
(British Museum.)

Copper also occurs in the remains at Knossos, and other sites in the island of Crete, which was the centre of an old civilization known as the Minoan: this copper dates back to about 3000 B.C. and may have come from the island of Cyprus. The Cretans probably learnt the use of metal from Egypt, with which they were in relation from the earliest times. The beautiful gold cup found at Vaphio, shown in Fig. 8, is regarded as of later Minoan origin, and the very advanced stage of pottery working in Minoan times is illustrated by the vase shown in Fig. 9.

FIG. 7.—LEAD STATUETTE, FOUND IN EGYPT, FIRST DYNASTY, 3400 B.C.
(British Museum.)

A later stage of the Minoan culture on the mainland was the so-called Mycenaean, represented by the finds of great stores of gold objects in tombs at Mycenae, and by remains at Tiryns, including blue copper glaze (*kyanos*). These belong to the period of 1500-1200 B.C., and the Mycenaean culture (which is that described by Homer) preceded the introduction of iron, which began with classical Greece. The use of iron came in with a new people, who destroyed the Minoan culture.

FIG. 8.—MINOAN GOLD CUP.
ABOUT 1500 B.C.

Troy was an outpost of a culture different from the Minoan, and probably extending through the north of the Balkans to the Danube

Basin and Hungary. Bronzes from the "second town" (2400-1900 B.C.) contain up to 11 per cent. of tin.

A great advance in metal working was the invention of bronze, an alloy of copper and tin. Bronze nearly always appears later than copper, and in several places at about the same time.

FIG. 9.—MINOAN POLYCHROME POTTERY WITH FLOWERS AND FOLIAGE FROM PALAIKASTRO, 2200 B.C.
(From Evans's *Palace of Minos at Knossos.*)

The earliest Egyptians were expert in working metals, as the remains of their craft show. There are representations of early metal working, such as the copper working shown in Fig. 10 and the goldsmiths shown in Fig. 11. (Fig. 10 was formerly thought to represent glass-blowing.)

The earliest Egyptian bronze is generally stated to be that found by Petrie in the Fourth to Sixth Dynasty remains at Medum, dating to about 3000 B.C., although a piece of true bronze from a tomb of the First Dynasty (*c.* 3400 B.C.) was described by Mosso. The source of the tin used in making these early bronzes is a problem, since tin is not found in many places. Some consider that this earliest tin came from Britain—the Islands of the Kassiterides (*kassiteros* was the Greek name for tin), perhaps the English coast of Cornwall, from whence tin was certainly shipped by the Phoenicians in later times, but a more probable theory * is that the tin came from mines in Drangiana in Persia (Iran) which are mentioned by Strabo (A.D. 7), although it is true that there is no tin there at the present day, nor does there seem to have been any for a very long time. The mines may have been worked out at an early period.

Egyptian and Mesopotamian bronzes sometimes contain lead in place of tin, and sometimes antimony (as do some early Chinese bronzes). A Sumerian vase of 2450 B.C. is of nearly pure antimony.

* Partington, *Scientia,* 1936, p. 197.

The metals iron, silver and lead were also known in Egypt, soon after copper and gold. in the Predynastic period, i.e. before King

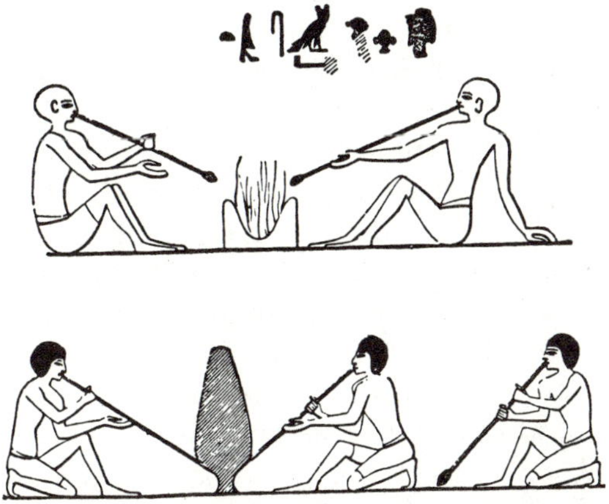

FIG. 10.—EGYPTIAN METAL WORKING.

Upper: BLOWING THE FIRE WITH CLAY-TIPPED REED BLOWPIPES; BENI HASAN, 1900 B.C. *Lower*: FASHIONING A COPPER VASE; THEBES, ABOUT 1550 B.C.

Menes (about 3400 B.C.). The early iron was very scarce and was probably (since it contains nickel) obtained from meteorites which originally fell upon the earth from outer space. This Predynastic

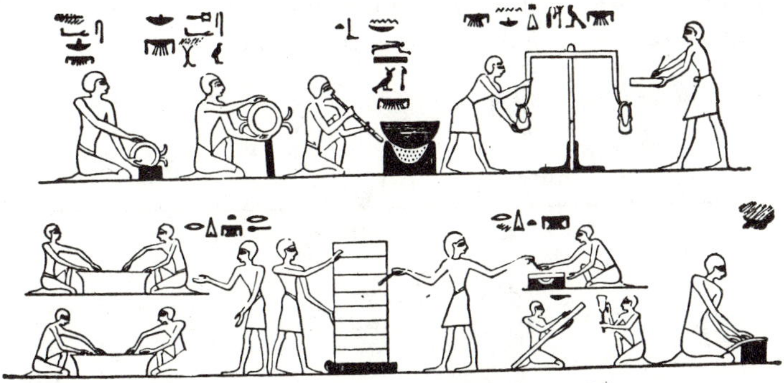

FIG. 11.—EGYPTIAN GOLDSMITHS WASHING, MELTING AND WEIGHING GOLD.

BENI HASAN, 1900 B.C.

iron was used as beads for jewellery, strung with beads of lapis lazuli, and a necklace of this kind was found in a Predynastic tomb by Petrie. Iron tools were found in the pyramid of Cheops (2900 B.C.), and do not contain nickel. Early Sumerian iron sometimes contains nickel and sometimes does not. Iron was used sparingly in Egypt about 2000 B.C. and generally after about 1500 B.C. This iron seems to have come from the Land of the Hittites, in Asia Minor, around the Black Sea. The Hittites were skilled in making iron, and an original letter of about 1250 B.C. has been found, written by a king of Egypt to the king of the Hittites asking for a supply of iron, and another from the king of the Hittites in reply, promising a *steel* dagger and asking for gold in exchange, " which in my brother's land is as common as dust "! The Assyrians from about 1400 B.C. made extensive use of iron for tools and weapons.

The metal *orichalcum*, mentioned by Plato (400 B.C.) as known but forgotten long before his day, may have been brass. Brass of about 1200 B.C. was found in Palestine ; it was well known in the Roman period, e.g. coins of A.D. 25. A passage in Strabo (XIII, i, 56), about the beginning of the Christian era, may describe the production of metallic zinc, and there was a well-developed brass industry in Cyprus. Actual specimens of zinc were found on the Island of Rhodes (500 B.C.), in Athens (4 to 2 centuries B.C.) and elsewhere in Europe.

Glass

Side by side with the working of metals, the Egyptians and the inhabitants of Mesopotamia perfected the arts of making glazed

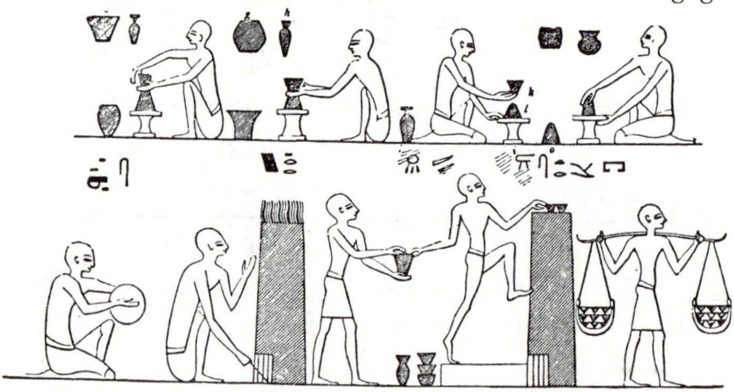

FIG. 12.—EGYPTIANS MAKING POTTERY, WITH FURNACE.
BENI HASAN, 1900 B.C.

pottery or other similar materials, and the production of glass. The Egyptian potter very soon learnt the use of the wheel for moulding the clay, and the vessels were baked in tall closed furnaces, not in open fires (Fig. 12). The very earliest pottery is unglazed and has a buff-coloured decoration (Fig. 13.)

Blue and green opaque glazes containing copper compounds were used at a very early period in Egypt (Fig. 14). A pale green true glass bead belonging to the Predynastic period (before 3400 B.C.) was found by Petrie at Naqada. A blue glass vase (Fig. 15) of Thothmes III (1550 B.C.) and some other early Egyptian

FIG. 13.—PREDYNASTIC EGYPTIAN POTTERY, WITH FIGURE OF BOAT (?), BEFORE 3400 B.C. (British Museum.)

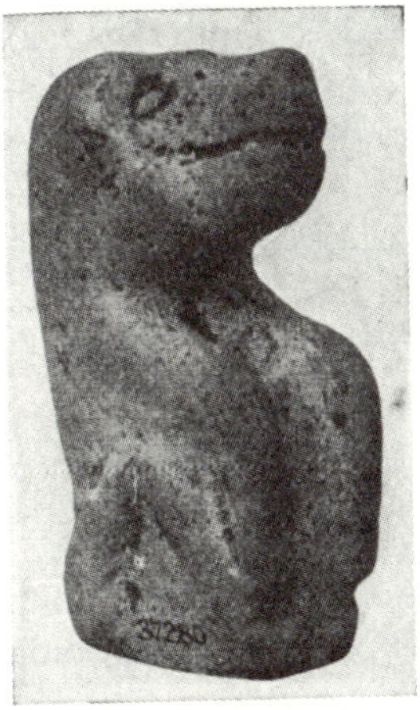

FIG. 14.—GLAZED FAIENCE FIGURE: EGYPT, FIRST DYNASTY, 3400 B.C. (British Museum.)

glass is in the British Museum. The manufacture of glass on the large scale apparently began in Egypt about 1370 B.C.; in the remains of a glass factory of this period at Tell el-Amarna, excavated by Petrie, nearly the whole process came to light. The alkali (*natron*, sodium carbonate found in Egyptian lakes near Alexandria, and incorrectly translated " nitre " in the English Bible) was melted with crushed quartz in crucibles. Nearly colourless glass was also made at an early date in Egypt.

Egyptian glass was exported to all parts of the Roman Empire, although in some places there were native glass factories, e.g. in Syria, and probably also in Britain, where remains of glass furnaces of the Roman period have been found near Warrington. Glass has been found on early Sumerian sites, and glass-making was also developed by the Assyrians; a glass bottle of King Sargon (700 B.C.) in the British Museum, formerly thought to be of Egyptian manufacture, is probably an Assyrian product, since there is independent evidence that the manufacture of coloured glass was highly developed in that country about this time. A blue glass called *uqnu*, containing copper, is described as an imitation of lapis lazuli in Assyrian tablets in the British Museum, dated about 650 B.C.*

FIG. 15.—TURQUOISE-BLUE OPAQUE GLASS VASE OF THOTHMES III, EGYPT, 1550 B.C. (British Museum.)

A definite compound, $CaO,CuO,4SiO_2$, of a deep blue colour ("Egyptian blue") was made in Egypt by heating silica with malachite and lime in the temperature range 830°-900°. It was applied with soda as a blue glaze on faience, and the blue glass is also coloured with copper. Some *early* specimens of Egyptian and Babylonian blue glass are coloured with cobalt.

Dyes

The blue dye indigo was obtained from the indigo plant by the Egyptians in the XVIII dynasty. The process of bringing the dye into solution by reduction must have been known to them: the Romans used indigo only as a pigment, called *vitrum* by Vitruvius.

The famous and valuable "purple of Tyre" was perhaps first made in Crete in very early times, and became celebrated in the Roman Empire, when it was a Phoenician product. It was obtained at great cost, by a process described by Pliny, from some marine molluscs. It is a dibromo-indigo and has been prepared artificially. Many other natural dyes were used in Egypt and Mesopotamia, and Pliny describes a process of dyeing the same piece of fabric in various colours in the same bath, which indicates

* R. Campbell Thompson, *A Dictionary of Assyrian Chemistry and Geology*, Oxford, 1936.

an early use of mordants, i.e. metallic compounds fixing the dye to the fabric, often in different colours according to the metallic base. The scarlet dye mentioned in the Bible was obtained from the kermes insect (hence the name " crimson ").

SUMMARY

The table gives a view of the occurrence of metals, and other materials requiring skill in applied chemistry for their preparation, in various periods in the oldest centres of civilization. Materials found only very rarely are enclosed in brackets.

B.C.	Egypt	Mesopotamia	Aegean, etc.
4000	PREDYNASTIC: gold, silver, lead, copper, (iron), glaze, (glass)	SUMER-AKKAD: *Lagash, Kish, Ur*: gold, silver, lead, (iron), copper, tin-bronze, glaze, (glass)	—
3400	I DYNASTY: (tin-bronze)		
3000		lead and antimony bronzes, (tin-bronze), metallic antimony	EARLY MINOAN: *Knossos*: copper, gold, silver, lead, tin-bronze, glaze
2500			purple dye
2000	XII DYNASTY: (tin), iron tools	tin-bronze common	*Mycenae*: glass
1700-1500	XVIII DYNASTY: tin-bronze common, tin, glass factories, cobalt glass, indigo, useful iron Ebers papyrus (medicine)	useful iron, steel weapons, tin oxide glaze, cobalt glass	THE HITTITES: iron working in the Black Sea region (" Chalybes ") *Tiryns*: lead-copper glass
1350-1200			*Palestine*: zinc brass THE PHOENICIANS: tin traffic ?
1000	Nubian iron industry, Assyrian iron common in Egypt	Assyrian iron common	IRON AGE: *Classical Greece*
650		Assyrian chemical tablets, imitation gems	

CHAPTER II

THE BEGINNINGS OF CHEMISTRY

The Four Elements

THE conceptions underlying the definitions of elements and compounds, although now almost obvious, were reached only after centuries of effort.*

The first clear expression of the idea of an element occurs in the teachings of the Greek philosophers.† Thales (640-546 B.C.) supposed that all things were formed of water; Anaximenes (560-500 B.C.) of air; Herakleitos (536-470 B.C.) of fire. Empedokles (490-430 B.C.) introduced the ideas of four "roots" of things: fire, air, water, and earth; and two forces, attraction and repulsion, which joined and separated them.

The name "elements" (*stoicheia*) was first used by Plato (427-347 B.C.), who assumed that things are produced from a formless primary matter, perhaps just space, taking on "forms". The minute particle of each element has a special shape: fire a tetrahedron, air an octahedron, water an icosahedron, and earth a cube, which are cut out of space by two kinds of right-angled triangles. The elements change into one another in definite ratios, by resolution into triangles and reassociation of these. Plato's dialogue *Timaeus* includes a discussion of the compositions of inorganic and organic bodies and is a rudimentary treatise on chemistry.

Aristotle (384-322 B.C.), who summarized the theories of earlier thinkers, developed the view that all substances were made of a primary matter, called *hulé*. On this, different forms could be impressed, much as a sculptor can make different statues from one block of marble, although Aristotle preferred to think of the form as evolving from within, as in organic growth. The form was called *eidos*. These forms can be removed and replaced by new ones, so

* Cf. Partington, *Chymia*, 1948, i, 109.
† Robin, *Greek Thought*, 1928.

that the idea of the transmutation of the elements arose. Aristotle took as the *fundamental properties* of matter hotness, coldness, moistness, and dryness. By combining these in pairs, he obtained what are called the *four elements*, fire, air, earth, and water, as shown in the diagram :

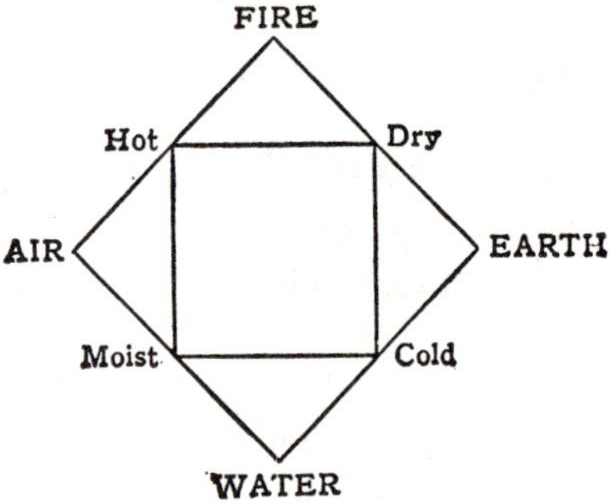

Thus, water is the type of moist and cold things, fire of hot and dry, and so on. To the four material elements a fifth, immaterial, one was added, which appears in later writings as the *quintessence*. This corresponds with the ether. The elements were supposed to settle out naturally into earth (below), water (the oceans), air (the atmosphere), fire and ether (the sky and heavenly bodies).

Aristotle defined an element or " simple body " as " one of those bodies into which other bodies can be decomposed and which itself is not capable of being divided into others ". He distinguished between mechanical mixture and solution, and chemical change with complete change of properties. He is the first Greek to mention mercury, as " silver water " (*chutos arguros*). He says that the vapour from sea-water evaporated in a vessel condenses as fresh water on the cool lid, and (wrongly) that wine would give water in a similar way, so that he knew a primitive method of distillation. He classified several chemical processes, giving them special names, thought (following Plato) that metals are mostly water, and listed the " homogeneous " parts of animals and plants, thus constituting a primitive organic chemistry. Plants and minerals were more fully dealt with by his pupil Theophrastos (372-288 B.C.).

The theory of the four material elements persisted until towards the end of the eighteenth century, and ether persisted as a medium for the transmission of light until the end of the nineteenth century. Many such fundamental ideas came from ancient Greece.

Galen of Pergamos (A.D. 129-199) applied Aristotle's theory to medicine. In a healthy body the four elements, which appear as four *humours* (blood, phlegm, bile and black bile) are in equilibrium (an idea going back to Anaxagoras, 450 B.C.). Disease is caused

FIG. 16.—ARISTOTLE, 384-322 B.C.

by the predominance of an element and is cured by the opposite element as shown in the diagram on p. 14.

Chemical Knowledge of the Classical Period

Several chemical substances, such as the oxides of copper, iron and zinc ; alum ; the sulphates of copper and iron ; sulphides of arsenic and mercury ; and vegetable and animal products, including dyes, were known in the Classical Period and, with some simple chemical operations such as the working of metal and alloys, gilding by means of solutions (amalgams) of gold in mercury, and testing gold and silver for purity, are described in the writings of Dioskurides (A.D. 60), Pliny the Elder (A.D. 23-79, the author of the famous *History of Nature* in 37 books),* and Galen. Even earlier than this we find Theophrastos (315 B.C.) describing some simple chemical operations, e.g. the manufacture of white lead (Greek, *psimuthion*): "lead is placed in an earthen vessel over sharp vinegar, and after it has acquired some thickness of a kind of rust, which it commonly does in about ten days, they open the vessels and scrape it off. They then place the lead over the vinegar again, repeating over and over again the same process of scraping it till it is wholly gone. What has been scraped off they then beat to powder and boil [with water] for a long time, and what at last settles to the bottom of the vessel is white lead" (*Treatise on Stones*). This is a fairly accurate account of the process.

Pliny thus describes the preparation of mercury: "They put *minium* [cinnabar from Spain] in an earthen vessel well luted over with clay, upon which there is set a pan of iron, and the same covered over the head with another pot, well cemented. Under the earthen pot a good fire is made and kept continually blown. And thus by circulation there will appear a dew or sweat in the uppermost vessel, proceeding from the vapours set free. When this is wiped off it will be as liquid as water but in colour will resemble silver." This description of a rudimentary sublimation or distillation process is imperfect, and was no doubt copied by Pliny from an author he did not clearly understand.

We might have expected that such processes would have aroused the interest of learned men, and although they did in fact provide exercises in making up *theories*, only a few seem to have made *experiments*, and the study of natural science was not held in high estimation in the Roman Empire, with the exception of Egypt. There the more refined arts of working precious metals seem to have been cultivated by the priests or, perhaps, by special temple workmen, for a very long time, and the practical traditions were handed down from one generation to another.

* K. C. Bailey, *The Elder Pliny's Chapters on Chemical Subjects*, 2 vols., 1929-32.

The Chemical Papyri

A collection of Egyptian recipes of this type appears to form the contents of the famous *Papyrus of Leyden*, discovered in 1828 in a tomb at Thebes in Egypt, and kept in the Leyden Museum. Another part of the same papyrus was sent to Stockholm, and is hence called the *Stockholm Papyrus*. The papyrus is written in Greek at a date probably round about A.D. 300 and contains recipes probably copied from much older Egyptian sources. It is possible that the work represents the note-book of a fraudulent goldsmith. These recipes in the *Leyden Papyrus* deal mainly with the production of imitations of an alloy of gold and silver which is called by its old Egyptian name *asem* (the Greek *elektron* and Roman *electrum*), which was regarded as a separate metal; and with the preparation of alloys and plated objects which would "serve for" the noble metals, and were even "better than the real" (an almost modern touch). The Stockholm part of the papyrus (Fig. 17) deals mostly with gems and valuable dyes and their imitation. There was no doubt in the mind of the writer that the imitations of noble metals, gems and expensive dyes were in no sense real. A quotation from the *Leyden papyrus* * will illustrate the point of view:

"One powders up gold and lead into a powder as fine as flour, 2 parts of lead for 1 of gold, and having mixed them, works them up with gum. One covers a copper ring with the mixture; then heats. One repeats several times until the object has taken *the colour*. It is difficult to detect the fraud, since the touchstone gives the mark of true gold. The heat consumes the lead but not the gold."

The recipe in the *Stockholm Papyrus* shown in Fig. 17 may be translated:

"Mix together and put into a pot 2 grams of malachite, 2 grams of azurite, 130 c.c. of the urine of a young boy, and 180 c.c. of solution of ox-gall. Put into the pot all the twenty-four pieces of stone,

* Complete translation in Berthelot, *Introduction à l'étude de la chimie des anciens et du moyen âge*, Paris, 1889; see also his *Origines de l'alchimie*, Paris, 1885; English translations of Leyden and Stockholm Papyri, Cayley, *J. Chem. Education*, 1926, iii, p. 1149; 1927, iv, p. 979.

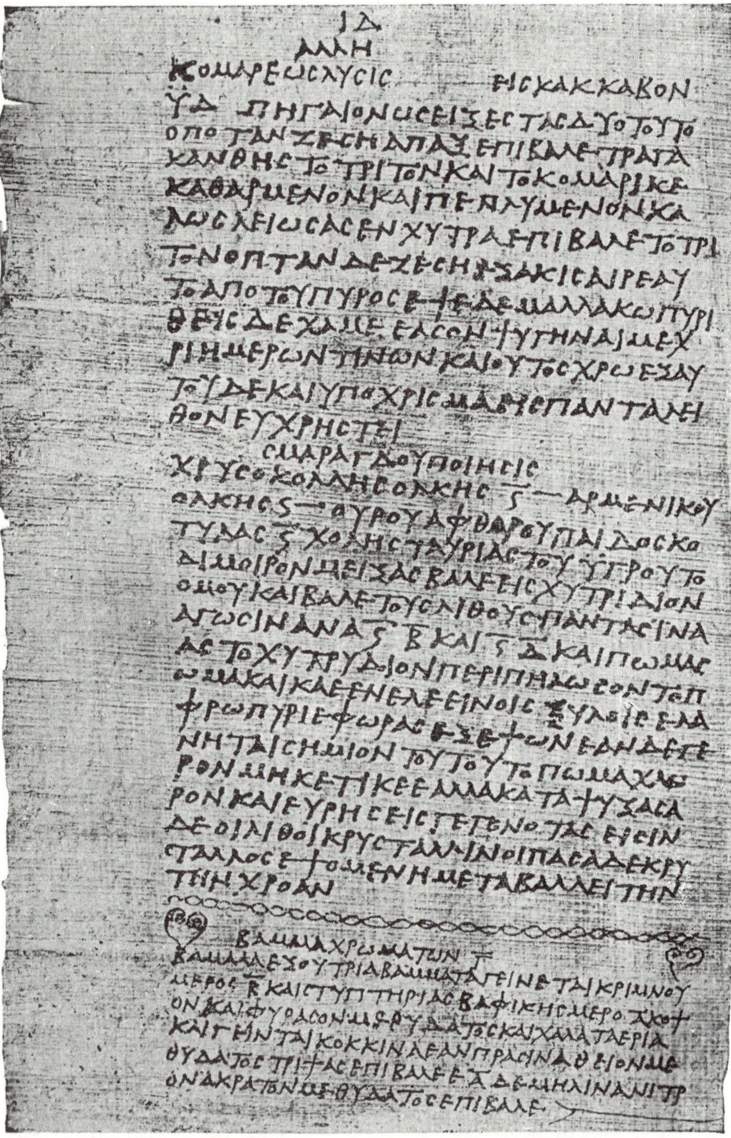

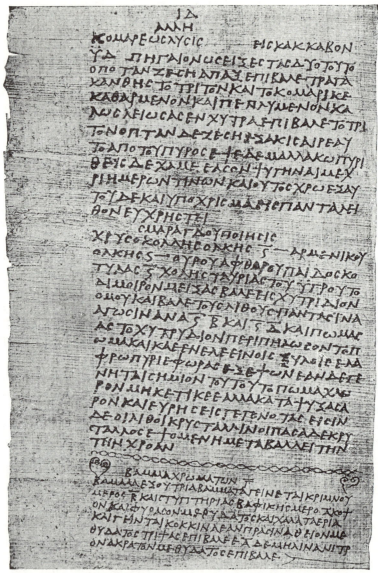

FIG. 17.—A PAGE OF THE STOCKHOLM PAPYRUS, A.D. 300.

THE RECIPE IN THE MIDDLE OF THE PAGE DEALS WITH THE "MAKING OF EMERALD" *i.e.* THE IMITATION.

each weighing 0·27 grams. Put the lid on the pot and lute it around with clay. Heat for six hours over a gentle fire of olive wood. When you see the lid has become green, do not heat any more but allow to cool and take out the stones, when you will find that they have become emeralds."

The recipe refers to the coloration of a stone, perhaps rock crystal, or zeolite or some other porous stone, with a solution of copper compounds in an ammoniacal liquid. Boyle refers to some such process as used in Italy in the seventeenth century.

Alexandria

It is generally agreed that chemistry had its origin about the beginning of the Christian era, in the city of Alexandria, in Egypt, which was founded by Alexander the Great at the mouth of the Nile in 331 B.C.

Alexandria had a mixed population of native Egyptians, Greeks, Syrians, and Jews, but was essentially Greek in culture. It contained a temple of the god Serapis, two Libraries and the Museum (or University) and, in later times, the Christian church of St. Mark and the famous Pharos or lighthouse, 500 ft. high. One library, said to contain 700,000 books, was accidentally destroyed by fire in 47 B.C., and the second followed it in A.D. 389.

The Museum was mainly interested in classical literature, philosophy, mathematics, and medicine. It included some famous names, and in the Roman period a voyage to Alexandria to study in its schools was regarded as essential to a medical man.

In Alexandria two streams of knowledge met and fused together in characteristic fashion. The ancient Egyptian industrial arts of metallurgy, dyeing and glass-making existed on one hand; and on the other were the philosophical speculations of ancient Greece, which had now undergone many changes, and had assimilated elements of Eastern mysticism. The teachings of Plato had become Neoplatonism, taught by the Egyptian philosopher Plotinos (204-270); this assumed that everything emanates from the One, passing, like a ray of light, through the stages of mind and soul, and finally becoming extinguished in the dark region of matter. The Gnostic sects also grew up in the early centuries of the Christian era, and, after Plotinos, Greek philosophy degenerated into mysticism, occultism, and magic. The technical arts, and the imitation and falsification of precious metals, gems and dyes, seem gradually to have assumed, under the influence of the theoretical views on the nature of matter, a new form, and the result was the "divine" or "sacred" art (θεῖα τέχνη; θεῖον also means sulphur) of making gold (*Chrusopoiïa*), or silver (*Arguropoiïa*). This contained the germ of chemistry, and during the first four centuries a considerable body of positive, practical chemical knowledge came into existence.

The Beginning of Chemistry

Up to the commencement of the Christian era we find no indication of the existence of chemistry in Europe or the Nearer East. Chemical operations were known to technologists, as we have seen, but the information was empirical. The name "chemistry" first occurs in an edict of the Emperor Diocletian in A.D. 296, given by Suidas (tenth century) from an older source, in which the books of the Egyptians (in Alexandria) on *chēmeia*, on making (i.e. imitating) gold and silver, are ordered to be burnt. The word appears in the Greek authors who report this as χημεία, but it is not a Greek word, and appears to have been derived from the native designation of Egypt, a country which Plutarch, in his treatise *On Isis and Osiris*, written about A.D. 100, says was called *chēmia* on account of the black colour of its soil. This statement is confirmed by the Egyptian inscriptions, where the hieroglyphic form of the word is used. The name probably meant "the Egyptian art", and never had the meaning of a "black art" as applied to magic. The name χημεία occurs also in a Greek manuscript now at St. Mark's in Venice, copied about A.D. 950, from a work by Zosimos of Panopolis (A.D. 300).

The art of chemistry was commonly ascribed to the mysterious Hermes Trismegistos, whose name is still preserved in the expression "hermetically sealed" (= "Hermes his seal" in the old books), who was invoked by the Gnostics as the source of all wisdom—as, in fact, his ancient Egyptian prototype, Thoth, was the god of learning and of the scribe.

There were Greek treatises on the divine art in existence as early as the first century A.D. The oldest of these is by Demokritos, perhaps Bolos the Demokritan, mentioned by Pliny, but certainly not the older famous philosopher Demokritos of Abdera, one of the founders of the atomic theory (see p. 165). The work of Demokritos contains recipes for alloys and imitations of dyes like those in the Leyden and Stockholm papyri. There is a commentary on it by Synesios. Practical operations and apparatus were invented about the same time by Maria the Jewess, who described the apparatus for distillation (the *ambix*) and for sublimation (the *kerotakis*): these are shown in Fig. 19. The most copious author was Zosimos (about A.D. 300), whose 28 books are addressed to his "sister" Theosebeia. Olympiodoros wrote a commentary on the work of Zosimos. The last author was Stephanos, a philosopher at the Byzantine court in the seventh century.

The Greek chemical treatises exist in the St. Mark's manuscript, and several later ones in Paris and elsewhere. A Latin translation of parts of them was published in 1572 (Fig. 18). Extracts from the Greek texts were published by Hoefer,* who clearly recognized their importance, and the texts almost in full, with a French translation, by Berthelot.†

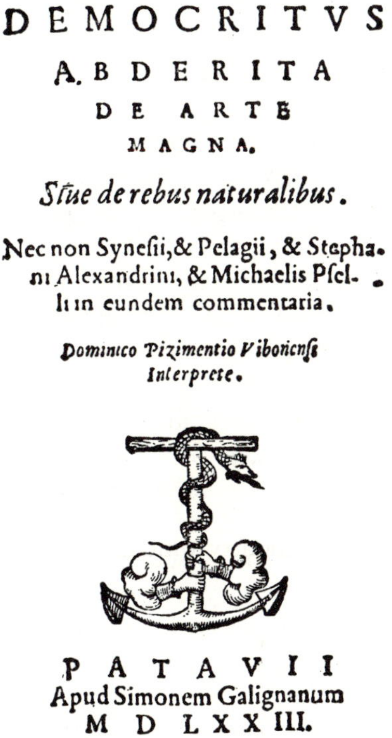
DEMOCRITVS
ABDERITA
DE ARTE
MAGNA.
Siue de rebus naturalibus.
Nec non Synesii, & Pelagii, & Stephani Alexandrini, & Michaelis Pselli in eundem commentaria.
Dominico Pizimentio Vibonensi Interprete.
PATAVII
Apud Simonem Galignanum
M D LXX III.

FIG. 18.—TITLE PAGE OF THE LATIN TRANSLATION OF THE PSEUDO-DEMOKRITOS.
(THE TRUE DATE OF PUBLICATION WAS 1572.)

The work of Zosimos appears to have been a kind of chemical encyclopaedia, summarizing many earlier treatises. It contains interesting descriptions and illustrations of chemical apparatus and

* *Histoire de la Chimie*, Paris, 1842.

† *Collection des anciens alchimistes grecs*, 3 vols., Paris, 1888-89; see also F. Sherwood Taylor, *Journ. Hellenic Studies*, 1930, vol. 50, p. 109; *Chemistry and Industry*, 1937, vol. 56, p. 38.

experiments, but also some mystical matter, including some remarkable " visions ", of psychological interest.

Alexandrian Chemistry

The treatises written in Greek in Egypt, at Alexandria, are the earliest known books on chemistry. They contain many technical expressions not found in our Greek dictionaries: one MS. in fact contains a small dictionary of such words. Sometimes strange names and expressions are used to conceal the meaning of the text, as when a solution of calcium polysulphide, made by boiling sulphur with milk of lime, is called the " divine water " (θεῖον ὕδωρ), or " bile of the serpent ": this reagent has a powerful action on metals and a strong smell of hydrogen sulphide—Zosimos says it is good to hold the nose when working with it! Such " cover names " were much used by the later alchemists. Some of the Greek names are technical terms, and a modern reader of an advanced book on chemistry who had access only to standard English dictionaries would be in a similar position to ours when we attempt to find the meaning of these curious Greek words by means of a literary vocabulary.

The treatises also contain much of what is sometimes described as " obscure mysticism ". This was, no doubt, the result of an effort on the part of the authors to make use of the newest and best philosophical ideas, viz. Neoplatonism, which seems to have been the point of departure for a new interest in science in Alexandria.

The study of astrology was connected with that of chemistry in the form of an association of the metals with the planets on a supposed basis of " sympathy ". This goes back to early Chaldaean sources but was developed by the Neoplatonists.

An old (tenth century) manuscript at St. Mark's, Venice, gives the following early list:

Metal		*Planet*		*Symbol*	
χρυσος	gold	Ηλιος	Sun	☉	gold
αργυρος	silver	Σεληνη	Moon	☾	silver
μολιβος	lead	Κρονος	Saturn	♄	lead
ηλεκτρος	electrum	Ζευς	Jupiter	♃	bronze
σιδηρος	iron	Αρης	Mars	♂	mixed metal
χαλκος	copper	Αφροδιτη	Venus	♀	tin
κασσιτηρος	tin	Ερμης	Mercury	☿	iron

The symbols are those of the planets. There is no mention of the metal mercury (known to Aristotle, 384-322 B.C.), the metal tin being assigned to the planet Hermes (Mercury). In later lists the symbol ☽ or ☿ was assigned to the metal mercury, and tin was given to Jupiter in exchange for electrum, which was recognized as a mixture of the metals gold and silver. It is noteworthy that the ascriptions vary in old lists: the last column gives a list reported by Celsus (quoted by Origen, *c.* A.D. 200) as the materials of the seven steps of the "ladder of Mithra". The list in the St. Mark's manuscript is the same as one reported from Proklos (A.D. fifth century) by Olympiodoros.

The Greek chemical treatises contain some interesting practical chemical information, which appears in them for the first time, and also many diagrams of chemical apparatus (Fig. 19). The operations are fusion, calcination, solution, filtration, crystallization, sublimation and especially distillation (not previously described); and methods of heating include the open fire, lamps, and the sand and water baths. Nearly all this practical knowledge has been ascribed by older writers on the history of chemistry to the Arabs, who really derived it from the very source we are now considering. The Arabic name *alchemy* is merely the Alexandrian-Greek *chēmeia* with the Arabic definite article *al* prefixed.

An important feature of the Alexandrian treatises is that the fraudulent processes described in the Papyrus of Leyden have now become real transmutations of base metals into gold. The process was to be effected by "changing the colour" of lead or mercury by means of various chemicals, and copper turned white by arsenic was regarded as a kind of silver. The transmuting agent was later called by the Arabs *aliksir* (elixir)—perhaps from the Greek ξήριον (*xerion*), a dusting powder or cosmetic—and by the European alchemists the *philosophers' stone*, or the *tincture* (i.e. an agent to change the colour). The idea was based on the marked effects of arsenic, mercury and sulphur in changing the colours of metals, and was quite a rational one at that time.

Zozimos distinguishes between what he calls *bodies*, by which he usually means metals, and *spirits* (πνεῦματα, *pneumata*), "certain substances that are, by reason of their peculiar nature, invisible", by which he understands the vapours of arsenic, sulphur and mercury, which exert a powerful action on metals. These spirits may be bound to bodies by affinity (συγγένεια, *sungeneia*) and also set free again by suitable processes.

FIG. 19.—ILLUSTRATIONS OF CHEMICAL APPARATUS.

COPIED FROM GREEK MSS. OF ZOSIMOS AND OTHERS IN THE BIBLIOTHÈQUE NATIONALE, PARIS. THE GREEK NAMES OF THE APPARATUS WILL NOT BE FOUND IN THE DICTIONARIES. *A*, *B*, *C* AND *F* REPRESENT APPARATUS FOR DISTILLATION, THE AMBIX, *later* CALLED ALEMBIC : IN THE MSS. THE LOWER PART IS CALLED *lopas*, THE UPPER *phiale*. THESE ARE SOMETIMES HEATED BY LAMPS (*phota*), SOMETIMES ON A SAND BATH (as in *F*). *D* IS A KEROTAKIS OR SUBLIMATION APPARATUS, *E* AN APPARATUS FOR HEATING A PHIAL IN A SAND BATH, *C* IS A COPPER STILL.

The spirits are also called by Zosimos, and by Arabic authors who copy him, " vapours " (νεφέλη, *nephelē*) or " smokes " (αἰθάλη, *aithalē*) : the word *tutia*, an old name for zinc oxide, is derived from the Persian *dudhā*, " smokes ", as it is given off as a white smoke on roasting zinc minerals with charcoal. Zosimos explains the burning of limestone to form quicklime as due to the taking up of a spirit from the fire. The Alexandrian chemists were very near to a recognition of gases.

The operation of transmutation is often compared with a colouring process, the four " primary " colours being black (associated with the primary matter, or sometimes lead), white (silver), yellow (electrum) and red (gold). The old philosopher Herakleitos (536-470 B.C.) says : " Nature strives towards opposites and brings harmony from them and not from likes . . . as the painter mixes the white, black, yellow and red colours and achieves likeness to the original."

One of the fundamental doctrines of the Greek-Egyptian alchemists was that " all is one " (ἓν τὸ πᾶν : *hen to pan*), represented by the symbol of a serpent coiled into a circle (representing the cosmos) with its tail in its mouth. Zosimos attributes this to an old philosopher, Chumes, who said : " the one is the all, through which all has come to be." This idea expressed the ultimate unity of matter, which was the great guiding principle of alchemy through the centuries and contains an important element of truth.

The origins of chemistry lay outside the development of classical learning and philosophy, and as the subject is not mentioned by Greek and Roman authors of the early centuries of the Christian era, it seems to have been unknown outside Egypt, the land of its birth, where its study was confined to small circles. One of its components is Greek philosophy, but its other and larger part is of diverse and rather uncertain origin. In its later developments, in the Byzantine period in Constantinople, it had lost its experimental character.

The later widespread popular interest in alchemy was a product of the Saracen culture, which rescued the older Egyptian and Greek learning from annihilation. The subject again became experimental, and since the first treatises to become well known in Europe were Latin translations of Arabic works, it was supposed that alchemy originated among the Arabs. The Greek treatises, in this period, were forgotten.

CHAPTER III

THE DIFFUSION OF ALCHEMY

Chemistry in Arabia

THE Arab conquests, including that of Egypt in A.D. 640, brought this vigorous and inquisitive people into contact with the remains of the Greek, or Hellenistic, civilisation; and learned Greeks, Syrians, and Persians living under their rule were active in the translation of Greek books into Arabic.* These books included works on philosophy (particularly those of Aristotle) and medicine, but also those on alchemy, which now appeared in Arabic translations. By the end of the seventh century several such works were in existence, but in the following century, through the new city of Baghdad, formed by the Abbasid Caliphate in 762, " a great stream of Greek and other ancient learning began to pour into the Mohammedan world and clothe itself in an Arabic dress."†

These translations often went through the intermediate stage of Syriac (i.e. theological Aramaic) made by Nestorian Christians—a sect which, cut off from the body of the Church of Alexandria for what was condemned as heresy, was especially instrumental in spreading the knowledge of the sciences to the Nearer and Far East. Some Syriac translations of alchemical works, especially of Zosimos, are extant.‡ Some illustrations from Syriac treatises and one from a late Arabic MS. are shown in Fig. 20. They may be compared with those of Fig. 19.

By this time chemistry had been known in Egypt for over five hundred years, but since much of this information reached Europe again in the form of Latin translations from the Arabic works,

* O'Leary, *How Greek Science passed to the Arabs*, 1948.

† E. G. Browne, *Arabian Medicine*, Cambridge, 1922.

‡ Berthelot, *La Chimie au moyen Âge*, 1893, vol. ii.

mostly produced in Spain from about A.D. 1100, it was long erroneously thought that the Arabs were the originators of chemistry.

Chemistry among the Arabs * was cultivated principally by: Jābir ibn Hayyān (about A.D. 720-813), who lived in Baghdad under the Caliph Harun al-Rashid, who appears in the *Arabian Nights*,

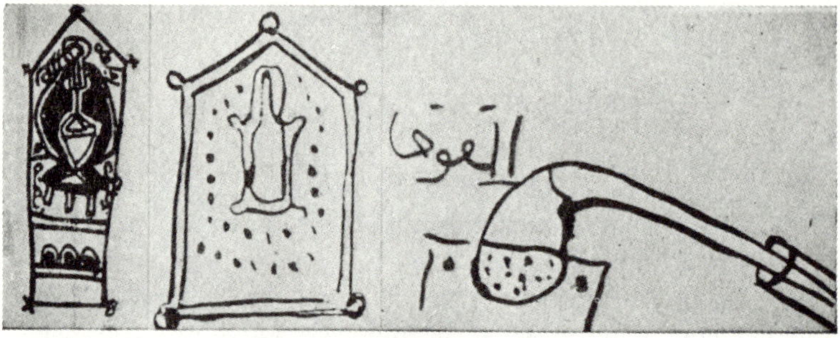

1. 2. 3.

4.

FIG. 20.—ILLUSTRATIONS OF CHEMICAL APPARATUS.
(1) APPARATUS FOR DIGESTION. (Syriac MS.)
(2) APPARATUS FOR DIGESTION IN SMALL CHAMBER. (Syriac MS.)
(3) RETORT AND RECEIVER. (Syriac MS.) A later addition.
(4) AN ILLUSTRATION FROM A PAGE OF A FAIRLY MODERN ARABIC CHEMICAL MS. IN THE BRITISH MUSEUM.

and al-Rāzī (866-925), commonly called Rhases, a Persian; a critic of alchemy was ibn Sina, generally called Avicenna (980-1036), born near Bokhara. Jābir was once thought to be the author of a treatise known in a Latin work attributed to Geber, but no Arabic original of this is known, and it is now supposed to have been composed at a much later date, perhaps about A.D. 1100. It is a

* Holmyard, *The Great Chemists*, 1928; Partington, *Ambix*, 1938, i, 192.

systematic treatise, giving clear accounts of the properties of metals and of chemical operations. The Arabic works of Jābir * contain some descriptions of experiments, but are largely mystical: the latest view † is that they are later than Jābir, and are religious propaganda of the sect of the Isma'īliyya.

Al-Rāzī (Rhases) was apparently a skilled practical chemist but is chiefly noteworthy as a physician. His book on alchemy, *Secret of Secrets*, was translated by Ruska (Berlin, 1937). It is curious that it contains such detailed descriptions of substances not mentioned in Rhases' genuine medical work, the *Continens*. He divided mineral bodies into six classes, an extension of an earlier classification by Zosimos:

1. *Bodies*, the metals.
2. *Spirits*, sulphur, arsenic, mercury and sal ammoniac.
3. *Stones*, marcasite, magnesia, etc.
4. *Vitriols* (known to Pliny).
5. *Boraces*, borax, natron (soda), plant ash.
6. *Salts*, common salt, kali (potash), " salt of eggs " (probably saltpetre, used in China for fireworks), etc.

Avicenna, who wrote a *Canon* of medicine long used in Latin translation, was the reputed author of a Latin work on alchemy called *De Anima*, but this is probably a later work compiled in Spain from Arabic sources. It was used by Roger Bacon and Albertus Magnus. In a genuine work, Avicenna expresses doubt about the possibility of transmutation, and says it is not in the power of alchemists to change the species of metals: they can only make imitations of gold and silver.

Jābir and al-Rāzī taught a new theory in chemistry, the germ of which is contained in Aristotle's *Meteorology*, viz. that metals are composed of mercury and sulphur, and are generated in the earth from these. This theory appears in an encyclopaedia written about A.D. 950 by the members of a secret society, the Brethren of Purity at Basra, a work largely compiled from earlier Syriac translations based on Greek sources. It is also possible that information from

* Berthelot, *La Chimie au moyen Âge*, 1893, vol. iii.

† P. Kraus, *Mémoires présentés à l'Institut d'Égypte*, Cairo, 1942-3, vols. xliv-xlv; Meyerhof, *Isis*, 1944, xxxv, 213; Holmyard, *Proc. Roy. Soc. Med.*, (*Hist. Med.*), 1923, xvi, 46; *Endeavour*, 1955, xiv, 17.

Mesopotamia and China was incorporated into Arabic alchemy, and may have passed directly to the Arabs by way of Harran.*

Hindu Chemistry

The Sanskrit *Vedas*, about 1000 B.C., mention five elements (not regarded as constituents of things): earth, water, air, ether (or space), and light. The *Rig-Veda* mentions gold, silver, copper, bronze, lead, and possibly iron and tin. The *Upanishads* (900-500 B.C.) definitely mention iron. Alexander entered India in 327 B.C. A coin of Euthydemos (235 B.C.), king of Bactria, consists of an alloy of copper with 20 per cent. of nickel. Indian wares reached Alexandria in the Roman period.

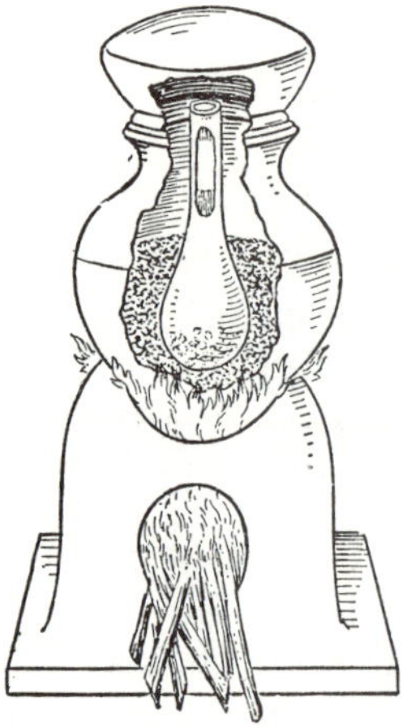

FIG. 21.—APPARATUS FROM INDIAN CHEMICAL MANUSCRIPT.

Old Indian medical works are the Bower Manuscript (4th century A.D.) and treatises attributed to Caraka (*c.* A.D. 100) and Suśruta (*c.* A.D. 200), the last two in the present form of the 6th century. They contain descriptions of metallic compounds; mercury is mentioned once in the Bower Manuscript and Caraka and twice in Suśruta. It is mentioned in the *Arthaśāstra* (4th century B.C. but interpolated), which has details of mining, metallurgy, medicine, pyrotechnics, poisons, fermented liquors and sugar. Brass is mentioned in Suśruta and zinc in Caraka and Suśruta. The preparation of zinc by distillation is in works of A.D. 1100-1150 and in detail in the *Rasaratnasamuccaya* (A.D. 1200), and may have originated in India. India produced good iron, the famous Delhi Pillar, weighing 6½ tons, of forged (not cast) iron, was made in A.D. 415. Steel, including *wootz*, occurs in graves of the 7th to 6th centuries B.C. The saltpetre industry was old in India. The mentions of alcohol in Caraka and Suśruta, and by Amara (6th-8th centuries), may be interpolations. Caraka and Suśruta describe solid caustic alkalis, potash being distinguished from soda, the "rusts" (oxides) of six metals, and impure chlorides and sulphates of iron and copper.

The Sāmkhya system of philosophy, ascribed to Kapila (*c.* 550

* Partington, *Nature*, 1929, vol. 120, p. 158.

B.C.?), has a primary matter (prakriti) and five subtle elements (tanmātras): sound, touch, colour, taste, and odour, from which proceed five coarse elements (mahābhūtas): ether, air, fire, water and earth, containing from one to five of the subtle elements in the above order. Each natural element contains varying proportions of all five coarse elements. Buddhism has six elements: earth, water, fire, air, ether and consciousness, symbolized by shapes, in section a cone (ether), crescent (wind), triangle (fire), circle (water), and square (earth); also atoms, which are not permanent but "momentary flashes of being".

The atomic theory occurs in the Vaiśeṣika system, attributed to Kaṇāda (500 B.C.?) and developed in Buddhist and Jainist works from the 2nd century B.C. The atom (aṇu, "small", later parām-aṇu, "absolutely small") is indestructible and spherical, six times smaller than the smallest mote in a sunbeam. Atoms have colour, taste, and smell (thus differing from the Greek atoms), and associate first in pairs, then in larger aggregates of pairs. Ether fills space and is the cause of sound.

Indian alchemy* (Rasasiddhi, "knowledge of mercury") dates from the 8th century A.D. when Buddhism changed to the Tantric form (also adopted in Tibet). In the Mercury system, taking prepared mercury was a means of conveyance beyond the series of rebirths, liberating its user whilst still alive. An early alchemist was Nāgārjuna (A.D. 700 or 850), other names are Patañjali, Narahari, Yasodhara, Gopal Krishna, and Vagbhata. The *Rasarṇava* (*c.* A.D. 1200) is supposed to mention mineral acids, which first appear clearly in works of the 16th and 17th centuries. The work of Śārṅgadhara, not later than the 13th century, deals with mercurial preparations and describes chemical operations. Albiruni (A.D. 973-1048) found that the Indians did not pay much attention to alchemy and could obtain very little information about it: "I only heard them speaking of sublimation, calcination, analysis, and the waxing of mica, which they call *talaka*, so I guess that they favour the mineral method."

* P. C. Rāy, *History of Hindu Chemistry*, 2 vols., Calcutta, 1902-09; B. Seal, *The Positive Sciences of the Ancient Hindus*, London, 1915; R. N. Bhagvat, *J. Chem. Education*, 1933, x, p. 659; Rāy, *ibid.*, 1948, xxv, 327; A. B. Keith, *Indian Logic and Atomism*, Oxford, 1921; Partington, *Annals of Science*, 1939, iv, 245.

Chemistry in China

The *I Ching*, " book of changes ", ascribed to Wen Wang (*c*. 1200 B.C.), contains an obscure " text " of eight symbols composed of long and short lines, and commentaries of later date. It introduces *yin* and *yang*, perhaps originally dark and light, or female and male, respectively, which later typified all kinds of dualities and entered scientific and medical theories. Whether they were regarded as material fluids or cosmic forces is uncertain. Later they were said to arise as " regulators " from a " great origin " (t'ai chi).

FIG. 22.—EXTRACTION OF ZINC IN CHINA.

The *Shu Ching*, " book of records ", of the Chou dynasty (722-221 B.C.) gives the " five things " or " five movers " (wu hsing) as water, fire, wood, metal and earth, usually called the " five elements ", which may go back to 350 B.C. They change into one another in a never-ending cycle and are sometimes said to come from *yin* and *yang*. Earth included all kinds of earth and stones, metal all the metals, wood all vegetables. The five elements were later related to regions, tastes, smells, colours, seasons, parts of the body, animals, planets, etc. The Greek four-element theory reached China with Buddhism from India early in the Christian era.

Bronze appeared about 1300 B.C., iron about 500 B.C. The Chinese had very efficient bellows, and cast iron was extensively worked in the Han dynasty (202 B.C.–A.D. 221); many objects (including cannon) were made of cast iron before it was used in the West. Early Chinese bronzes contain zinc; pai t'ung or paktong, a copper-nickel alloy, was smelted from a mixed ore about 250 B.C. Brass (t'ou shi) was made from copper and zinc ore at least as early as the 7th century, and in the 8th century A.D. by exposing copper plates to zinc vapour. The production of metallic zinc by heating the ore with coal in crucibles, when the metal condensed on the lids, is described and illustrated (see Fig. 22) in the *T'ien kong k'ai wu* (1637), and Chinese zinc was imported to Europe in the 17th century. Mercury (hung) was known about 150 B.C., mercury compounds are described in the *Pên Ts'ao Kang Mu* (1578), which also says distillation of alcohol was first practised in the Mongol dynasty (1260-1368). Paper was probably first made about A.D. 100, porcelain about A.D. 600. Saltpetre was known fairly early in China and gunpowder was made for fireworks before it was known in the West; its use may go back to A.D. 1150.

Alchemy in China * arose with later developments of Taoism, a philosophy introduced by Lao Tzŭ (*c.* 350 B.C.) but greatly modified by his later disciples. The *tao* was for the early school "the first and last cause of existence". A life regulated on the principles of Taoism could be prolonged.

Later Taoism degenerated into occultism, beginning under Liu Nan, or Huai-nan Tzŭ, a prince of the court of the Emperor Wu-ti (140-86 B.C.), which was frequented by magicians and preparers of the *tan*, or elixir of life, the production of which was now one of the principal objects of Taoism. It would enable the possessor to prevent death, to rise to heaven, to assume other forms, and generally to perform miracles. Early Chinese alchemy is more concerned with the "medicine of immortality" than with the transmutation of metals.

The earliest Chinese text dealing exclusively with alchemy, by Wei Po Yang (*c.* A.D. 140), has much about *yin* and *yang* and the

* A. Forke, *World Conception of the Chinese*, 1925; O. S. Johnson, *Chinese Alchemy*, 1928; Waley, *Bull. School Oriental Studies*, 1930, vi, 1; T. L. Davis *et al.*, *Isis*, 1932, xviii, 210; 1933, xix, 524; 1938, xxviii, 73; 1939, xxx, 236; *Proc. Amer. Acad. Arts and Sci.*, 1935, lxx, 219; 1939-40, lxxiii, 97, 371; 1941, lxxiv, 287; *Harvard J. Asiat. Stud.*, 1942, vii, 126; 1945, ix, 23; 1946, x, 186; W. H. Barnes, *J. Chem. Education*, 1934, xi, 655; 1936, xiii, 453; M. Chikashige, *Alchemy and other Chemical Achievements of the Ancient Orient*, Tokyo, 1936; Dubs, *Isis*, 1947-8, xxxviii, 62; Spooner and Wang, *ibid.*, 235.

tao, and also some chemical operations, such as crystallization, rather picturesquely described.*

The most celebrated alchemist of China was the Taoist, Ko Hung, in the fourth century A.D., whose treatises on alchemy are extant. His ideas are said to show a close resemblance to the Yoga system of India. He was chiefly concerned with the elixir of long life, the *chin tan*, by taking which, and gold, the body is preserved from decay. When vegetable matter is burnt it becomes ashes, but when the *tan sha* (cinnabar) is subjected to heat it produces mercury. When it suffers further changes this becomes cinnabar again. Ko Hung says that if the *tan* be placed over a hot fire, "gold will be instantaneously produced; the production of gold is a sign of the completion of the elixir".

Alchemy in Europe

The Greek treatises on the "divine art", which contain the beginnings of chemistry, were unknown in Europe during the Middle Ages, and the information on alchemy arrived with translations of Arabic works made in Spain—the point of contact of the Saracen and European cultures and the principal focus for the transmission of Arabic learning to Europe. Michael Scot, who was translating in Toledo in 1217 and was later in the service of the Emperor Frederick II, wrote on alchemy, but a *De Alchimia* attributed to him is probably spurious; it gives recipes for transmutation.† The earliest Latin translation of an Arabic work on alchemy was made by Robert of Chester in A.D. 1140, and some of the great schoolmen became interested in the subject.

Albertus Magnus (1193-1280) found that alchemy was a pretended science. Roger Bacon (1214-1292) believed in alchemy. After the adverse criticism of Albertus Magnus the schoolmen lost interest in the subject, which was then cultivated mainly by "artists", or "adepts", who wandered over Europe in search of wealthy patrons. Large numbers of books on alchemy written in the period 1250-1500 are mostly quite unintelligible. The alchemist is a favourite figure in art (Fig. 23) and literature. He is criticized in Chaucer (1345-1400), and in Ben Jonson's play, *The Alchemist* (1610).

* Partington, *Nature*, 1935, vol. 136, p. 287.

† J. W. Brown, *Michael Scot*, 1897; D. W. Singer, *Isis*, 1929, xiii, 5; Thomson, *Osiris*, 1938, v, 523.

Experiments on the supposed transmutation included the roasting of the sub-metallic mineral galena (lead sulphide) in air, when lead was formed, with a strong smell of sulphur ; and the produc-

FIG. 23.—THE ALCHEMIST. (*National Gallery.*)
Part of a picture by Ostade.

tion of a small button of silver when the lead was burnt off by heating on a cupel or dish made of bone-ash. Also, if iron pyrites, a yellow mineral looking somewhat like gold, was melted with lead, and the lead cupelled, a minute amount of gold was left. Both the silver and gold, of course, pre-existed in the minerals. A steel knife-blade dipped into a solution of blue vitriol (copper sulphate) apparently became converted into copper.

The later history of alchemy is often that of fraud. One method of effecting transmutation was to stir the materials in the crucible with a hollow iron rod filled with gold powder and stopped with wax. Another deception was to take a nail, half of iron and half of gold, and cover it with black ink. It was then dipped into a liquid and stirred, when the black was washed away and the part of the nail which dipped in the liquid was apparently turned into gold. Still another was to take a coin made from a white alloy of silver and gold and dip it in nitric acid, when the silver was dissolved and half the coin apparently converted into gold. Some of these coins are still in museums and have been analysed.

There are several interesting old accounts of transmutations supposed to have been carried out in the presence of witnesses of the highest reputation for honesty.*

Early European Writers on Alchemy

Albert the Great (Albertus Magnus)† was born in Bavaria in 1193 (or 1206) of a noble family, studied in Paris, and entered the Dominican Order. He lectured at Cologne and Paris and was bishop of Ratisbon in 1260-62, relinquishing his office with the consent of the Pope so as to be able to devote himself to study. He died at Cologne in 1280. Albert composed a large number of works—the latest edition is in thirty-eight volumes—mostly dealing with theology, physics and natural history, in which he made extensive use of Aristotle but added some original material. His *De Mineralibus* includes sections on chemistry and alchemy, which he calls " a beggarly union of genius and fire ". He knew the frauds of the alchemists and says he had alchemical gold tested (*ego expiriri feci*), and although it stood six or seven ignitions in the fire it was at length consumed and turned into dust. He quotes the dictum of the genuine Avicenna (p. 29) that species cannot be transmuted, but was puzzled by the apparent inconsistency of this and the statements in the work *De Anima*, attributed to Avicenna but probably spurious, in which transmutation is described as possible. A *De Alchimia* attributed to Albert, which describes fused caustic soda, etc., is probably interpolated but may be based on a genuine work.

Thomas Aquinas (1225-1274), the famous Dominican pupil of Albert, discusses in his genuine works whether payments in alchemical gold are legal, and says they are if the gold is genuine,

* Thomson, *History of Chemistry*.

† Partington, *Ambix*, 1937, i, 3.

as he thought could be the case. Alchemical works attributed to Thomas are probably spurious.

Vincent of Beauvais (1190?-1264), another Dominican, has a long section on alchemy in his encyclopaedia called *Speculum Naturale*, mostly derived—like the knowledge of Albertus Magnus—from Latin translations of Arabic works. Albert, Thomas and Vincent made no original contributions to chemistry, but their study of the subject is significant as showing the great interest in alchemy among the prominent schoolmen and clerics of their time.

Roger Bacon

Roger Bacon was born at Ilchester, in Somerset, probably in 1214. When quite young he went to study in Oxford, where he later took the degree of M.A. (he is said not to have had the degree of doctor of divinity, although he was afterwards called *Doctor Mirabilis*). He afterwards, perhaps about 1236, studied and lectured in the University of Paris, and he was thoroughly educated in scholastic philosophy. He joined the Franciscan Order about 1247. His later history is very obscure: he was probably in Oxford for some part of the period 1247-57, and after 1268, but he spent much time in Paris, where he became keenly interested in experimental science as a result of meeting Peter of Maricourt, the author of a treatise on the magnet. His criticism of members of his Order, and of famous Dominicans like Albertus Magnus and Thomas Aquinas, brought Bacon into disgrace, and he is generally said to have been imprisoned from 1277 until 1292, in which year, the chronicler John Rous says, he died: "the noble doctor Roger Bacon was buried at the Grey Friars at Oxford A.D. 1292, on the feast of St. Barnabas the Apostle" (11th June): other accounts say he died in 1294. Leland, writing about 1550, was ashamed to say (*pudet dicere*) that Bacon's works were lying neglected and mutilated in libraries, and some are still unpublished.

In 1267 Bacon says he had expended in the last ten years a sum of over 2000 livres (nearly £10,000) on books, apparatus, assistants, and "forming friendships with the wise", and it is said that practical alchemy "was carried out by him in places more private, sometimes in the suburbs [of Oxford]". Borrichius, in the seventeenth century, was shown what was supposed to have been Bacon's house there.

Bacon divides alchemy into (*a*) *speculative*, dealing with the generation of things from the elements and all kinds of metals, minerals, salts, etc., " of which Aristotle and the Latin authors are ignorant " (he is thinking of the Saracens as the source of this knowledge) ; (*b*) *operative*, teaching how to make things (including

FIG. 24.—ROGER BACON (?) WITH A PUPIL.

gold) better by art than in nature, and also powerful medicines by sublimation, distillation, etc. Bacon emphasized, long before Paracelsus, that medicine should make use of remedies provided by chemistry, and he realized that chemistry is a science intermediate between physics (in the Aristotelian sense) and biology.

There is a section on alchemy in the *Opus tertium*, sent to the Pope in 1268 as part of a great encyclopaedia composed by Bacon. Unfortunately, the Pope died in the same year and his successor was an enemy of Bacon.

Bacon in his works describes gunpowder, which became known in Europe (perhaps from China by way of the Arabs) in his time.

He gives its composition as seven parts of saltpetre, five of wood charcoal and five of sulphur, which contains too little saltpetre to be good. Several alchemical works attributed to Bacon were printed;* one is a summary of the *De Anima* attributed to Avicenna (p. 29). There seems little doubt that Bacon made many chemical experiments and that he fully recognized the importance of the subject.

Arnald of Villanova

Arnald of Villanova (born either in France or Spain) (1240-1311) was a physician in Montpellier; he is the author of several medical treatises, and some alchemical works are also attributed to him. He describes the distillation of spirit of wine, which he used medicinally, and emphasises the utility of chemical remedies in medicine. Arnald is said to have made artificial gold for Pope Boniface VIII at Avignon. His main alchemical works, which are regarded by some modern critics as genuine, are a *Rosarium philosophorum* and a *Flos florum*, extant in old manuscripts.

Raymund Lully

Raymund Lully (Ramon Lull), born in Majorca about 1232, was a learned man and a missionary enthusiast (*Doctor illuminatissimus*, as he is called) who suffered martyrdom at Bugia on the North African coast in 1316. He was heterodox and (unlike Thomas Aquinas and, recently, Albert the Great) has not been canonized by the Roman Church. His genuine works (mostly in Catalan) criticize fraudulent alchemy, but there is a large alchemical literature attributed to him which has given rise to much discussion. It was probably written by members of his school, the Lullists, soon after his death. He is said to have performed a transmutation in 1332 (i.e. several years after his death!) for the king in St. Catherine's Church near the Tower of London. In the alchemical works attributed to Lully (e.g. the *De Secretis Naturae* and the *Testamentum*) the preparation of nearly anhydrous alcohol by rectification and dehydration over potassium carbonate (salt of tartar) is described. They also describe the preparation of nitric acid and aqua regia. The first discovery of the mineral acids is quite obscure. The preparation of nitric acid by distilling a mixture of saltpetre

* *De Arte Chymiæ*, Frankfurt, 1603; *Essays*, edit. A. G. Little, Oxford, 1914.

and copperas (*corprossa*), i.e. crystalline ferrous sulphate, is described by the Franciscan Vital du Four (d. 1327), who quotes no authorities later than 1150. The acid was unknown to Roger Bacon and was probably discovered after his time, but whether by Muslim or European alchemists is not known. Sulphuric and hydrochloric acids were apparently not known until the sixteenth century (pp. 47, 55).

Technical Treatises

The technical recipes and traditions represented by the early Leyden and Stockholm papyri (p. 17) found a continuation in such works * as the *Compositiones ad tingenda* (*c.* A.D. 800), the *Mappæ clavicula* (in two forms, 10th and early 12th centuries), and the *Diversarum artium schedula* of Theophilus the Priest (*c.* 1075). These deal with pigments, etc. The 12th century *Mappæ clavicula* has a recipe for alcohol which is one of the earliest known. Of a different character is the Book of Fires (*Liber ignium*) of Mark the Greek (Marcus Græcus), known in 13th-14th century MSS., which describes incendiary mixtures and gunpowder, and seems to be based on Arabic (not Greek) sources. The "Greek fire" which had been used in Constantinople was, apparently, not a mixture containing saltpetre, but an incendiary containing light petroleum.

* See Berthelot, *La Chimie au moyen Âge*, 1893, i; Hime, *Origin of Artillery*, 1915; Partington, *Isis*, 1934, xxii, 136; T. L. Davis and Chao Yün-Ts'ung, *Proc. Amer. Acad. of Arts and Sciences*, 1943, lxxv, 95.

CHAPTER IV

IATROCHEMISTRY, OR CHEMISTRY IN THE SERVICE OF MEDICINE

Iatrochemistry

ALTHOUGH not one of the alchemists ever succeeded in transmuting a base metal into gold or in preparing the elixir of life, the great amount of work done with these ends in view bore fruit in other directions, and led to the discovery of many new substances, such as alcohol, the mineral acids, and many metallic salts. In the sixteenth and seventeenth centuries another school of chemists arose, called the Iatrochemists, i.e. the Medical Chemists, who attempted to apply chemistry to the preparation of medicines and to the explanation of processes in the living body. Paracelsus was the founder of this sect; he believed in the philosophers' stone and in the elixir of life.

Paracelsus

Theophrastus Bombast von Hohenheim,* commonly called Paracelsus, was born in 1493 at Einsiedeln in Switzerland, his father being a German licentiate in medicine who had migrated there; he taught his son medicine, and some mineralogy and chemistry. Paracelsus then wandered about Europe making a study of mining, medicine and alchemy. He took the degree of M.D. at Ferrara in Italy and in 1527-8 he was appointed professor of medicine at Basel. He lectured in German and was unsparing in his criticism of both ancient and contemporary physicians. There is a story told by Boerhaave that in his introductory lecture he burnt the works of Galen and Avicenna in a brass pan with sulphur and nitre and expressed the hope that their authors were in like circumstances. A dispute with a canon, who refused to pay a large fee for three small pills (*tres murini stercoris pilulas*) which cured him of

* Titley, *Ambix*, 1938, i, 166; Partington, *Nature*, 1941, cxlviii, 332.

gout, led to the disgrace of Paracelsus, and he was compelled to leave Basel. The rest of his life was spent in wandering about Europe. His failure to bring to fruition his ambition to reform medicine, the contempt shown for his teachings by the medical men of his time and by most of his students, a scurrilous attack upon him posted in public places in Basel, and the necessity for his flight in disgrace, left an indelible stamp on his character. He is said to have been often intoxicated and to have spent a large part of his time in taverns in the company of peasants, sleeping in his clothes on the floor. Oporinus, his factotum, says he dictated his works when drunk, and it is difficult to find a more probable explanation for the confused, self-contradictory and obscure style of most of them. Paracelsus died in 1541, Boerhaave says " as he lived, in an inn at Saltzburg, at the sign of the white horse, on a bench, in the chimney corner ", but others say he died in hospital.

Boerhaave says Paracelsus's fame was due to his being a good surgeon and a competent physician, to his understanding the use of metallic remedies, to his marvellous cures with opium, and especially to his use of preparations of mercury to cure new diseases which resisted all the old remedies. " These five concurring circumstances take in his whole merit, and were the matter of all his glory ; the rest was empty smoke, and idle ostentation."

Paracelsus's writings are full of mystical ideas. He believed in astrology and associated different parts of the body with the planets, e.g. the heart with the Sun, the brain with the Moon, the liver with Jupiter, etc. He believed that digestion was caused by an independent spiritual being, called Archeus, in the stomach.

Paracelsus was essentially a reformer of medicine and his contributions to chemistry are trifling. He opposed the teachings of Galen and Avicenna, whose worn-out systems were still the mainstay of the physicians of his day, and he directed attention to the great utility of a knowledge of chemistry in medicine and pharmacy. In medicine he used tinctures, essences, extracts of plants, etc., hoping to extract the active principle from the mass of inert material. His theory of tartar is important : he believed that many diseases were caused by morbid deposits formed in the body, as tartar is deposited from wine on standing.

Paracelsus was impressed by the industry of the medical chemists, who did not over-dress and wear gold rings and chains like the physicians, but worked patiently in the laboratory, wearing leather aprons, soiling their hands with coals and dirt and not

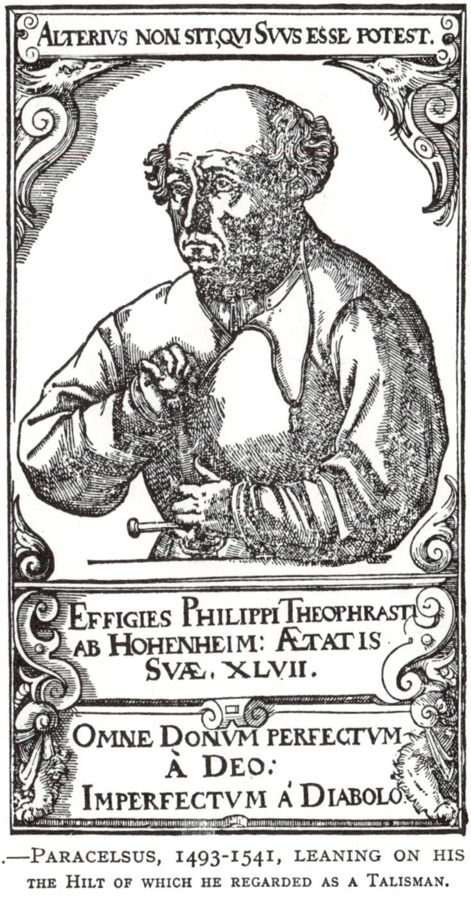

FIG. 25.—PARACELSUS, 1493-1541, LEANING ON HIS SWORD, THE HILT OF WHICH HE REGARDED AS A TALISMAN.

boasting of their skill. " They leave such things alone and busy themselves with working with their fires and learning the steps of alchemy, which are distillation, solution, putrefaction, extraction, calcination, reverberation, sublimation, fixation, separation, reduction, coagulation, tinction, and the like."

In theory he believed in the four elements, but thought they appeared in bodies as the *three principles*: salt, sulphur and mercury. Salt was the principle of fixity and incombustibility, mercury of fusibility and volatility, sulphur of inflammability. The last two had long been recognized by the alchemists, but Paracelsus seems to have been the first to add salt, making up the *tria prima*, which he compared with body, soul and spirit. He thought everything had its own particular kind of salt, sulphur and mercury, " yet these sulphurs, salts and mercuries are only three things." Paracelsus was, apparently, the first to use the name *alcohol* for strong spirit of wine, and the first in Europe to mention zinc, which he calls a " bastard metal ". His immediate followers, who made his theories more precise and added to the stock of chemical remedies by new mineral preparations, included Gerhard Dorn, Oswald Croll, who in his *Basilica Chymica* (1609) describes silver chloride as *luna cornea* (horn silver), and Hadrian Mynsicht, who describes tartar emetic (potassium antimonyl tartrate, $K(SbO)C_4H_4O_6$).

Van Helmont

Johann Baptista van Helmont was born in Brussels in 1579, and died on December 30, 1644, either in Brussels or Vilvorde (near Brussels). He was descended from a noble and ancient family: his mother was Marie de Stassert, and he belonged to the family of Mérode through his wife, Margaret van Ranst. He studied arts at Louvain until 1594 but took no degree, since he considered academic honours a mere vanity. He then went to the Jesuits' School at Louvain, but, still dissatisfied, he turned to the mystics and studied the works of Thomas à Kempis and Johann Tauler. He then took up medicine and read the works of Hippocrates, Galen, Avicenna and a great number of contemporary authors, from which he says he " noted all that seemed certain and incontrovertible, but was dismayed on reading my notes to find that the pains I had bestowed and the years I had spent were altogether fruitless." He says he gave away to students books worth 200 crowns, but wished afterwards that he had burnt them. He took the degree of M.D. at Louvain in 1609 after ten years of study and travel. In 1609, also, he married, and he says: " God has given me a pious and noble wife. I retired with her to Vilvorde and there for seven years I dedicated myself to pyrotechny [i.e. Chemistry] and to the relief of the poor."

Boerhaave was told that Helmont was "wholly taken up in chemical operations night and day" and that "he was scarce known in his neighbourhood; that he did not apply himself to practice; nor scarce ever stirr'd out of doors." He was a very

FIG. 26.—JOHANN BAPTISTA VAN HELMONT, 1579-1644.
HE BELONGED TO THE NOBILITY (HIS COATS OF ARMS ARE SEEN IN THE BACKGROUND) BUT DEVOTED HIS LIFE TO STUDY AND EXPERIMENT.

influential and highly respected man, with a great reputation, although Boyle, who constantly quotes him as an authority, remarks that Helmont was "an author more considerable for his experiments than many learned men are pleased to think him". Helmont's harsh although deserved criticism of the conventional medicine of his time made him many enemies and retarded the general acceptance of his views.

Van Helmont's works were published in 1648 (and in later editions) as *Ortus medicinae.* An English translation, *Oriatrike* *or Physick Refined*, appeared in 1662.* The position of van Helmont in the history of Chemistry is much more important than has usually been supposed. Boyle, in particular, was much influenced by him, and he had expressed doubts as to the earlier theories of the elements before Boyle.

Van Helmont was proud of his claim to be called a chemist, and calls himself (e.g. in the introduction to his *De lithiasi*) " Philosophus per ignem." He represents the transition from alchemy to chemistry. He made a careful study of the chemical as well as the medical writings of Paracelsus (which he later found full of errors) and carried out a large number of chemical experiments in his house at Vilvorde. He believed in alchemy and gives a very circumstantial account of the transmutation of nearly 2000 times its weight of mercury into gold, by means of a quarter of a grain of the philosophers' stone given him by a stranger. This was a heavy red powder, glittering like powdered glass and smelling of saffron; it was enclosed in wax and projected on the mercury heated to the melting-point of lead, when the metal grew thick and, on raising the fire, melted into pure gold. Helmont did not believe that the philosophers' stone was also the elixir of life, as Paracelsus assumed. He says that by means of the alkahest, or universal solvent, he had converted vegetables or oak charcoal into water. He calls the alkahest *ignis aqua*, and it was probably nitric acid. In some of the examples which he gives of converting bodies into water, he neutralizes acids with chalk and distils off water.

An important feature of van Helmont's chemical work is its quantitative character; he made extensive use of the balance, expressed clearly the law of indestructibility of matter, and emphasized that metals when dissolved in acids are not destroyed but can be recovered again by suitable means. He also realized that when one metal precipitates another from a solution of a salt, there is no transmutation as Paracelsus thought. When silver is dissolved in aqua fortis it is not destroyed, but is concealed in the clear liquid as salt is contained in a solution in water, and can be recovered in its original form. Dissolved copper is precipitated by iron, which takes its place, and copper similarly precipitates silver. " Nothing is made of nothing, therefore weight is made of another body of

* On van Helmont, see Partington, *Annals of Science*, 1936, i, 359-84.

equal weight in which there is a transmutation as it were of the matter." When mercury is boiled with oil of vitriol it forms a white precipitate like snow which on washing with water turns yellow, and on revivification gives the original weight of mercury. He describes the preparation of blue vitriol by concentrating mine-water; by lixiviating roasted pyrites exposed to air; by throwing sulphur on melted copper and putting the mass (cuprous sulphide) in rain-water; and by boiling copper plates with oil of vitriol, when a black mass is obtained which is dissolved to a blue solution in water. Copper vitriol yields little or no acid on distillation, but when common (iron) vitriol is distilled by a strong fire in a coated glass retort it yields a very acid oil of vitriol—a clear distinction between copper and iron vitriols, not made by Paracelsus.

Van Helmont was well acquainted with sulphuric acid, made both from the distillation of vitriol and by burning sulphur under a bell. He describes the preparation of nitric acid by distilling equal parts of saltpetre (*salispetræ*), vitriol and alum, first dried and then mixed together, and apparently knew that it converted sulphur into sulphuric acid (*salispetræ spiritus elevat sulfur humidum et embryonatum vitrioli*). He mentions aqua regia, made from nitric acid and sal ammoniac, and the gas (chlorine and nitrosyl chloride) evolved from it. He describes the distillation of spirit of sea-salt (*spiritus salis marini*), i.e. hydrochloric acid, from salt and dried potter's clay. Fixed alkali is not present (as such) in plants but is produced by combustion. Van Helmont uses the name *sal salsum* for a neutral salt. When strong spirit of vitriol is poured on salt of tartar, heat is produced, and when sugar of lead is calcined the residue takes fire when exposed to the air—van Helmont thought because an alkali in it took up moisture from the air.

The volatile red oil obtained by repeatedly distilling sulphur (with ammonia?) was probably ammonium sulphide, usually credited to Beguin.* Van Helmont knew that silver chloride is soluble in ammonia; he mentions a fixed salt (potassium arsenate) formed when white arsenic is fused with saltpetre, and prescribes burnt sponge (containing iodides) for goitre.

* Jean Beguin, *Tyrocinium Chymicum*, 1610; Patterson, *Annals of Science*, 1937, ii, 243.

Van Helmont on Gas

Van Helmont says that flame, which is only burning smoke (*non est nisi accensa fuligo*),* perishes at once in a closed vessel, and charcoal may be heated continuously in a closed vessel without wasting. Yet if 62 lb. of oak charcoal contain 1 lb. of ashes, the remaining 61 lb. are " wild spirit " (*spiritus silvestre*) which cannot escape from the shut vessel. (These ideas contain the germ of Stahl's theory of combustion.) " I call this spirit, hitherto unknown, by the new name of gas, which can neither be retained in vessels nor reduced to a visible form, unless the seed is first extinguished " (*Hunc spiritum, incognitum hactenus, novo nomine Gas voco, qui nec vasis cogi, nec in corpus visibile reduci, nisi extincto prius semine, potest*). The last part of this famous definition he explains by saying that the gas of flame is not yet water (the fundamental element) because, " although the fire has consumed the seminal forces of the burning body, yet some primitive fermentive differentiations of the body remain, which being at last consumed and extinguished, the gas returns to the element of water."

Flame is ignited smoke and smoke is gas. Van Helmont describes an experiment of burning a candle in air in a cupping-glass over water, when the water rises and the flame goes out: the suction is caused by consumption of part of the air. " There is in the air something that is less than a body, which fills up the vacuities in the air and is wholly annihilated by fire." The contraction is due to the pressing together of the empty spaces in the air by the smoke from the burning candle, the air having been " created to be a receptacle of exhalations ". The air in mines, saturated with exhalations from minerals, extinguishes a flame. All this shows, he says, that a vacuum, which Aristotle thought impossible, is " something quite ordinary ".

The name *gas sylvestre* (*sylvestris*, " of the wood ") is given by van Helmont to the " wild spirit ", " untameable gas ", which

* Although van Helmont is usually credited with this definition, e.g. by Roscoe and Schorlemmer, *Treatise on Chemistry*, 1905, i, 811, it is due to Aristotle: *De cœlo*, iii, 4; *De gen. et corr.*, ii, 4; *Meteor.*, iv, 9, who also frequently uses the word *φλογιστά*. Kopp, *Beiträge*, iii, 84, incorrectly attributes this definition of flame to Albertus Magnus: see Partington, *Nature*, 1935, cxxxv, 916.

breaks vessels and escapes into the air. If nitric acid is poured on sal ammoniac in a glass vessel which is closed by cement or by melting the glass, a gas is produced which bursts the vessel: "the vessel is filled with plentiful exhalation (yet an invisible one) and however it may be feigned to be stronger than iron, yet it straightway dangerously leapeth asunder into broken pieces." This explosive property explains the effects of gunpowder. The name gas is almost certainly derived by van Helmont from the Greek word *chaos* (*non longe a Chao veterum secretum*). Juncker's derivation from *Gäscht*, "froth", and a favourite derivation from *Geist* (spirit), are probably incorrect.

Van Helmont was the first clearly to realize the production of gas in various chemical processes. He says more than once that he was the "inventor" of gas, which Paracelsus was ignorant of, and there is no doubt that Paracelsus had no such ideas on gases as he. Van Helmont distinguished gases from condensable vapours and from air, and from one another. He says that gas is composed of invisible atoms which can come together by intense cold and condense to minute liquid drops. He recognized that gas may be contained in bodies in a fixed form (*spiritus concretus, et corporis more coagulatus*), set free again by heat, fermentation, or chemical reaction.

The kinds of gas mentioned by van Helmont are: (i) The poisonous gas, extinguishing a candle flame, which collects in mines and in the Grotto del Cane, i.e. carbon dioxide. (ii) The *gas carbonum*, formed by burning charcoal and other combustibles, which is usually carbon dioxide but sometimes carbon monoxide, since van Helmont says he was nearly poisoned at the age of 65 by the fumes of burning charcoal, and gives the symptoms of carbon monoxide poisoning. (iii) The gas forming in cellars, especially from fermenting wine (carbon dioxide). Grapes can be dried to raisins if the skin remains whole, but if the skin is broken they ferment and evolve *gas sylvestre*, which makes them appear to boil and is contained in wines which have been closed up in casks before the fermentation is ended, making them effervescent. Since the fresh grape on distillation is reduced by art to elementary water, but gives rise to gas in presence of a ferment, it follows that gas itself is water. (iv) Gas formed by effervescence of sulphuric acid and salt of tartar, or distilled vinegar and calcium carbonate (carbon dioxide). (v) A poisonous red gas formed when aqua fortis (*chrysulca*) acts on metals such as silver. This was nitric oxide, which Juncker in 1730 still called *gas sylvestre*. (vi) The gas evolved from aqua fortis and sal ammoniac in the cold (chlorine and nitrosyl chloride). (vii) The gas evolved in bubbles from Spa water, which then deposits an ochry sediment (carbon dioxide). (viii) The gas evolved in eructations (*gas ventosum*), i.e. carbon dioxide, sharply distinguished from inflammable intestinal gas, i.e. (ix) *gas pingue*, which is inflammable, is evolved in putrefaction, and is contained in intestinal gas which he (as did Albertus Magnus) knew was inflammable (*transmissus per flammam candelæ, transvolando accenditur, ac flammam diversicolorem, Iridis instar exprimit*) (hydrogen, methane, with fetid impurities). (x) A gas, different from (ix), which inflates the tympanum (? in gas gangrene). (xi) A combustible gas (*gas pingue, siccum, fuliginosum, endemicum*) formed on dry distillation of organic matter (a mixture of hydrogen, methane and carbon monoxide). (xii) A sulphurous or acid gas (i.e. sulphur dioxide) which flies off from burning sulphur, which is a material wholly fatty and combustible (*totum sit pingue et φλογιστὸν*) : this gas when formed in a vessel filled with air extinguishes a candle-flame, and it can be condensed in a bell-jar into a juice (sulphuric acid). (xiii) A *gas sylvestre* from fused saltpetre and charcoal (carbon dioxide). Van Helmont missed the oxygen from heated saltpetre, although he says that when saltpetre is strongly heated it gives off a little acid water and leaves salt of tartar (really potassium nitrite or oxide). (xiv) Gunpowder when inflamed evolves gas which bursts vessels, yet the charcoal, sulphur and saltpetre when heated separately do not explode : the detonation of the mixture is due to a mutual antipathy by which they try to destroy one another. (xv) An ethereal or vital gas, a kind of vital spirit of a gaseous nature, which is the reason why other gases act so swiftly and powerfully on the body.

In respiration, the air mingles in the lungs with the venous blood, which would otherwise coagulate ; it mixes with the sulphur of the blood, and is exhaled together with watery vapour in an unperceivable gas. From the arterial blood no dregs or filth are expelled, but venous blood, in wasting itself by the guidance of heat, produces a gas, as water does a vapour, and this gas is subsequently of necessity expelled. Van Helmont criticizes Galen for teaching that the object of respiration is refrigeration, and says its purpose is to maintain animal heat, by a ferment in the left ventricle of the heart changing the arterial blood into a vital spirit. The friction together of saline and sulphurous particles in the blood, caused by the beating of the heart, produces heat and a " formal " light in the blood. This theory is very like that adopted by Willis and Mayow (p. 83).

Van Helmont on the Elements

Van Helmont rejects the theory of the four elements and three principles as taught by Paracelsus, and the " heathen " theory of a primary matter of Aristotle. He asserts that the true elements are air and water. Neither of his two primary elements (air and water) is convertible into the other and an element cannot be reduced to a simpler state. The other two so-called elements, viz. fire and earth, do not deserve the title, since fire is not a form of matter at all and earth can be formed from water. He points out that water, with heaven and earth, was formed on the first day in the account of Creation in Genesis. He describes the famous " tree experiment " to prove that " all vegetables proceed out of the element of water only ":

> " I took an Earthen Vessel, in which I put 200 pounds of Earth that had been dried in a Furnace, which I moystened with Rain-water, and I implanted therein the Trunk or Stem of a Willow Tree, weighing five pounds; and at length, five years being finished, the Tree sprung from thence, did weigh 169 pounds, and about three ounces: But I moystened the Earthen Vessel with Rain-water, or distilled water (alwayes when there was need) and it was large and implanted into the Earth, and least the dust that flew about should be co-mingled with the Earth, I covered the lip or mouth of the Vessel, with an Iron-Plate covered with Tin, and easily passable with many holes. I computed not the weight of the leaves that fell off in the four Autumnes. At length, I again dried the Earth of the Vessel, and there were found the same 200 pounds, wanting about two

ounces. Therefore 164 pounds of Wood, Barks, and Roots, arose out of water onely."

The conclusion is mainly correct, since the tree is largely water (about 50 per cent. of fresh willow wood is free water), but it is an irony of fate that van Helmont did not know the part played by the carbon dioxide in the air, since, as has been shown, he was the first to realize the existence of this gas, to which he gave a special name. In the idea, but not the performance, of this famous experiment, van Helmont had been anticipated by a century and a half in a work of Nicolaus of Cusa.

As further proofs of his thesis, van Helmont says that spirit of wine, carefully dephlegmated (dehydrated) with salt of tartar, gives only water on combustion, and that fish are nourished and their fatty matter produced from the water in which they swim. He establishes links between materials to prove that they are formed from water; for example, since wood was shown to be formed from water in the tree experiment, all the products obtained from wood, such as charcoal and ash, must also consist of water. If gold is to be formed from water, this will involve a compression to one-sixteenth the volume, which is quite possible to Nature, although water has no pores. Grain by fermentation is converted into beer, which still leaves a solid residue on evaporation. But beer can undergo a further fermentation, becoming sour and consuming its dregs, and finally it returns of its own accord into water.

Earth is not an element, but is formed from water. For if sand is fused with excess of alkali it forms a glass. If this glass is exposed to the air it liquefies to water, and if sufficient aqua fortis is added to saturate the alkali (*quantum saturando alcali sufficit*) the sand settles out again of the same weight as was used to make the glass. Fire—which is clearly distinguished from light—is not an element, cannot form a material constituent of bodies, and is "a positive death of things, a singular creature, second to no other", which can pierce glass. Air cannot be condensed to water, as is proved by an experiment with the air-gun, in which compressed air remains elastic and can propel a ball through a board.

Van Helmont on the Stone

Of all Helmont's works, that on urinary calculi (*De lithiasi*) is said by Boerhaave to be "incomparable, and the best", and it also contains the greatest number of chemical experiments, which must have occupied him for a long period of time. He gives a fairly accurate description of the formation of tartar in wine-casks, and says tartar is not contained in food and does not cause the disease of the stone. The stone, called *duelech* by Paracelsus, is not tartar, since it does not dissolve in boiling water. By mixing spirit of urine (ammonium carbonate solution) with spirit of wine he observed the formation of a white precipitate, afterwards called *offa Helmontii*. Van Helmont isolated from urine two fixed salts, one of them common salt, which he asserted was that taken with the food, and another of different crystalline form—probably microcosmic salt.

Van Helmont on Ferments

Van Helmont says the name of ferment was unknown before, except for the leaven used for making bread, whereas there is no change or transmutation brought about by the sleeping affinity of matter except by the work of the ferment. The two chief beginnings of bodies (*prima initia*) are water and ferment or seminal origin; the ferment is an indwelling formative energy, "hardly 1/8200 part of a body", which disposes the material of water so that a seed is produced and life, and the mass develops into a stone, metal, plant or animal. There are specific ferments in the stomach, the liver, and other parts of the body, which bring about digestions and other physiological changes.

Van Helmont's ideas on ferments, although rather crude and undeveloped, were in the right direction and in many ways resemble the modern theory of enzymes.* The acid of the gastric juice is necessary for digestion, but an excess of acid causes discomfort and illness, since it cannot be neutralized into salt by the alkali of the gall in the duodenum. Van Helmont compared the

* Sir M. Foster, *Lectures on the History of Physiology*, Cambridge, 1901, p. 135; W. M. Bayliss, *Principles of General Physiology*, 1915, p. 307; Partington, *Everyday Chemistry*, 1929, pp. 508, 551, 555.

interactions of various juices in the organs with chemical reactions between liquids outside the body. He showed by experiment that salt can pass with water through a bladder (by osmosis), and explains how the digested food (chyle) can pass through the walls of the intestines into the veins.

Van Helmont distinguished six fermentations of the food in passing through the body. The stomach and spleen produce an acid liquor (which van Helmont said had been tasted in the saliva of birds) which carries out the first digestion. The mass passes through the pylorus into the duodenum, where it is neutralized by the gall (*fel*) of the gall-bladder, the second digestion. The third digestion takes place in the mesentery, to which the gall-bladder sends the prepared fluid. The fourth digestion occurs in the heart, where the red blood becomes more yellow by the addition of vital spirit; the fifth digestion consists in the conversion of arterial blood into vital spirit, and occurs mainly in the brain; and the sixth digestion consists in the elaboration of the nutritive principle in each separate member from the blood, by a separate ferment.

Sylvius

Franciscus Sylvius de le Boë (1614-1672), professor of medicine at Leyden, persuaded the curators of the university to build him a "Laboratorium, as they call it", which seems to have been the first university chemical laboratory. He represents the culmination of Iatrochemistry (of which he is sometimes mistakenly called the founder). He was largely a theorist. He taught that the functions of the living organism were mainly determined by chemical activities ("effervescences"), particularly by the real or imaginary acidic or alkaline characters of the body fluids—a precursor of the modern cult of *p*H—, that an excess of one of these constituents gave rise to a disturbance in the chemical processes taking place in the body, and that the removal of the excess or the supplementing of the defect could effect a cure of the disease—a survival of the old humoral pathology (p. 15).

Agricola

Georg Agricola (1494-1555), a German physician, wrote several works on mineralogy and metallurgy, the most famous being his *De Re Metallica*, first published in 1556. His works are written in a quiet, dignified style and are very practical; they describe all the mining and metallurgical processes known in his time, and are illustrated with curious cuts (Fig. 27). He mentions bismuth. Agricola's *De Re Metallica* was one of the first treatises on applied chemistry: a smaller work (*Pirotechnia*, 1540) on similar lines had been written in Italian by Biringuccio: it contains accounts of casting bells and cannon.

FIG. 27.—A SALT WORKS.
FROM AGRICOLA'S *De Re Metallica*, 1556.

Basil Valentine

Basil Valentine is supposed to have been a Benedictine monk of Erfurt who wrote about 1470, but the works extant under his name were probably written about 1600 by their "editor", Thölde, a salt manufacturer of Halle. The best known is the *Triumphal Chariot of Antimony*, first published in German in 1604 (*Triumph-Wagen Antimonii*). A work called *Haligraphia*, published under his own name by Thölde in 1603, contains material afterwards published in a work, *Letztes Testament*, attributed to Basil Valentine. The works of Basil Valentine describe the preparation of many compounds of antimony, and also the mineral acids, e.g. sulphuric acid, made by deflagrating a mixture of stibnite and sulphur with nitre under a glass bell, and hydrochloric acid by distilling a mixture of common salt and copperas. They mention the *tria prima* (p. 44).

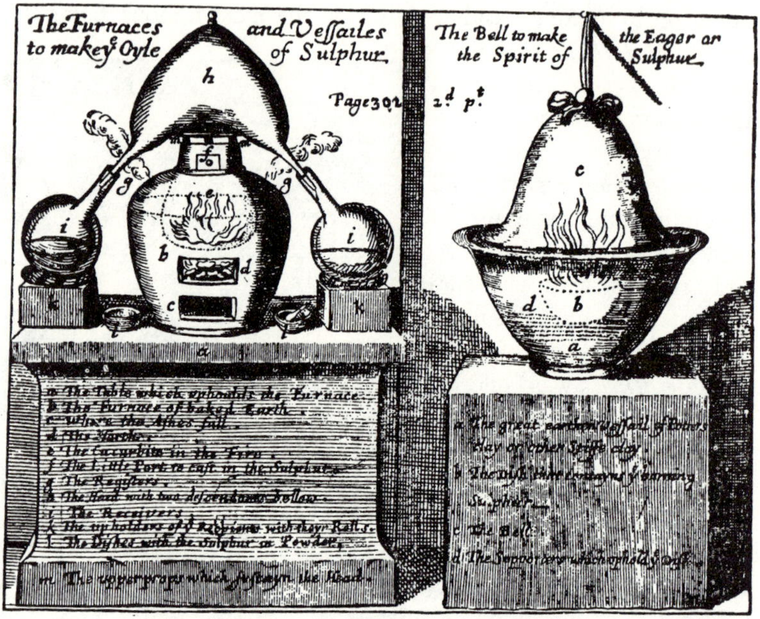

FIG. 28.—PREPARATION OF SULPHURIC ACID (OIL OR SPIRIT OF SULPHUR) ACCORDING TO LE FÉVRE (1660).

The alchemists John and Isaac of Holland (mentioned by Ben Jonson) are of doubtful date but probably also after Paracelsus.

Libavius

Andreas Libavius (1540?-1616) was a German schoolmaster who had an excellent knowledge of chemistry and wrote the first text-book on the subject (*Alchemia*, 1597), and other works. He criticized some of the absurdities of Paracelsus but still believed in the possibility of transmutation. Libavius describes zinc, lead nitrate, and the preparation of anhydrous stannic chloride from tin and corrosive sublimate: it was long known as "Libavius's fuming liquor." He also describes reactions and analytical tests, particularly for use with mineral waters, and dry reactions in assaying metallic ores.

Glauber

Johann Rudolph Glauber (1604-1670), born in Bavaria, was essentially a practical chemist although he wrote a large number of works, most of which were translated into English by Packe and

Deutsches Museum, Munich.

FIG. 29.—J. R. GLAUBER, 1604-1670.

published in folio in 1689. His most important work is the *New Philosophical Furnaces* (Amsterdam, 1646-49) in which he describes the preparation of " spirit of salt " by distilling salt with green vitriol and alum or with clay. In 1658 he described the preparation from salt and oil of vitriol. The residue from the preparation (sodium sulphate) he called " miraculous salt " (*sal mirabile*), but

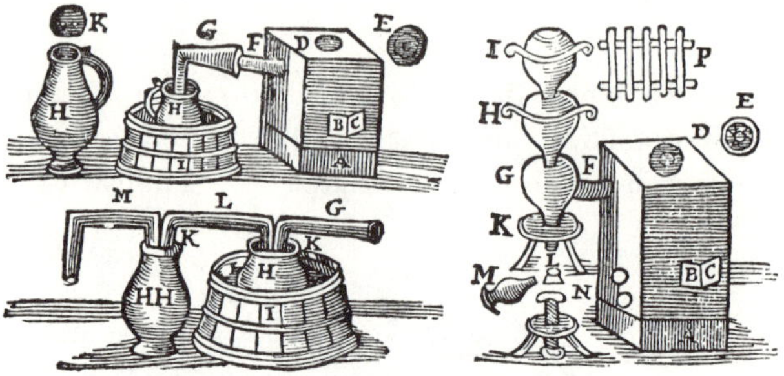

FIG. 30.—APPARATUS USED BY GLAUBER.

ON THE RIGHT IS A FURNACE FOR DISTILLATION, WITH RECEIVERS G. H, I. ON THE LEFT IS A RECEIVER COOLED IN A TUB OF WATER, AND BELOW IS SHOWN THE METHOD OF CONNECTING TWO RECEIVERS TOGETHER.

FIG. 31.—BEEHIVE OVEN FOR RECOVERY OF WOOD TAR (GLAUBER, 1657).

NOTE THE AIR CONDENSER AND TAR COLLECTOR.

it was usually called "Glauber's salt." He also described the distillation of wood in closed ovens with the production of acid and spirit. Glauber had a fairly clear idea that salts consist of acid and base. He mentions the curious precipitates ("chemical garden") formed by metal salts with soluble glass (sodium silicate) and recognized that silica is contained in them. He explains clearly some cases of double decomposition, e.g. in the preparation of "butter of antimony" ($SbCl_3$) from stibnite (Sb_2S_3) and corrosive sublimate ($HgCl_2$), when the antimony and mercury exchange acid radicals. He prepared "butter of arsenic" ($AsCl_3$) similarly, by using white arsenic or orpiment in place of stibnite. He had fairly correct ideas on affinity, explaining, for example, that when sal ammoniac is heated with zinc oxide, the latter combines with the acid because of its greater affinity and lets the ammonia go free.

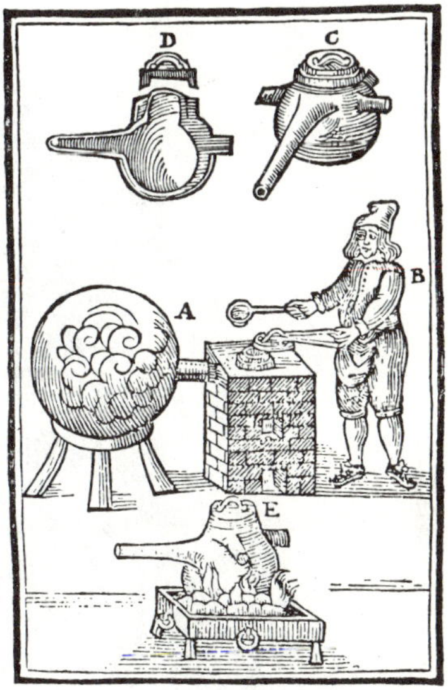

FIG. 32.—PREPARATION OF OIL OF VITRIOL BY GLAUBER BY DISTILLING FERROUS SULPHATE.

Lemery

Nicolas Lemery (1645-1715) was a French chemist who, in common with Le Févre and Willis, admitted five principles: three active (mercury or spirit, sulphur or oil, and salt) and two passive (water or phlegm, and earth). His text-book, *Cours de Chymie*, first published in 1675, went through a large number of editions and was translated into Latin and most modern languages. He classifies substances into three groups according to the three "Kingdoms of Nature", viz. mineral, vegetable and animal. The book is very practical, although Lemery makes use in it of a form of Descartes' corpuscular theory which supposed that the properties of substances depend mainly on the shapes of their particles. Acids have sharp spiky particles, which prick the tongue, and their salts form sharp crystals. In precipitation reactions, the spikes of the acid

FIG. 33.—N. LEMERY, 1645-1715.

particles break off in the pores of metal corpuscles, e.g. of silver, and are carried down in the precipitate. Metals dissolve in acids because the points of the acid tear apart the particles in the mass of metal.

Tachenius

Otto Tachenius, a German who lived for some time in Venice, in his book *Hippocrates Chimicus* (1666) gave a clear definition of a salt: "all salts are composed of two parts, of acid and alkali" (*omnia salsa in duas dividuntur partes in alcali nimirum et acidum*). He recognized that silica is an acid, since it combines with alkalis, that acids differ in strength, a stronger acid displacing a weaker from its salts, and that soap is a salt of an oily acid. He remarks that lead increases by one-tenth of its weight when calcined to red lead and regains its original weight on reduction. He thought the increase was due to absorption of an acid from the smoke and flame of the wood fire. Tachenius gives several wet reactions and devised a rudimentary system of qualitative analysis.

Kunckel

Johann Kunckel (1630-1703) was a laborious and skilled practical chemist who was for a time an alchemist in the service of the

FIG. 34.—J. KUNCKEL, 1630-1703.

Elector John George of Saxony in Dresden, and of Frederick William of Brandenburg (the " Grosse Kürfürst ") in Berlin. He was a very honest man and firmly believed in the possibility of transmutation, but he never says he achieved it. He published several works during his life, but his most famous book is his *Laboratorium Chymicum*, published in 1716 after his death. Kunckel denied the presence of sulphur in metals, but thought these contained mercury. He remarks on the increase in weight of antimony on calcination, which he at first put down to the fixation of igneous corpuscles (1677), but afterwards to the matter becoming denser and expelling the air between its pores (1716). He describes the preparation of gold ruby glass and wrote a treatise on glass manufacture (*Ars Vitraria*, in German, 1679), based on an earlier work by an Italian, Neri. Kunckel was an independent discoverer of phosphorus, which was first obtained by Brand of Hamburg about 1674. Kunckel in 1675 tried to buy the secret from Brand, but the latter sold it to Kunckel's friend Krafft. Kunckel then found the process (distilling evaporated urine) for himself early in 1676. Krafft in 1677 exhibited the specimen of phosphorus he obtained from Brand to Boyle, and gave him a hint of the method of preparation, so that Boyle was able in 1680 to publish the method for the first time.*

We have now traced the development of Chemistry from the earliest period to the time where it began to take shape as a separate science. "The purpose of chemistry", it has been said, "seems to have changed much from time to time. At one time chemistry might have been called a theory of life, and at another time a department of metallurgy; at one time a study of combustion, and at another time an aid to medicine; at one time an attempt to define a single word, the word *element*, and at another time the quest for the unchanging basis of all phenomena. Chemistry has appeared to be sometimes a handicraft, sometimes a philosophy, sometimes a mystery, and sometimes a science."†

* Partington, *Science Progress*, 1936, xxx, 402.

† M. M. Pattison Muir, *A History of Chemical Theories and Laws*, New York, 1907, p. vii.

Summary of the Early History of Chemistry

140 B.C. Alchemy said to have begun in China.

A.D. 0-50. The first treatises on the "divine art" appeared in Alexandria, in Egypt, containing the earliest chemistry.

300. Ko Hung, the most celebrated Chinese alchemist.

300. Papyri of Leyden and Stockholm, summarizing earlier technical information on metallurgy, dyeing, imitation of precious stones, etc

300. Zosimos, in Egypt, describes many chemical operations: solution, filtration, fusion, sublimation, distillation, etc., and several chemical substances and reactions. Belief in transmutation of metals arose

640. Egypt conquered by the Arabs, who later caused translations to be made of the Egyptian books on chemistry (which were written in Greek). The subject was especially studied by Jabir ibn Hayyan (720-813), Rhases (866-925) and Avicenna (980-1036). The idea that metals were composed of mercury and sulphur was introduced Arabic chemistry is largely a continuation of that of Egypt.

800-900. Hindu chemistry resembled that of the Arabs.

1144. Alchemy appeared in Europe by way of translations made in Spain from Arabic works. Roger Bacon and Albertus Magnus wrote on it about 1250.

1493-1541. Paracelsus, the founder of Iatrochemistry, or chemistry applied to the service of medicine. Three principles: salt, sulphur and mercury.

1494-1555. Agricola: wrote *De Re Metallica* (1556), on mining and metallurgy.

1579-1644. Van Helmont: invented the name *gas* and described carbon dioxide as *gas sylvestre*. Criticized the older theories of the elements. Regarded water as the fundamental element ("tree experiment"). Quantitative experiments.

1540-1616. Libavius: first text-book of Chemistry, *Alchemia*, (1597). Qualitative analysis. Discovered stannic chloride.

1600? "Basil Valentine"; *Triumphal Chariot of Antimony*, probably really by Thölde; antimony compounds; sulphuric acid.

1604-1670. Glauber: mineral acids, salts (especially *sal mirabile*, "Glauber's salt"). Anhydrous chlorides of arsenic and metals Ideas on affinity. *New Philosophical Furnaces* (1646-49).

1614-1672. Sylvius: "effervescence" of acid and alkali in the body.

1645-1715. Lemery: text-book, *Cours de Chymie* (1675): corpuscular theory, corpuscles with spikes and pores. Division of substances into mineral, vegetable and animal.

1650. Tachenius: every salt composed of acid and alkali: *Hippocrates Chimicus* (1666).

1630-1703. Kunckel: gold ruby glass; phosphorus; *Laboratorium Chymicum* (1716).

CHAPTER V

EARLY STUDIES ON COMBUSTION AND THE NATURE OF THE ATMOSPHERE

Combustion and the Calcination of Metals

THERE are two kinds of chemical change which, since they were investigated side by side, and depend on the same cause, may conveniently be described together. These are combustion, and the calcination of metals.

The alchemists attached great importance to the effects of heat on substances, and their writings describe many types of furnaces and experiments made with them (Fig. 35). The metals, except gold and silver, were found to change when heated in open crucibles, and to leave a dross, which was called a calx (Latin *calx*, lime). It was noticed in the sixteenth century that this calx is heavier than the metal: the explanations usually given were that some kind of "soul" escaped from the metal, or that the matter became denser, or that some kind of acid was absorbed from the fire, or that fire possessed weight and was absorbed by the metal in forming the calx.*

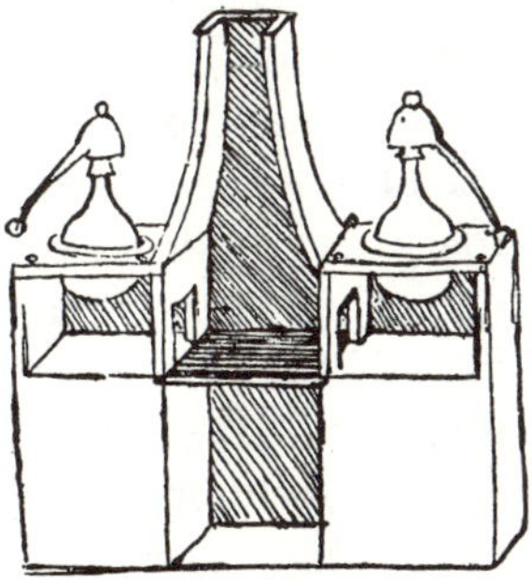

FIG. 35.—DRAWING OF A CHEMIST'S FURNACE, WITH TWO ALEMBICS (cf. FIG. 19), MADE BY THE FAMOUS ARTIST, LEONARDO DA VINCI (1452-1519). From the *Codice Atlantico*, Milan.

The further study of combustion and calcination was taken up by Boyle, Hooke and Mayow, who have been called the "Oxford Chemists", since they all worked at some time in their lives at Oxford, and Hooke and Mayow were graduates of the University.

* Cf. J. C. Gregory, *Combustion from Heracleitos to Lavoisier*, London, 1934; Kopp, *Geschichte der Chemie*, vol. iii, p. 119 f.

Boyle

Robert Boyle was born in Lismore Castle, Ireland, in 1627 and died in 1691 in London. His father Richard Boyle was first Earl of Cork. Robert was educated at Eton and travelled on the Continent, returning home in 1644, when, after the death of his father and the loss of his fortune, he retired to the family house at Stalbridge in Dorsetshire, where he lived very simply. In 1654 he moved to Oxford and, in conjunction with his assistant Robert Hooke, he worked for a time in lodgings next to University College in the High Street, on experiments with the air-pump, on combustion, etc. In 1668 he moved to London, where he lived with his sister, Lady Ranelagh, and set up a laboratory at the back of the house in Pall Mall. Boyle was one of the original Fellows of the Royal Society, founded in 1644-45, which received its charter from Charles II in 1662. He was chosen president in 1680, but declined to take office and Wren was appointed.

Boyle had poor health and was fond of dosing himself and his friends with recipes from very miscellaneous sources: it is related that " he had divers sorts of cloaks to put on when he went abroad, according to the temperature of the air, and in this he governed himself by his thermometer ".* He avoided honours and affairs, preferring to spend his life in a quiet and dignified study of science, including its applications to metallurgy, medicine and the manufacture of chemicals, dyes and glass.†

Boyle's works are very voluminous; they were collected and published in 1744 by Birch (five vols. folio; or six vols. quarto, 1772): a handier but abridged edition is that edited by Shaw (three vols. quarto, 1725). Many of the separate works appeared simultaneously in English and Latin and were well known on the

* Birch, Life of Boyle, *Works*, 1744, i, 86; Picton, *The Story of Chemistry*, 1889, 127.

† Autobiography of Boyle, *Works*, ed. by Birch, 1744, vol. i; Flora Masson, *Robert Boyle*, 1914; R. Gunther, *Early Science in Oxford*, Oxford, 1923 f., vol. i; Thorpe, *Essays*, 1902, 1 f.; Ramsay, *Gases of the Atmosphere*, 1915; Tilden, *Famous Chemists*, 1921, 1 f.; Agnes M. Clerke, *Dict. Nat. Biogr.*, 1908, ii, 1026; L. T. More, *The Life and Works of the Hon. Robert Boyle*, 1944.

Continent. He describes all his experiments clearly, in contrast to many of his contemporaries, such as Glauber.

Boyle has been called the founder of modern chemistry for three reasons: (1) he realized that chemistry is worthy of study for its own sake and not merely as an aid to medicine or as alchemy—

FIG. 36.—ROBERT BOYLE, 1627-1691.

although he believed in the possibility of the latter; (2) he introduced a rigorous experimental method into chemistry; and (3) he gave a clear definition of an element and showed by experiment that the four elements of Aristotle and the three principles of the alchemists (mercury, sulphur and salt) did not deserve to be called elements or principles at all, since none of them could be extracted from bodies, e.g. metals. In some ways he was anticipated by

FIG. 37.

OTTO VON GUERICKE'S AIR PUMP, SHOWN ON THE LEFT BELOW. ON THE RIGHT BELOW IS THE GLOBE FOR WEIGHING AIR. ON THE RIGHT AT THE TOP ARE THE FAMOUS "MAGDEBURG HEMISPHERES".

(Engraved title of von Guericke's book, Amsterdam, 1672.)

van Helmont, whose works he studied with care and to whom he frequently refers as an authority.*

Boyle's views on the elements are mostly set out in his book, *The Sceptical Chymist: or Chymico-Physical Doubts & Paradoxes, touching the Spagyrist's Principles commonly call'd Hypostatical, as they are wont to be Propos'd and Defended by the Generality of Alchymists*, London, 1661, which appeared in a second (anonymous) edition in 1680. It has been reprinted in the *Everyman Series*.

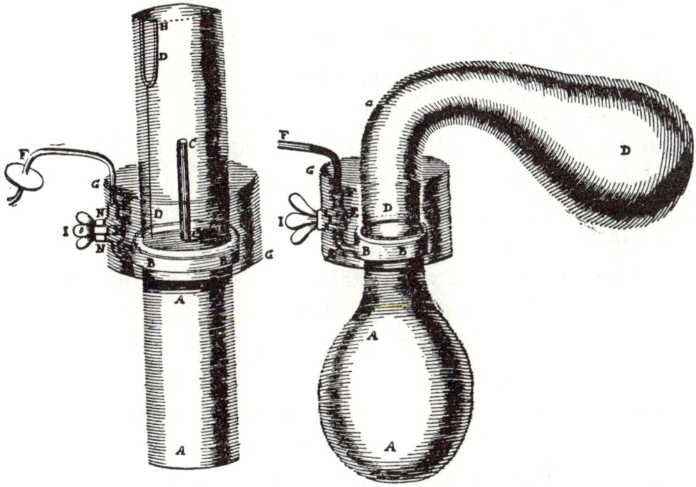

FIG. 38.—BOYLE'S APPARATUS FOR DISTILLATION UNDER REDUCED PRESSURE.

Boyle was a good experimenter and improved much of the apparatus in use in his time. He made many experiments on the effects of reduced pressure by means of the air pump, which was adapted from the one invented by Otto von Guericke, the Burgomaster of Magdeburg, in 1654 (Fig. 37). Among other things Boyle describes distillation under reduced pressure and the apparatus for carrying out the process (Fig. 38). The *Sceptical Chymist* does not give an adequate impression of his experimental skill, for which his other works must be consulted.

* Some of the above ideas are also clearly stated by Joachim Junge, of Lübeck (1587-1657), in his *Doxoscopiae Physicae Minores*, written about 1630 and published in 1662.

The *Sceptical Chymist* is written in a good, though rather prolix, style, enlivened with touches of humour, as when the alchemists are compared with " the Navigators of Solomon's Tarshish Fleet, who brought home . . . not only Gold, and Silver, and Ivory, but Apes and Peacocks too ", since their theories " either like Peacock's feathers make a great shew, but are neither solid nor useful ; or else, like Apes, if they have some appearance of being rational, are blemish'd with some absurdity or other " which makes them appear ridiculous.*

Boyle on the Elements

In the *Sceptical Chymist*, Boyle says that salt, sulphur and mercury cannot be extracted from gold. " I can easily enough sublime Gold into the form of red Chrystalls of a considerable length ; and many other ways may Gold be disguis'd, and help to constitute Bodies of very different Natures both from It and from one another, and nevertheless be afterwards reduc'd to the self-same Numerical, Yellow, Fixt, Ponderous and Malleable Gold it was before its commixture." The same is true of mercury. Hence " the Corpuscle of Gold and Mercury, though they may not be primary Concretions of the most minute Particles of matter, but confessedly mixt Bodies, are able to concurre plentifully to the composition of several very differing Bodies, without losing their own Nature or Texture, or having their cohesion violated by the divorce of their associated parts or ingredients." † In copper nitrate also, the copper is only disguised : the nitric acid may be separated by heat and the copper by precipitation.‡ It must be remembered that this idea had already been expressed by van Helmont (p. 46).

Boyle's definition of an element is as follows :

" I mean by Elements, as those Chymists that speak plainest do by their Principles, certain Primitive and Simple, or perfectly unmingled bodies ; which not being made of any other bodies, or of one another, are the Ingredients of which all those call'd perfectly mixt Bodies are immediately compounded, and into which they are ultimately resolved." § By " perfectly mixt bodies " he means chemical compounds as distinguished from mechanical

* *Sceptical Chymist*, 1661, p. 429.
† *ibid.*, p. 40.
‡ *ibid.*, p. 154.
§ *ibid.*, p. 350.

mixtures. Also, "Elements and Principles" mean "those primitive and simple Bodies of which the mixt ones are said to be composed, and into which they are ultimately resolved".* He argues at some length that the action of fire upon bodies, previously used as a method of resolution into elements, is unsatisfactory for this purpose.

Boyle was not so clear and dogmatic in his discussion of the chemical elements as could have been wished for his time. He still seems to regard the different elements as being made up of some primary matter, and the varying properties of the elements might be due to the different shapes and motions of the particles of the primary matter:

"The greatest part of the affections of matter, and consequently of the Phænomena of nature, seems to depend upon the motion and the contrivance of the small parts of Bodies",† so that "there is no great need that Nature should alwaies have Elements before hand, whereof to make such Bodies as we call mixts",‡ and "the difference of Bodies may depend meerly upon that of the schemes whereinto their Common matter is put . . . so that according as the small parts of matter recede from each other, or work upon each other . . . a Body of this or that denomination is produced." § If Boyle had made up a table of bodies which he regarded as satisfying his definition of an element, such as gold, copper, and sulphur, some chemists would no doubt have followed him, but as it was he left the matter in too indefinite a form and nowhere says what he considered to be elements.

It will be noted that Boyle believed in the atomic theory. He also considered chemical combination to occur between the elementary particles and had some good ideas on chemical affinity. He says:

"There are Clusters wherein the Particles stick not so close together, but that they may meet with Corpuscles of another Denomination, which are dispos'd to be more closely United with some of them, than they were among themselves." ||

* *Sceptical Chymist*, 1661, p. 16. † *ibid.*, p. 333.

‡ *ibid.*, p. 411. § *ibid.*, p. 422.

|| *ibid.*, p. 153; cf. *Origin of Forms and Qualities*, Birch, 1744, ii, 471.

Boyle repeated van Helmont's experiment proving that water can be converted into earth (p. 51), using instead of a willow tree a vegetable marrow, which took less time to grow, but he concluded that the substance of the marrow came mostly from the dust particles floating in the air.* Whereas van Helmont thought a gas could not be collected in a vessel (p. 48), Boyle showed that this was possible by collecting the gas (hydrogen) from iron nails in a bottle filled up with dilute oil of vitriol and inverted in a dish of the same liquid. He says : " The cavity of the glass was possessed by the air, since by its spring it was able to hinder the liquor from regaining its former place."† He also used dilute nitric acid and so obtained nitric oxide (see p. 83).

Boyle's Law

Boyle's interest in the effect of pressure on air led him in 1662 to discover the law named after him, that the volume of a gas is inversely proportional to the pressure. He proved this experimentally both for pressures greater than atmospheric and pressures less than atmospheric, in the first case using the well known U-tube arrangement with mercury, and in the second case a straight glass tube containing some air confined over mercury, which could be raised in a vessel of mercury.‡ This is sometimes called Mariotte's law, but Mariotte does not state it until 1679 and does not claim it as new. To explain the compressibility of air, Boyle compared its particles with small coiled springs, fleeces of wool, or little sponges, the lower " springs " in a column of air being compressed by the weight of those above them. (Mariotte uses the same analogies.) He also remarks that the pressure could be explained as due to the motion of the particles, as assumed by Descartes : " The restless agitation of that celestial matter [the ether] wherein the particles swim, so whirls them round each other that each corpuscle endeavours to beat off all others from coming within the little sphere requisite to its motion about its

* *Sceptical Chymist*, 1661, p. 107.

† *New Experiments Physico-Mechanical touching the Spring of the Air*, 1660; in his *Continuation* of these experiments, 1682, he calls the gas a "factitious air."

‡ *A Defence of the Doctrine touching the Spring and Weight of the Air*. 1662.

own centre "—a good statement of the famous Theory of Vortices of Descartes.

Boyle concluded * that air was composed of at least three parts:

(i) Vapours from water and living animals.

(ii) A very subtle emanation from the earth's magnetism, producing the sensation of light.

(iii) A fluid, compressible and dilatable, having weight and able to refract light.

Boyle's Experiments on Combustion

Boyle's experiments on combustion were in the first place carried out in conjunction with Hooke, who was his assistant at Oxford, and in 1659 had devised a modification of the air-pump invented in 1654 by Otto von Guericke (Fig. 37).

In his *New Experiments touching the Relation betwixt Flame and Air* (1673),† Boyle described experiments on combustion in an exhausted receiver. The method of experimenting was to put a red-hot iron plate under a bell jar, the air in which could be exhausted by the air pump, and then to drop the combustible material on the hot plate by means of a contrivance fitted in the neck of the bell jar. With sulphur wrapped in paper and lowered on to the plate, fumes were seen but no combustion. On the admission of air, however, Boyle saw " divers little flashes, as it were, which disclosed themselves by their blue colour to be sulphurous flames." He describes the extinction of a candle flame and the flame of burning hydrogen, and was inclined to think that a flame could not exist without air. The experiment with hydrogen is interesting:

" A very volatile and saline piercing liquor [hydrochloric acid] being dropped upon filings of steel, the mixture grew hot, and emitted out of the vial it was contained in, very fetid steams, which would kindle at the flame of a candle, and continue to burn a good while; where we convey'd it into our receiver, and upon the first exsuction of air it flam'd brisker than before; and likewise upon the second and third; but after it went out, it would not be kindled again, tho' the air was let in upon it." ‡

* *General History of the Air*, 1692. † *Works*, ed. Birch, iii, 250.

‡ Text as abridged by Boulton, *The Works of the Honourable Robert Boyle, Esq., Epitomiz'd*, 4 vols. 8°, London, 1699-1700, vol. iii, p. 224.

The combustion of hydrogen evolved from iron and dilute sulphuric acid was later described by Lemery (1700): the statement that Paracelsus knew of the inflammability of hydrogen is incorrect, but it is mentioned in a work of Turquet de Mayerne, who died at Chelsea in 1655.*

Boyle, however, obtained very curious results with gunpowder. When dropped upon the heated plate in vacuum the powder burned slowly but did not explode: "We saw a pretty broad blue flame like brimstone, which lasted so long as we could not but wonder at it." A ring of gunpowder burned in a vacuum only at the place where it was heated by a burning glass, but on admitting air the whole went off with a flash ("Expt. V: About an endeavour to fire gunpowder in vacuo with the sun beams"). Gunpowder in a goose quill would also burn under water. Fulminating gold would also explode with a flash when dropped on the hot plate in a vacuum. Boyle at first thought that some air might be mixed with the nitre crystals used to make the gunpowder, but he obtained the same result with powder made from nitre crystallized in a vacuum. He came to the conclusion that substances could burn mixed with nitre even in absence of air, and that nitre on heating gives "agitated vapours which emulate air". The fulminating gold was made with nitric acid.

Boyle remarked (1681-2) that the glow gradually disappears when phosphorus is shut up in a glass tube, and hence he concluded that "the air had some vital substance preyed upon or else tamed by the fumes of the phosphorus." †

Boyle's Experiments on Calcination

In 1673 Boyle published his *New Experiments to make Fire and Flame Stable and Ponderable* ‡ in which he describes the increase in weight of metals on calcination in air. This fact was known before his time, and is mentioned by Cardan and others in the sixteenth century (see p. 65). Boyle found that 8 oz. of block tin when heated

* *Opera medica*, ed. by Browne, London, 1700-3, ii, 5,150.

† *New Experiments, and Observations, made upon the Icy Noctiluca*, 1681/2.

‡ *Works*, ed. Birch, iii, 340. This and the *Additional Experiments about Arresting and Weighing Igneous Corpuscles*, are parts of Boyle's *Essays of Effluviums*. See McKie, *Science Progress*, 1934, xxix, 253.

in an open flask increased in weight by 18 grains. He tried putting the tin in a retort, weighing, then sealing the neck and heating, but owing to the expansion of the air the retort burst "with a noise like the report of a gun." He then heated 2 oz. of tin in an open retort and sealed the neck when as much air as possible had been driven out by heat. After heating so as to calcine the tin, the retort was cooled and the neck cut off, when Boyle "heard the outward air rush in, because when the vessel was sealed the air within it was highly rarefied." He thus missed the absorption of air by his method of working, and concluded that the increase in weight, 12 grains, was "gained by the operation of fire on the metal." Boyle thought the igneous corpuscles from the fire had passed through the glass and been absorbed by the metal, and hence concluded that fire had weight. He made the important observation that the density of the calx is less than that of the metal.

The theory adopted by Boyle to explain the increase in weight of metals on calcination, viz. the fixation of ponderable particles of fire (*Sceptical Chymist*, 1661, p. 212), was also adopted by Becher,* by Lemery and others, but was criticized in 1679 by Chérubin d'Orléans (François Lasseré), who objected that Boyle should have weighed his retort before opening it, and argued that ponderable matter could not pass through glass. Boyle refers to Chérubin in a letter in which he makes the surprising statement that he had once experimented by weighing before opening the retort and also obtained an increase in weight.† The untenability of Boyle's theory was proved by Kunckel, who showed that a mass of iron had the same weight when red hot as when cold,‡ and similar experiments were made by Boerhaave.§

In his *Mechanical Origin or Production of Fixedness* (1675) Boyle says red precipitate *per se* is formed when mercury is calcined in air, yet "with a greater and competent degree of heat . . . would, without the help of any volatilizing additament, be easily reduced into running mercury again. Chemists and physicians who agree in

* *Physicæ subterraneæ*, Frankfurt, 1669, p. 195.

† *Works*, ed. Birch, v, 233; D. McKie, *Science Progress*, 1934, xxix, 253; 1936, xxxi, 55; *Annals of Science*, 1936, i, 269; *Ambix*, 1938, i, 143.

‡ *Laboratorium Chymicum*, 1716, 31.

§ *Elementa Chemiae*, Leyden, 1732, i, 259, 362.

supposing this precipitate to be made without any additament, will, perchance, scarce be able to give a more likely account of the consistency and degree of fixity, that is obtained in the mercury ; in which since no body is added to it, there appears not to be wrought any but a mechanical change . . . though, I confess, I have not been without suspicions, that in philosophical strictness this precipitate may not be made *per se*, but that some penetrating igneous particles, especially saline, may have associated themselves with the mercurial corpuscles ".*

Boyle's Miscellaneous Experiments

In 1664 (*Experiments . . . Touching Colours*) and in 1675 (*Reflections upon the Hypothesis of Alcali and Acidum*) Boyle gives some general properties of acids and alkalis, such as the sour taste of acids, their action as solvents, their precipitation of sulphur from liver of sulphur, their action on vegetable colours (indicators) such as tournsol, juice of violets and decoctions of cochineal and Brazil wood, the colours being restored by alkalis, and their reaction with alkalis with disappearance of the characteristic properties of each and the formation of a neutral salt.

Boyle describes many tests and was one of the founders of qualitative analysis. He mentions the green colour imparted to a flame by copper salts, the white fumes produced with ammonia and nitric and hydrochloric acids (already mentioned by Kunckel), tne black colour produced with iron salts by tincture of galls (well known to Pliny), the white precipitate formed by calcium salts with sulphuric acid, and the precipitation of silver salts by chlorides. Most of these were known before.

Boyle obtained anhydrous cuprous chloride (" resin of copper ") and bismuth chloride by heating copper and bismuth (" tin glass "), respectively, with corrosive sublimate, and mentions the precipitation of a solution of bismuth in nitric acid by water. He showed that the volatile alkali (ammonium carbonate) on distillation with lime becomes more volatile and loses the property of effervescing with acids : † " alcalizate spirit of urine drawn from some kinds

* *Works*, ed. Birch, iii, 620.

† *Memoirs for the Natural History of Human Blood*, 1684 ; *Of the Reconcileableness of Specific Medicines to the Corpuscular Philosophy*, 1685 ; *Reflections upon the Hypothesis of Alcali and Acidum*, 1675 ; Birch, iii, 605.

of quicklime, being mixed with oil of vitriol moderately strong, would produce an intense heat, whilst it produced either no manifest bubbles at all, or scarce any, though the urinous spirit was strong, and in other trials operated like an alcali." He thought that the alkali of plant ashes was not present in the plant but produced by combustion. He showed that oil of vitriol gives sulphur when distilled with turpentine,* but does not decide if sulphur is a constituent of the acid.

Boyle made many determinations of the specific gravities of liquids and solids, and used freezing mixtures (e.g. nitric acid and snow) in 1665.† He describes the distillation of sugar of lead, and thought the liquid (containing acetone) came from the vinegar which left one of its parts combined with the lead, although others had called it the "sulphur" of the lead,‡ and he also mentions the precipitation of dyes by alum and potash, and sugar of lead. §

In his work on phosphorus, || Boyle established the following important results: (1) the phosphorus glows only in presence of air; (2) a very small quantity of phosphorus (1 in 500,000 parts of water) can be detected by the glow; (3) an acid is produced which differs from phosphoric acid in giving little flashes of light on heating (phosphine from phosphorous acid); (4) the glow is exhibited by solutions of phosphorus in olive and some other oils, but oils of mace and aniseed prevent it; (5) after long exposure to phosphorus, the air acquires a strong odour (ozone) distinct from the visible fumes.

Boyle's views on the elements, as expressed in the *Sceptical Chymist*, had a great influence on chemical thought: they are not often referred to specifically by chemists who followed him immediately, but their indirect implications are clear, although the followers of Becher and Stahl, the Phlogistic Chemists, still retain the four elements and ideas of the older times.

* *Sceptical Chymist*, 1661, p. 218.

† *New Experiments touching Cold*, 1665.

‡ *Sceptical Chymist*, 1661 pp. 155, 231.

§ *Experiments Touching Colours*, 1664.

|| *New Experiments and Observations made upon the Icy Noctiluca*, 1681/2.

Hooke (1635-1703)

Although Boyle had some idea of the part played by the air in combustion, his statements are not clear and the credit for putting forward the first rational theory of combustion must be given to his assistant, Robert Hooke. Born in the Isle of Wight, and originally intended for the Church, his constitution proved too weakly and he took up scientific investigation. He studied at Christ Church, Oxford, and was in succession assistant to Willis and Boyle, curator of experiments to the Royal Society (1662) and, after the Great Fire, a surveyor of London, in which capacity he amassed a considerable sum of money which was found locked up in an iron chest after his death. He was reputed to be miserly and rather cynical: * it is said he never went to bed or undressed for two years before his death. He claimed to have anticipated Newton in the discovery of the law of gravitation and is known for his discovery, in 1660, of Hooke's law, which he first stated in 1676 in the form of an anagram : ceiiinosssttuu = ut tensio sic vis, which he published in an intelligible form in 1678.

In 1665 (again in 1667) Hooke published his *Micrographia: or some Physiological Descriptions of Minute Bodies made by Magnifying Glasses. With Observations and Inquiries thereupon*, dedicated to Charles II, with an interesting preface and a large number of plates of things seen under the microscope. It also contains some observations on charcoal and on the sparks from flint and steel, and a theory of combustion.†

Hooke's work on combustion apparently had its origin in a statement in Boyle's *Sceptical Chymist*, ‡ that if charcoal is strongly heated in a closed retort it does not disappear, but the *caput mortuum* (residue) becomes black on cooling. If, however, air is admitted, the charcoal burns away and crumbles down to a white ash. Boyle, therefore, distinguished between the effects of heating substances in closed vessels and with exposure to air.

Hooke gives no details of any experiments in the *Micrographia*, but on the basis of his unpublished observations he puts forward

* But cf. Andrade, *Nature*, 1935, cxxxvi, 358, 603.

† *Alembic Club Reprint* No. 5; Lysaght, *Ambix*, 1937, i, 93.

‡ *Everyman* edit., p. 43; see van Helmont, p. 48.

in twelve propositions a theory of combustion in which he says, among other things: (1) " Air is the universal dissolvent of all sulphureous bodies; (3) this action of dissolution produces a very great heat and that which we call fire; (5) this dissolution is made by a substance inherent, and mixt with the air, that is like, if not the very same, with that which is fixt in saltpeter; (6) in this dissolution of bodies by the air, a certain part is united and mixt, or dissolv'd and turn'd into the air, and made to fly up and down with it." Hooke did not succeed in isolating this constituent common to air and nitre, which in 1682 he called " Nitrous Air ". Boyle also speaks vaguely of a " volatile nitre " in the air.*

In some experiments made in 1678–9, by Hooke,† a piece of charcoal weighing 128 grains lost only $1\frac{1}{2}$ grains when kept red-hot for two hours in an iron box filled with sand; a box containing charcoal was arranged so that the air in it could be circulated by bellows, when after a while the fire went out, the air being " satiated ". Hooke also observed the brilliant combustion of pieces of charcoal and sulphur dropped on the surface of fused nitre.

In another work, *Lampas* (1677), Hooke gives an extension of a description of the structure of a candle-flame which he had referred to in the *Micrographia*. He says " that transient shining body which we call flame " is " nothing but the part of the oyl rarified and raised by heat into the form of a vapour or smoak ", acted upon by air, which " by its dissolving property preyeth upon those parts of it that are outwards . . . producing the light which we observe; but those parts which rise from the wick which are in the middle are not turned to shining flame till they rise to the top of the cone, where the free air can reach and so dissolve them ". This can be seen by holding a thin piece of glass or mica across the flame, when it is observed that " all the middle of the cone of flame neither shines nor burns, but only the outer superficies thereof that is contiguous to the free and unsatiated air ".

Hooke thus supposed that air, in virtue of the presence in it of something which also exists in saltpetre, *dissolves* combustible

* *Suspicions about some hidden Qualities of Air*, 1674.
† Birch *History of the Royal Society*, London, 1756-7, iii, 460 f., 465, 469; cf. McKie, *Discovery*, 1935, 200; Lysaght, *Ambix*, 1937, i, 93.

bodies *as such*, and that the heat evolved is a heat of solution, analogous to that given out when some substances, e.g. potassium carbonate and sulphuric acid, dissolve in water. An advance on Hooke's experiments was made by Mayow.

Mayow (1641-1679)

John Mayow was born in Morval, near Looe, in Cornwall, entered Wadham College, Oxford, in 1658, was elected a Fellow of All Souls

FIG. 39.—JOHN MAYOW, 1641–1679.

College in 1660, and took the degree of D.C.L. in 1670 at Oxford, but became a medical practitioner in Bath and London.* Most of his chemical work seems to have been done in Oxford. In 1673 he completed his *Tractatus quinque Medico-Physici*, which was pub-

* Hartog, *Dict. Nat. Biogr.*, 1894, xxxvii, 175; *ibid.*, 1909, xiii, 175; Gunther, *Early Science in Oxford*, 1923, vol. i; T. S. Patterson, *Isis*, 1931, xv, 47; McKie, *Nature*, 1941, cxlviii, 728; *Phil. Mag.*, 1942, xxxiii, 51; Partington, *Isis*, 1956.

lished in Oxford in 1674.* In this he puts forward theories of combustion and respiration supported by ingenious and novel experiments.† Mayow concluded that air contains two kinds of particles, one of which he called igneo-aerial or nitro-aerial particles, or nitro-aerial spirit (*spiritus nitro-aereus*), which are taken out of the air in combustion and respiration, and another kind of particles which remain in a diminished volume after these processes.

All combustible bodies contain " sulphureous particles " and the heat evolved in combustion is due to the violent collision of sulphureous and nitro-aerial particles (a theory different from Hooke's, p. 80). Animal heat is due to the interaction of sulphureous particles in the blood with nitro-aerial particles in the inspired air. Willis (*Exercitatio de Sanguis Accensione*, 1670) had a similar theory, but assumed the existence of particles of nitre in the air.

Gunpowder will burn in a vacuum or under water (see p. 74) and the nitro-aerial particles are then supplied by the nitre; they are contained in the acid (nitric acid) part of the nitre. Sulphur does not contain an acid; the sulphuric acid formed on its combustion, or by the oxidation of pyrites in air, is formed from sulphur and nitro-aerial particles.

Mayow remarks that when powdered metallic antimony is calcined on a marble slab by means of a burning glass, the residual calx is heavier than the original metal, in spite of the loss of abundant fumes—an experiment reported by Hamerus Poppius (1618) and Le Févre (1660) and explained as due to the fixation of light. Mayow says the nitro-aerial particles of the atmosphere are fixed in the metal, and the product is identical with that obtained by the action of nitric acid on antimony and heating the solid formed.

Mayow proved that only part of the air is concerned in combustion and respiration, which he clearly recognized as essentially similar processes, by some ingenious experiments. He burnt a candle in a glass globe inverted over water (Fig. 40), equalizing the levels of the water inside and outside the globe by means of a glass siphon, which was then quickly withdrawn. The water rose inside the globe, showing that some air had disappeared. When the

* Extracts transl. by F. G. Donnan in Ostwald's *Klassiker*, No. 125; translated in full in *Alembic Club Reprint*, No. 17.

† H. B. Dixon, *B. A. Report*, 1894, 594.

candle was extinguished, a large bulk of air was left, but this would not support the combustion of sulphur or camphor lying on a small shelf inside the globe, when they were heated by a burning-glass. To ensure that light from the burning-glass could reach the camphor, Mayow kept part of the glass free from soot deposited in the first combustion by a paper patch which was pulled off by a thread attached to it and passing outside the flask.

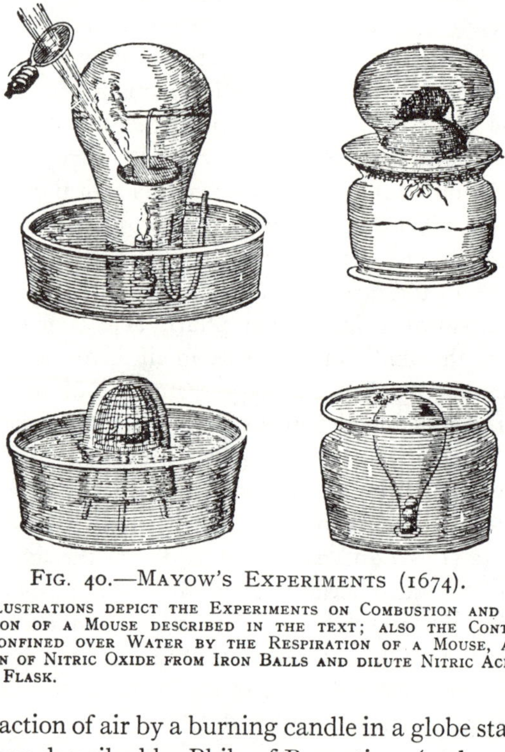

FIG. 40.—MAYOW'S EXPERIMENTS (1674).

THE ILLUSTRATIONS DEPICT THE EXPERIMENTS ON COMBUSTION AND ON THE RESPIRATION OF A MOUSE DESCRIBED IN THE TEXT; ALSO THE CONTRACTION OF AIR CONFINED OVER WATER BY THE RESPIRATION OF A MOUSE, AND THE COLLECTION OF NITRIC OXIDE FROM IRON BALLS AND DILUTE NITRIC ACID IN AN INVERTED FLASK.

The contraction of air by a burning candle in a globe standing over water had been described by Philo of Byzantium (2nd century B.C.), Robert Fludd (1617), and van Helmont (p. 48), but they did not equalize the water levels and some air may have bubbled out owing to heating. Scheele (p. 105) later pressed the mouth of the flask into a cake of soft wax, but Mayow's siphon was neater.

Mayow collected nitric oxide by the action of dilute nitric acid on steel balls in an inverted flask (Fig. 40) (see p. 72), and showed by experiment that this gas obeys Boyle's law (p. 72). By generating nitric oxide in a flask of air over water he found that a quarter of the air is removed by condensing with the gas and the solution of the product in the water. (This represents the oxygen, which combines with the nitric oxide. Van Helmont had found that the gas from nitric acid and metals becomes red on contact with air.)

If a mouse was kept in a vessel of air closed by a bladder (Fig. 40), the contraction of the air was perceptible by the bulging inwards of the membrane owing to the pressure outside. A mouse lived only half as long when kept in a vessel together with a burning candle as a mouse in the vessel without the candle. A candle caused a contraction of one thirtieth of the air by its combustion, but a mouse caused a contraction of one fourteenth, and a mouse could live for a time in air in which a candle had burnt to extinction.

Mayow found that fresh arterial blood effervesced in a vacuum and evolved a gas, but would not do this after standing. He stated that the blood was heated by respiration, not cooled as had previously been supposed, and that the heating occurs in the muscles.* Lavoisier † incorrectly thought the heat was evolved in the lungs. Although Lower ‡ found that blood acquires a florid colour in the lungs by insufflation (as Hooke had noticed in 1667), and suggested that blood absorbs a " nitro-aerial spirit " from the air, he describes no experiments on combustion.§

Mayow had very clear views on affinity. He says: " When the acid spirit of salt [HCl] is coagulated with a volatile salt [ammonia] . ., although the mixed salts [acid and base] seem to be destroyed, yet they may be separated from each other with their forces unimpaired, as takes place when sal armoniac . . . is distilled with salt of tartar [K_2CO_3]. . . . And the reason of this is, that the acid spirit of salt is capable of entering into closer union with any fixed salt than

* Bayliss, *General Physiology*, London, 1915, 606, quotes several passages from Mayow on respiration and says: " Mayow rightly held that combustion went on in the muscles themselves, although he was incorrect in his statement that it took place in the blood also." The relevant passage, not correctly translated in the *Alembic Club Reprint*, is from the *Tractatus quinque*, 1674, p. 152: " quanquam calor iste in animalibus, per exercitia violenta excitatus, etiam ab effervescentiâ particularum nitro-aerearum et salino-sulphurearum in partibus motricibus ortâ, partim provenit, ut alibi ostendetur."

† *Œuvres*, Paris, 1862, ii, 232. ‡ *Tractatus de corde*, 1669, 61, 166 f.

§ Sir M. Foster, *Lectures on the History of Physiology*, Cambridge, 1901, 181 f.

it is with a volatile salt, so that it immediately leaves the volatile salts that it may be combined more intimately with the fixed salt." He also gives a very clear account of the displacement of the more volatile nitric acid from nitre by sulphuric acid, on lines long afterwards elaborated by Berthollet (p. 157). Mayow says:

" No doubt it is because the volatile acid salt [HNO_3] of the nitre has been expelled from the society of the alkaline salt by the more fixed vitriolic acid that the acid of nitre, now liberated from union with the alkaline salt, ascends under a heat no greater than is required for the rectification of the spirit of nitre. . . . It is a corroboration of this view that the mass left in the retort . . . closely resembles vitriolated tartar [K_2SO_4], and can be properly substituted for it."

Jean Rey

We should not leave the path of true discovery opened out by Boyle, Hooke and Mayow for the jungle of the Theory of Phlogiston without a word for Jean Rey, a physician of Périgord, who, long before them, had published in 1630 an essay: *Sur la Recerche de la cause pour laquelle l'Estain & le Plomb augmentent de poids quand on les calcine*, republished by Gobet in 1777.* Rey says, in his euphuistic style, that he " devoted several hours " to the consideration of the increase in weight of tin and lead on calcination [which was well known in his time], and came to the conclusion that it is due to " thickened air " which " mixes itself among the *calx* " (*c'est l'air qui se mesle parmi la chaux de l'estain & du plomb qu'on calcine, qui l'augmente de poids*). Rey says he made some experiments on the calcination of tin. He nowhere says the formation of the calx is due to a union of the air, in whole or in part, with the *metal*, and that the increase in weight is due to this union. His ideas are interesting but crude, and their importance has been over-estimated. Rey points out clearly that the increase in weight does not exceed a certain amount: " Nature in her inscrutable wisdom has set limits which she never oversteps (*s'est ici mise des barres qu'elle ne franchit jamais*)."

* *Alembic Club Reprint* No. 11; McKie, *The Essays of Jean Rey*, 1951.

The Theory of Phlogiston

Johann Joachim Becher, who was born at Speyer in 1635 and was for a time in England, where he is said to have died in 1682, published in 1669 a book called *Physicæ subterraneæ*, in which he states that the constituents of bodies are air, water, and three earths, one of which is *inflammable* (*terra pinguis*), the second *mercurial*, the third fusible or *vitreous*. These correspond with the sulphur, mercury, and salt of the alchemists. On combustion, the " fatty earth " burns away.

FIG. 41.—J. J. BECHER, 1635-1682.

Georg Ernst Stahl, born in Anspach in 1660, studied medicine and lectured on chemistry at Jena, and in 1687 became physician to the Duke of Saxe-Weimar. In 1694 he became professor of medicine and chemistry in the new university of Halle, but he left in 1716 to become physician to the King of Prussia in Berlin, where he died in 1734. He was morose in disposition. Stahl's writings are often obscure and some are printed in a mixture of Latin and German, interspersed with alchemical symbols provided with Latin case-endings. Although in his youth Stahl believed in alchemy, he afterwards issued a warning against its frauds, and as a proof that metals do not " ripen " into gold in the earth, he mentions British tin, which is still the same as when exploited by the Phoenicians : " a peculiar kind of addled egg which will not be hard-boiled ".

In 1703 Becher's treatise was republished, with a long commentary, by Stahl, who in his lectures and text-book (*Fundamenta*

FIG. 42.—G. E. STAHL, 1660-1734.

Chymiae, 1723) popularized Becher's view in an improved form. Becher's *terra pinguis* he renamed *phlogiston* (which he writes in Greek, φλογιστὸν), a word which had been used in the same sense by Hapelius (Raphael Eglin) (1606), Sennert (1619), and van

Helmont (p. 48). It is " the matter and principle of fire, not fire itself ", escapes from burning bodies with a rapid whirling motion, and is contained in all combustible bodies and also in metals (which can be burnt to calces). The burnt product may be restored to the original substance by supplying phlogiston from any material containing it, such as oil, wax, charcoal, or soot (which is nearly pure phlogiston). Zinc on heating to redness burns with a brilliant flame, hence phlogiston (ϕ) escapes. The white residue is calx of zinc. If it is heated to redness with charcoal (rich in phlogiston) zinc distils off. Hence, calx of zinc $+\phi=$ zinc. Similarly with other metals. If phosphorus is burned, it produces an acid matter, and much heat and light are evolved. Hence, phosphorus $=$ acid $+\phi$. If the acid is heated with charcoal, phlogiston is absorbed and phosphorus is reproduced.

Stahl in 1697 gave the following " proof " that sulphur is a compound of sulphuric acid (an element) and phlogiston. Sulphur burns with a flame (due to the escape of phlogiston) and forms sulphuric acid (Stahl drew attention to the intermediate formation of sulphurous acid): sulphur $=$ sulphuric acid $+\phi$. If we could replace phlogiston into sulphuric acid we should get sulphur. To prevent volatilization of the acid on heating, it is first " fixed " with potash, and the salt (potassium sulphate) is heated with charcoal (rich in phlogiston). A dark brown mass is formed, identical with the " liver of sulphur " obtained by fusing potash with sulphur:

$$\text{(sulphuric acid + potash)} + \phi = \text{liver of sulphur;}$$
$$\text{sulphur + potash} = \text{liver of sulphur.}$$

It readily follows from these experiments that:

$$\text{sulphuric acid} + \phi = \text{sulphur.}$$

Phlogiston was material, sometimes the matter of fire, sometimes a dry earthy substance (soot), sometimes a fatty principle (in sulphur, oils, fats and resins), and sometimes invisible particles emitted by a burning candle. It is contained in animal, vegetable and mineral bodies, and is the same in all. It can be transferred from one body to another. It is the cause of metallic properties, of colours (as in soot and Prussian blue), of odours (as in sulphur compounds, essential oils, etc.). Salts are compounds of acid and base. The bases are different (Stahl distinguished potash from soda) but all acids are modifications of a universal acid, sulphuric acid; nitric acid is a compound of sulphuric acid and phlogiston. Fermentation is slow combustion. Alcohol is a compound of water and phlogiston; on burning, phlogiston escapes and water remains.

Stahl inverted the true theory of combustion and calcination; adding phlogiston was really removing oxygen, and removing phlogiston was adding oxygen. He neglected the quantitative aspects of chemical changes, disregarded what was known of gases, and paid little attention to the atomic theory. His theory of acids was incorrect. On the other hand, he turned away from alchemy and the three alchemical principles (salt, sulphur and mercury), his theory linked together a large number of facts into a coherent body of false doctrine, suggested new experiments, and led to discoveries.

As time went on, various modifications of the phlogiston theory were proposed. Cavendish (1766) suggested that inflammable air (hydrogen) is phlogiston; Baumé (1777) that the inflammable principle is composed of the matter of fire united with an earth in varying proportions, and metals on calcination lose phlogiston and absorb pure fire or fire less charged with earthy principle. Macquer (1779) identified phlogiston with the matter of light; metals and combustibles on combustion lose it, at the same time combining with air or its purest part. The difficulty that a calx becomes lighter when reduced to a metal by taking up phlogiston, when considered at all, was explained in various ways. Stahl * said that addition of phlogiston reduces weight (*per accessionem enim partium inflammabilium levius fit concretum*); Scheffer (1757) that metals gain or lose weight as phlogiston is removed from, or added to, them. Chardenon (1764) distinguished between specific gravity (density) and absolute gravity (weight); compounds with elements of less gravity than air tend to rise. Guyton de Morveau (1772) said phlogiston is lighter than air or the rarest medium, and reduces the weight of bodies in that medium. Venel (about 1750) attributed negative weight to phlogiston, a theory adopted by Gren (1786) but

* *Fundamenta Chymiae*, 1747, Pars iii, p. 375; Partington and McKie, "Historical Studies on the Phlogiston Theory", Parts i-iv, in *Annals of Science*, 1937, ii, 361 (the levity of phlogiston); 1938, iii, 1 (the negative weight of phlogiston), 337 (light and heat in combustion), 1939, iv, 113 (last phases of the theory); Partington, *Scientia*, 1938, lxiv, 121.

later abandoned by him after criticisms by J. T. Meyer (1790) and Hindenburg (1790).

After the foundation of Lavoisier's oxygen theory of combustion (see p. 130), an attempt was made to retain some features of the phlogiston theory in providing an explanation of the heat and light developed in combustion. Lubbock (1784), Gadolin (1788), and Richter (1791) supposed that a combustible might contain a material base united with phlogiston (ϕ), oxygen gas a material substrate combined with the matter of heat (*caloric*). On combustion, the two material bases combined by attraction and at the same time the phlogiston and caloric united by affinity to give the fire and light set free. It was only after the idea of energy was accepted (about 1850) that the need for some such explanation of the heat and light of combustion ceased to be felt.

CHAPTER VI

DISCOVERY OF GASES

The Discovery of Gases

WE can easily understand why the discovery of gases and the investigation of their properties were rather late in the study of chemistry. Van Helmont, who invented the name gas about 1630 and described at least two gases—*gas sylvestre* (carbon dioxide) and *gas pingue* (impure hydrogen, or perhaps marsh gas)—was of the opinion that a gas cannot be contained in a vessel. Robert Boyle was probably the first to collect a gas. He also knew that hydrogen was inflammable (see pp. 72, 73).

The manipulation of air over water was described by Moitrel d'Element in 1719. The Rev. Stephen Hales in 1727 published his *Vegetable Staticks*, describing several experiments on gases, but he contented himself with measuring their volumes without studying their properties, and so missed the discovery of individual gases.

" Fixed air " (carbon dioxide) was rediscovered by Joseph Black in 1754, and this gas as well as the inflammable air (hydrogen) discovered by Boyle were carefully investigated in 1766 by Cavendish, who describes several methods for the manipulation of gases, and was the first to collect a soluble gas over mercury.

Joseph Priestley, a nonconformist minister of Leeds, devoted his spare time to the study of science. He invented from 1770 much of the apparatus now used in the manipulation of gases. He discovered a number of new gases, and in addition to collecting gases over water, he collected soluble gases (ammonia, hydrochloric acid, sulphur dioxide) over mercury.

Hales

Stephen Hales, who had studied at Corpus Christi College, Cambridge, was vicar of Teddington in Middlesex—he refused a canonry of Windsor so that he could devote himself to scientific experiments and parochial duties—and his principal work is botanical. In his two books, *Vegetable Staticks* (1727) and *Haemastaticks* (1733), however, he has some interesting chemical observations. His chemical work—probably under Newton's influence—is entirely quantitative, and is a good example of

Fig. 43.—Stephen Hales, 1677-1761.

the poor results obtained when the qualitative chemical characters of the substances investigated are neglected, but his experiments inspired both Black and Priestley. Hales attempted to determine the *amount* of "air" which could be extracted from various substances by heating in a gun-barrel and collecting over water (Fig. 44). Among the materials he used were coal (which would give coal gas), red lead and saltpetre (which would give oxygen), iron filings and dilute sulphuric acid (which would give hydrogen), and iron filings and dilute nitric acid (which would give nitric oxide). The various gases were all " air " to Hales; they were measured and then thrown away. In some experiments on respiration he found that he could breathe a fixed volume of air much longer if it were passed, between exhalation and inhalation, by means of valves

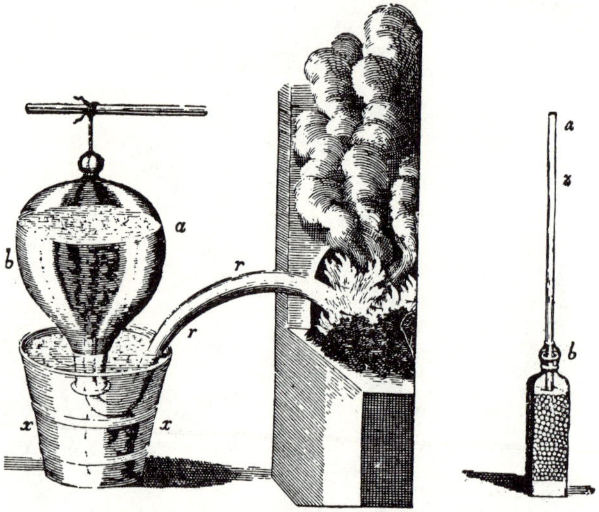

FIG. 44.—APPARATUS USED BY HALES.

ON THE LEFT IS THE GUN BARREL (*r*) HEATED IN A FIRE. THE GAS IS COLLECTED IN THE GLOBE (*ab*) OVER WATER IN THE TUB (*xx*). ON THE RIGHT IS A BOTTLE FILLED WITH FERMENTING PEAS OVER MERCURY. THE PRESSURE OF THE GAS EVOLVED DRIVES THE MERCURY INTO THE VERTICAL TUBE (*ab*), WHICH IS FIRMLY FIXED INTO THE BOTTLE AND DIPS INTO THE MERCURY ON WHICH THE PEAS FLOAT.

through rags wetted with potash, but he did not pursue this experiment, " fearing that I might thereby some way injure my lungs ". He found that much air is absorbed when phosphorus is burnt in air in a receiver.

Hales came to the vague conclusion * that " air abounds in animal, vegetable and mineral substances " and that its presence " leavens " them. He says† that plants probably draw through their

* *Vegetable Staticks*, 1727, 313 f. Cf. Ramsay, *Gases of the Atmosphere*, 1915; Clara M. Taylor, *The Discovery of the Nature of the Air* (*Classics of Scientific Method*), 1923, 33 f.; A. E. Clark-Kennedy, *Stephen Hales*, Cambridge, 1929.

† *Vegetable Staticks*, 1727, 325.

leaves some part of their nourishment from the air (which Priestley afterwards said was his opinion also)*, and he emphasized that " elasticity is not an essential immutable property of air particles ; but they are, as we see, easily changed from an elastick to a fixt state, by the strong attraction of the acid, sulphureous and saline particles which abound in the air ", and are also capable of once more resuming the elastic state. He regarded air as an element.

Black

Joseph Black was born in Bordeaux, where his father, of Scots descent but born in Belfast, had a wine business. Joseph was educated in medicine at Glasgow University (1746), where he came under the excellent influence of Dr. Cullen, professor of medicine and lecturer in chemistry and a practising physician,† in whose laboratory Black worked. In 1750 or 1751 Black moved to Edinburgh, where in June 1754 he presented his inaugural dissertation : *De Humore Acido a Cibis orto, et Magnesia Alba* (" On the acid humour arising from food, and magnesia alba ").‡ This has an interesting appendix of chemical experiments with an explanation of the relation between mild and caustic alkalis. It was read in an extended version, in English, to the Philosophical Society of Edinburgh in June 1755 and published in 1756 § as *Experiments upon Magnesia alba, Quicklime, and some other Alcaline Substances.* The work was begun in 1750. Black succeeded Cullen as professor of anatomy and lecturer on chemistry in Glasgow in 1756, but exchanged the chair of anatomy with the professor of medicine. He succeeded Cullen as professor of chemistry in Edinburgh in 1766, and occupied that position until his peaceful death, seated in his chair, in 1799. His last course of lectures was given in 1796-7. Black was very popular as a lecturer, taking great pains both with his courses and with his lecture experiments.

* Rutt, *Memoirs of Priestley,* 1832, i, 299.

† Dobbin, *Annals of Science,* 1936, i, 138.

‡ Transl. by Crum Brown, *J. Chemical Education,* 1935, xii, 225, 268.

§ *Essays and Observations, Physical and Literary, read before a Society in Edinburgh, and published by them,* Edinburgh, 1756, ii, 157-225 ; reprinted with Cullen's *Memoir on the Cold Produced by Evaporation,* Edinburgh, 1777, 1782 ; *Alembic Club Reprint* No. 1.

One of his pupils was Benjamin Rush (1745-1813) afterwards, from 1799 at Philadelphia, the first professor of chemistry in America.

Black had little regard for hypotheses unless they were founded on experiment: he welcomed Lavoisier's new views, and taught them for this reason. Robison says: "many were induced, by the

FIG. 45—JOSEPH BLACK, 1728-1799.

report of his students, to attend his courses, without having any particular relish for chemical knowledge." His lectures were taken down by students and manuscript copies are still obtainable: * some lectures were published anonymously in London in 1770 as: *An Enquiry into the General Effects of Heat, with Observations on the Theory of Mixture*; a more complete account was published by Robison in Edinburgh in 1803 as: *Lectures on the Elements of*

* McKie, *Annals of Science*, 1936, i, 101.

Chemistry, delivered in the University of Edinburgh; by the late Joseph Black, M.D., now published from his Manuscripts, 2 vols. quarto.

Black made important experiments on latent heats of fusion and evaporation, which he explained in his lectures from about 1758 (fully in 1761), and he recognized about 1760 that bodies have different capacities for heat or specific heats. His views on latent heat were of very great service to James Watt, then instrument-maker at Glasgow, in his invention in 1765 of the improved condensing steam-engine.*

Black's Researches on the Alkalis

Black showed that when *magnesia alba* (basic carbonate, $xMgCO_3,yMg(OH)_2,zH_2O$) is heated a gas is evolved, which he called *fixed air*: it was identical with van Helmont's *gas sylvestre* (p. 48). The residue of *calcined magnesia* (MgO) is lighter than the magnesia alba and more alkaline. By weighing the *magnesia alba* in a glass retort and heating, Black found that, "of the volatile parts in that powder, a small proportion only is water; the rest cannot, it seems, be contained in vessels, under a visible form .. the volatile matter lost in the calcination of magnesia is mostly air, and hence the calcined magnesia does not emit air, or make an effervescence, when mixed with acids." Thus:

magnesia alba = calcined magnesia + water + fixed air;
magnesia alba + acid = magnesia salt + fixed air;
calcined magnesia + acid = magnesia salt.

Black mentions that Hales had "proved that alkaline salts contain a large quantity of fixed air, which they emit, in great abundance when joined to a pure acid". Black also made experiments on the loss in weight when *magnesia alba* was treated with acid in a flask.

Black now turned his attention to limestone and quicklime, and the mild and caustic alkalis. He found that limestone effervesces with acids, giving off fixed air. On heating, it forms quicklime and

* Ramsay, *Life and Letters of Joseph Black*, 1918; Thomson, *History of Chemistry*, ii, 313; Lord Brougham, *Works*, 1872, i, 1 f., 477—Brougham attended Black's lectures; McKie and Heathcote, *The Discovery of Specific and Latent Heats*, 1935; Tilden, *Famous Chemists*, 1921, 22; Agnes M. Clerke, *Dict. Nat. Biogr.*, 1908, ii, 571.

(in contrast to *magnesia alba*) only a trace of water : the considerable loss in weight must be due to the escape of fixed air :

limestone = quicklime + fixed air. - - (1)

There were three alkalis recognized in Black's time : vegetable (potash), marine (soda) and volatile (ammonia), and a mild and caustic form of each. The mild form (carbonate) was known to be converted into the caustic form (hydroxide) by treatment with slaked lime. Duhamel in 1736 and Marggraf in 1757 had distinguished by chemical tests between potash and soda, and Duhamel in 1747 had shown that there is a loss in weight when limestone is burnt to quicklime, which is slowly recovered when the quicklime is exposed to the air. Black found that if a given weight of limestone is converted into quicklime according to equation (1), the quicklime slaked with water, and the slaked lime boiled with a solution of mild alkali (potassium carbonate), the alkali becomes caustic and the original weight of limestone is recovered :

quicklime + mild alkali = limestone + caustic alkali. - (2)

Equations (1) and (2) show that :

mild alkali = caustic alkali + fixed air. - (3)

" If quicklime is mixed with a dissolved alkali it shows an attraction for fixed air superior to that of the alkali. It robs this salt of its air and thereby becomes mild itself, while the alkali is consequently rendered more corrosive, or discovers its natural degree of acrimony, or strong attraction for water ; which attraction was less perceivable, as long as it was saturated with air. And the volatile alkali [ammonium carbonate *], when deprived of its air, besides this attraction for various bodies, discovers likewise its natural degree of volatility, which was formerly somewhat repressed, by the air adhering to it, in the same manner as it is repressed by the addition of an acid."

Also : " as the calcareous earths and alkalis attract acids strongly and can be saturated with them, so they also attract fixed air and in their ordinary state are saturated with it : and when we mix an acid with an alkali or earth, the air is set at liberty . . .

* Commercial ammonium carbonate is mainly a mixture of bicarbonate NH_4HCO_3, and carbamate, $NH_2{\cdot}CO{\cdot}ONH_4$, with some normal carbonate, $(NH_4)_2CO_3$.

because the alkaline body attracts it more weakly than it does the acid, and because the acid and air cannot both be joined to the same body at the same time."

When limestone is dissolved in acid, fixed air is evolved with effervescence and a salt is formed. On addition of mild alkali to the solution, the original weight of limestone is reprecipitated and there is no effervescence because the fixed air of the mild alkali joins to the lime to produce limestone. Slaked lime does not contain any parts more caustic than the rest, and " as any part of it can be dissolved in water, the whole of it is also capable of being dissolved ". Caustic alkali contains no lime or a mere trace which had been dissolved by the water.

" Quicklime does not attract air when in its most ordinary form, but is capable of being joined to one particular species only . . . to this I have given the name of fixed air, and perhaps very improperly; but I thought it better to use a word already familiar in philosophy, than to invent a new name before we be more fully acquainted with the nature and properties of this substance." It will have been noted that Black's definition of fixed air, given on p. 95, is identical with van Helmont's definition of gas (p. 48) and was borrowed by Black from that author.

Black's theory was opposed by the German chemist Johann Friedrich Meyer in 1764 in a book with a characteristic title: *Chymischen Versuche zur näheren Erkenntniss des ungelöschten Kalchs, der elastichen und elektrischen Materie, des allerreinsten Feuerwesens, und der ursprünglichen allgemeinen Säure.* In this Meyer explains that limestone on burning takes up an oily acid (*acidum pingue*) from the fire, and this is transferred to the mild alkali when it becomes caustic. Since limestone and alkali always effervesce with acids, it follows that quicklime and caustic alkali, which do not effervesce with acids, are already saturated with *acidum pingue.* Black's theory was defended by Jacquin (1769) and accepted by Wenzel (1777). Meyer's theory was supported by Wiegleb (1767), Cranz (1770), and Smeth (1772). Lavoisier (*Opuscules Physiques et Chimiques*, 1774) praised Meyer's work highly and gave Jacquin credit for much of Black's.

Lord Brougham says Black in his lectures showed the experiment of decanting fixed air from a jar over a candle; he also showed

FIG. 46.—A CHEMICAL LABORATORY IN 1765.
(From W. Lewis, *Commercium Philosophico-Technicum ; or the Philosophical Commerce of the Arts*, London, 1765.)

that it is contained in exhaled air by blowing through a tube into lime water, which became turbid, and that it is formed by blowing air over red-hot charcoal. Macbride * made some further experiments on fixed air, and Bergman in 1774 published a long dissertation on it, calling it the "aerial acid". Priestley in 1772 says,† "It is not improbable but that fixed air may be of the nature of an acid, though of a weak and peculiar sort. Mr. Bergman of Upsal, who honoured me with a letter upon the subject . . . says that it changes the blue juice of tournsole into red." Black had already distinctly said that fixed air behaves as an acid.‡ He had ascertained its effects on animals, its production by respiration and fermentation, and by the burning of charcoal, and inferred its presence in small quantities in the atmosphere, in 1757.§

Cavendish

Henry Cavendish (Nice 1731–London 1810)—the prefix "Honourable" is a mistake, although he was the grandson of the second Duke of Devonshire—had an excellent education and devoted himself to scientific research, his town mansion being fitted up as a laboratory. He was extremely shy and eccentric, and refused to sit for his portrait, the only one in existence being sketched surreptitiously by the artist. Lord Brougham says Cavendish "probably uttered fewer words in the course of his life than any man who lived to fourscore years." Cavendish died very wealthy, although he was not miserly.‖

Besides his chemical work, Cavendish made fundamental researches in electricity and heat, most of which he did not publish. He was the first to determine the electrical conductivity of salt solutions, and he differentiated between electrical quantity and intensity. He gave an experimental proof of the inverse square law by showing that there is no charge inside a hollow conductor. He experimented on latent and specific heat—perhaps with some

* *Experimental Essays*, London, 1764.

† *Phil. Trans.*, 1772, 153. ‡ *Alembic Club Reprint* No. 1, p. 22.

§ *Lectures on Chemistry*, 1803, ii, 87 f.; McKie, *Annals of Science*, 1936, i, 101.

‖ G. Wilson, *Life of Cavendish*, 1851; Lord Brougham, *Lives of Philosophers of the Time of George III*, Edinburgh, 1872; *Scientific Papers of Cavendish*, 2 vols., Cambridge, 1921; Tilden, *Famous Chemists* 1921, 41.

knowledge of Black's experiments and theories—and rejected the theory of the material nature of heat, believing it to consist in internal motion of the particles of bodies. He calls this " Sir Isaac

FIG. 47.—HENRY CAVENDISH, 1731-1810
(From the portrait in the British Museum.)

Newton's opinion ", but it is really Francis Bacon's.* Cavendish does not seem to have made any chemical experiments after 1785.

Cavendish's Experiments on Gases

Cavendish's first publication on chemistry was his paper *On Factitious Airs* (1766),† divided into three parts. The name "factitious air" had been used by Boyle. Cavendish says: "By factitious air, I

* Gregory, *Ambix*, 1938, ii, 93.
† *Phil. Trans.*, 1766, 141 ; *Scientific Papers*, 1921, ii, 77.

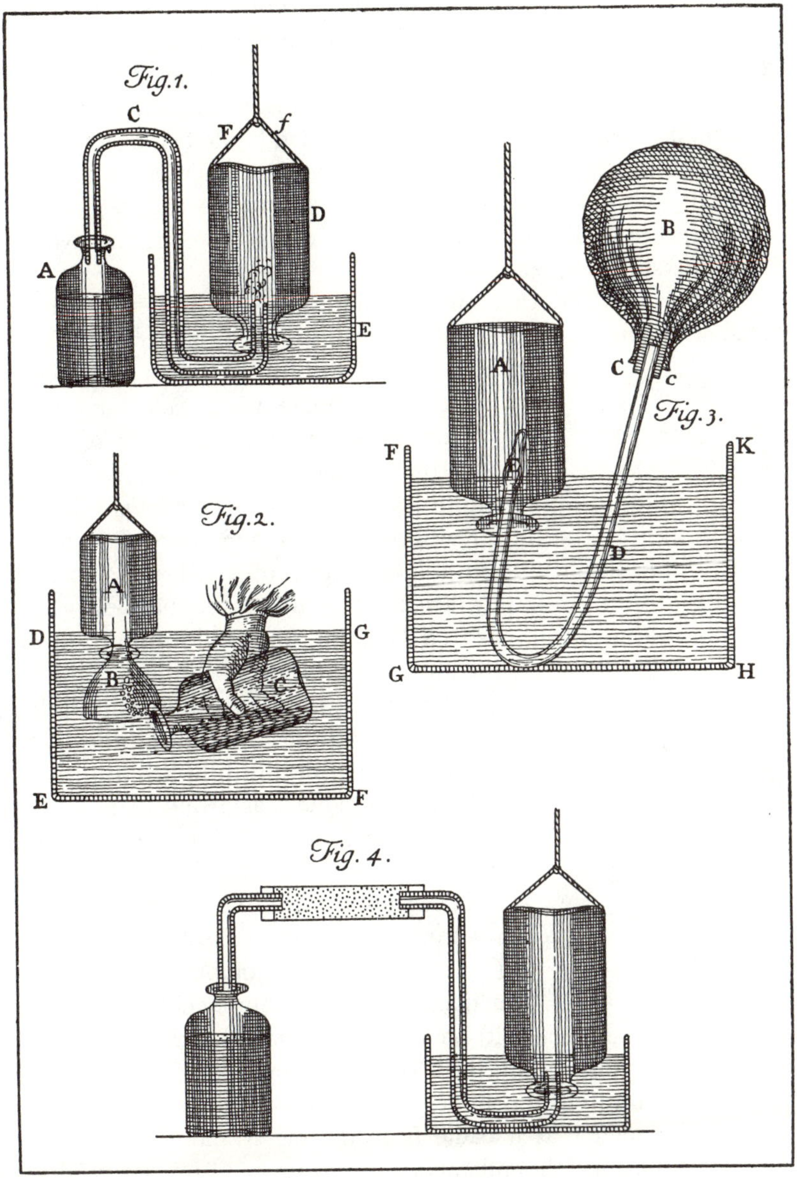

FIG. 48.—CAVENDISH'S APPARATUS FOR MANIPULATION OF GASES.

mean in general any kind of air which is contained in other bodies in an unelastic state, and is produced from thence by art," and he gives Black's fixed air as an example. He describes the manipulation of gases, including collection in bottles filled with water in a pneumatic trough without shelf, the inverted bottle being supported by cords; the transfer of gas from one bottle to another by pouring upwards through a funnel under water; and the transfer of gas from a jar to a bladder by means of a siphon tube and the hydrostatic pressure (Fig. 48).

Cavendish found that air containing one-ninth its volume of fixed air extinguishes a candle, that water absorbs rather more than its own volume of fixed air at room temperature and more when cold, and that boiling expels it, and that spirit of wine absorbs $2\frac{1}{4}$ times its volume of fixed air. He noted its solubility in alkalis. Fixed air was collected over mercury in a cylinder. The weight of fixed air expelled by acids from limestone, marble, pearl-ash (K_2CO_3), etc., was determined with the apparatus shown in Fig. 49, the acid being first put in and the bottle and tube weighed. The weighed solid was then put in and the drying tube fixed in with a lute of almond paste and glue. The drying tube contained shreds of filtering paper.

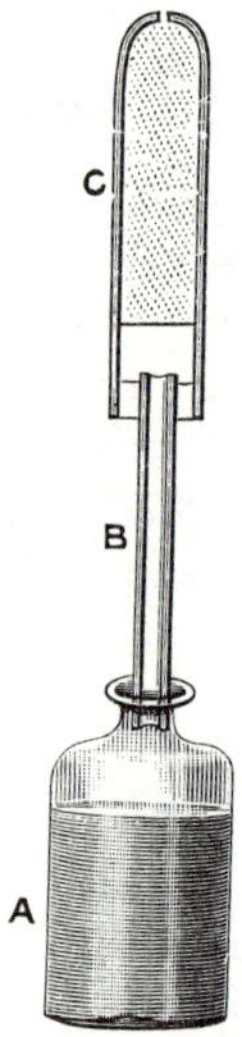

FIG. 49.—CAVENDISH'S APPARATUS FOR FINDING THE WEIGHT OF GAS EVOLVED IN A REACTION.

In his experiments on inflammable air (hydrogen), already collected by Boyle (p. 72), Cavendish obtained it by the action of dilute sulphuric acid and hydrochloric acid (spirit of salt) on zinc, iron and tin. He found that "the air is the same and of the same amount whichever acid is used to dissolve the same weight of any particular metal", and hence concluded that it came from the metal and not from the acid. He calls it "the inflammable air from metals". Cavendish was a believer in the theory of phlogiston and would represent the action of an acid on a metal as:

$$\underbrace{\text{calx} + \phi}_{\text{metal}} + \text{acid} = \underbrace{\text{calx} + \text{acid}}_{\text{salt}} + \underbrace{\phi}_{\text{inflammable air}}.$$

Cavendish says " their phlogiston [i.e. that in the metals] flies off, without having its nature changed by the acid, and forms the inflammable air ". The gas probably did not come from the acid, since an inflammable air is also evolved in putrefaction and the dry distillation of vegetable matter, which (like metals) is rich in phlogiston. With hot concentrated sulphuric acid, the metals evolved " sulphurous vapours " (SO_2), which he regarded as a compound of sulphuric acid and phlogiston ; with nitric acid, red fumes were evolved, which he regarded as a compound of nitric acid and phlogiston.

The inflammable air differed from fixed air in being insoluble in water and alkali. Cavendish found that it formed an explosive mixture with air, the most violent detonation being with a mixture of three volumes of inflammable air with seven volumes of common air (he did not try the ratio 2 : 5). He determined its density in two ways. In the first he filled a weighed bladder of 80 fluid oz. capacity (by measurement of the circumference) with the air and found it lost 41 grains, whence he calculated that inflammable air was either 7 or $10\frac{8}{19}$ times lighter than common air, according to the figure taken from previous experimenters for the density of common air. (The correct figure is 14·4). In the second method he used the apparatus of Fig. 49, finding the loss in weight when a known weight of metal is dissolved in acid. The drying tube in this case was filled with pearl-ash which would keep back both moisture and acid spray, any carbon dioxide evolved from the pearl-ash by the acid being fixed as potassium bicarbonate. In this way Cavendish found inflammable air to be eleven times lighter than common air.

By heating copper wire with spirit of salt until all the air was expelled from the vessel, Cavendish found that " on a sudden, without any sensible alteration of the heat, the water rushed violently . . . into the bottle and filled it almost entirely full ". He remarks that the vapour formed from the acid " loses its elasticity as soon as it comes in contact with the water ". This gave Priestley the hint which enabled him to collect the " acid air " (HCl) over mercury (p. 115).

Cavendish in 1781 found that the method of measuring the "goodness" of air by mixing with nitric oxide over water, which had given Priestley very variable results, can be made to give reliable results if standardized. He reports * that "during the last half of the year 1781, I tried the air of near 60 different days . . . but found no difference that I could be sure of, though the wind and weather on those days were very various ; some of them being very fair and clear, others very wet, and others very foggy ". In other trials of the air of London and at Kensington—then "the country" —he found the same result. Although Cavendish will not permit himself to consider air as a mixture of dephlogisticated air (oxygen) and phlogisticated air (nitrogen), "as Mr. Scheele and La Voisier suppose", his results correspond with 20·83 per cent. by volume of oxygen and 79·17 of nitrogen, very near the modern values. Cavendish was also the first, in 1784, to analyse air brought from the upper atmosphere by a balloon, finding it practically the same as that at ground level.†

Cavendish anticipated Scheele in the discovery in 1764 of arsenic acid, which he prepared by boiling white arsenic (arsenious oxide) with nitric acid, but he did not publish his results. He also discovered the true nature of cream of tartar (potassium hydrogen tartrate) and its relation to soluble tartar (normal potassium tartrate) obtained from cream of tartar and potash. His work on equivalents and that on the composition of water and of nitric acid are described elsewhere (pp. 159, 136). It will be seen that Cavendish had followed van Helmont, Boyle, and Black in careful *quantitative* experiments in chemistry long before Lavoisier's.

Scheele

Carl Wilhelm Scheele was born in 1742 at Stralsund, then the chief town of Swedish Pomerania. He was apprenticed in 1757 to an apothecary in Gothenburg, where he began to study chemistry. He occupied positions in pharmacies in Malmö, Stockholm, Uppsala (where he became acquainted with Bergman) and Köping, where he died in 1786 at an early age. Scheele was a man of great modesty and his circumstances were often poor. He worked with very simple apparatus and in periods of scanty leisure, in a cold

* *Phil. Trans.*, 1783, 106 f. † *Scientific Papers*, 1921, ii, 22.

and uncomfortable laboratory, yet he made a great number of chemical discoveries of the very first rank, those on combustion being published in his *Chemische Abhandlung von der Luft und dem Feuer* (*Chemical Treatise on Air and Fire*).* Some of these experiments were made before the autumn of 1770, and nearly all prior to 1773, but the book, although sent to the printer in 1775, did not appear until 1777, when many of Scheele's discoveries had been made independently, and published, by Priestley in England. Scheele's priority was only established in 1892, from his original laboratory notes.†

Scheele's Experiments on Air

In his first set of experiments " on air and fire " Scheele noticed the contraction of a confined volume of air standing in contact with various materials. He used, for example, a solution of liver of sulphur (*hepar sulphuris*), a solution of sulphur in lime-water, linseed oil, and iron filings moistened with water, all of which, he observes, are rich in phlogiston, or, as he often called it, the *inflammable substance.* In all cases there was a loss of air. After a few days about one-fourth of the air was absorbed and the residual gas extinguished a taper.

The inflammable substance was not contained in the residual gas, which differed from common air. For, if this gas had been formed by the union of common air with phlogiston, and contraction, it should be denser than common air. But: " a very thin flask which was filled with this air, and most accurately weighed, not only did not counterpoise an equal volume of ordinary air, but was even somewhat lighter ". Thus, " the *air is composed of two fluids*, differing from each other, the one of which does not manifest in the least the property of attracting phlogiston, whilst the other, which composes between the third and fourth part of the whole mass of the air, is peculiarly disposed to such attraction ". These two constituents of common air Scheele called *Foul Air* (" verdorbene Luft ") and *Fire Air* (" Feuerluft ") : they were afterwards named nitrogen and oxygen, respectively.

* *Collected Papers of Carl Wilhelm Scheele*, transl. by L. Dobbin, London, 1931 ; *Alembic Club Reprints* Nos. 8 (oxygen) and 13 (chlorine) ; Thorpe, *Essays*, 1902, 60 ; Tilden, *Famous Chemists*, 1921, 53.

† Nordenskjöld, *Scheele, Nachgelassene Briefe*, Stockholm, 1892.

Scheele next placed a little phosphorus in a thin flask, corked the latter, and warmed it until the phosphorus took fire. A white cloud was produced, which attached itself to the sides of the flask in white flowers of "dry acid of phosphorus". On opening the flask under

FIG. 50.—C. W. SCHEELE, 1742-1786
(From a posthumous portrait by Falander.)

water, the latter rushed in, and occupied a little less than one-third of the flask. By allowing phosphorus to stand for six weeks in air in the same flask, until it no longer glowed, again some of the air was lost.

Scheele then burned a hydrogen flame under a glass globe standing over water (Fig. 51). The water at once began to rise, until it filled one-fourth of the flask, when the flame went out. Scheele assumed that the inflammable substance (hydrogen) had combined with the fire air, and since he was unable to find the product of the combination (he missed the dew deposited on the flask, since he used hot water) he assumed that it was *heat*, which had escaped through the glass: heat = fire air + ϕ. He therefore tried to decompose heat and set free the fire air by several methods and in this way he made an immense step forward, because he was able to obtain pure fire air, which he found was a colourless gas in which a taper burned with great brilliancy.

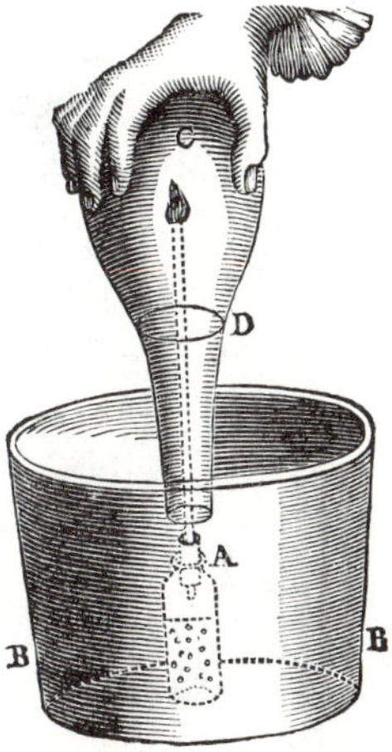

FIG. 51.—COMBUSTION OF HYDROGEN IN AIR: SCHEELE'S EXPERIMENT.

THE HYDROGEN WAS GENERATED IN THE BOTTLE A, IMMERSED IN A TUB OF HOT WATER B, AND THE HYDROGEN FLAME C BURNT INSIDE THE INVERTED FLASK D. THE WATER ROSE IN THE FLASK.

The Discovery of Oxygen by Scheele

In order to decompose heat into its supposed constituents, fire air and phlogiston, Scheele argued that he must present to heat some substance having a greater attraction for phlogiston than is exhibited by fire air. He chose nitric acid, which readily acts upon metals, taking out their phlogiston and forming red fumes (nitric acid + ϕ). In order to subject it to heat, the nitric acid was fixed by combination with potash (cf. Stahl's experiment, p. 88) and was set free again at a higher temperature by distilling the resulting nitre with oil of vitriol (sulphuric acid) in a retort. Red fumes came off, which were absorbed in milk of lime contained in a bladder attached to the neck of the retort (Fig. 52). The bladder gradually filled with a colourless gas, in which a taper burned with a flame of dazzling brilliance. This was the fire air, and thus oxygen was at last discovered. Its existence was made probable by the experiments of Hooke and Mayow, but it was first obtained, as an expected result in a carefully planned experiment, by Scheele.

Scheele prepared fire air (oxygen) in several other ways. He heated calx of mercury (*mercurius calcinatus per se*: mercuric oxide), which he supposed absorbed phlogiston from the heat, setting free the fire air:

$$\text{calx of mercury} + \underbrace{(\phi + \text{fire air})}_{\text{heat}} = \underbrace{(\text{calx of mercury} + \phi)}_{\text{mercury}} + \text{fire air.}$$

He also obtained fire air by heating "black manganese" (manganese dioxide) with sulphuric or arsenic acid; by strongly heating nitre alone; by heating mercurous and mercuric nitrates; and by heating silver or mercurous carbonate, the *aerial acid* (carbon

FIG. 52.—SCHEELE'S ISOLATION OF FIRE AIR.

dioxide) simultaneously produced being absorbed by bringing the gas in contact with an alkali:

$$\text{silver carbonate} = \text{silver} + \text{fire air} + \text{aerial acid.}$$

Scheele found that fire air is *completely* absorbed by moist liver of sulphur. When he burned phosphorus in a thin flask of it, the flask burst on cooling. With a thicker flask, the cork could not be taken out under water, but could be pushed in, when water rushed in and filled the flask. A hydrogen flame continued burning in the gas until seven-eighths were absorbed.

When fire air was added to the foul air left after combustion of hydrogen, etc., in air, so as to restore the original volume, the mixture had all the properties of ordinary air, e.g. it left the same residue after standing over liver of sulphur. Scheele also noticed that fire air is partly dissolved out of common air when this stands over water which had been boiled. A candle burns more brightly in the air expelled from the water by boiling than in common air.

Scheele found that a rat and flies died in confined volumes of air. After the aerial acid (CO_2) was separated by lime water the air had contracted and the residue extinguished a flame. Similar results were found with sprouting peas. A large bee confined in a bottle of fire air, together with some honey, in contact with lime water, was dead after a week, the lime water was milky and almost filled the bottle. He breathed air in a bladder, separated the aerial acid, and found that the contracted residue extinguished a flame.

Other Discoveries by Scheele

Besides his experiments on combustion, leading to the discovery of oxygen, Scheele made a number of other most important discoveries.

I. Chlorine, manganese and baryta in an investigation on " black magnesia " (native manganese dioxide) in 1774 (see p. 185).

II. Silicon fluoride and hydrofluoric acid from fluorspar (1771, 1780, 1786).

III. Phosphorus from bone ash, and phosphoric acid by the action of nitric acid on phosphorus (1774, 1777 ; phosphoric acid was discovered in bone ash in 1770-71 by Gahn).

IV. Arsenic acid (1775), molybdic acid (1778), tungstic acid (1781). Distinction between molybdenite (MoS_2) and graphite (1779). Arsenic hydride (1775), copper arsenite (" Scheele's green ") (1778).

V. Several organic acids : tartaric (1770), mucic (1780), lactic (1780), uric (1780), prussic (1782-83), oxalic (1776, 1784-5), citric (1784), malic (1785), gallic and pyrogallic (1786) ; also glycerol (1783-84), murexide (1780), several esters (1782), aldehyde (1782), and casein (1780).

VI. Action of light on silver salts (1777).

VII. Hydrogen sulphide (" stinking sulphurous air ") and crude hydrogen persulphide (1777).

VIII. Distinction between nitrous and nitric acids (1767) ; nitric oxide, nitrous acid (N_2O_3), nitrogen dioxide, and nitric acid contain decreasing amounts of phlogiston (increasing amounts of oxygen) (1777 ; *Collected Papers*, 1931, 103).

IX. Formation of cyanide by the action of ammonia on a mixture of charcoal or graphite and potassium carbonate (1782).

Priestley

Joseph Priestley * was born in 1733 at Fieldhead, near Leeds, the son of a cloth dresser, and died in America in 1804. As a boy he was sickly and his education was neglected for a time. He then studied for the nonconformist ministry and entered the Dissenting Academy at Daventry. He acquired a knowledge of Hebrew, Greek and Latin which, with his extensive theological learning, enabled him to meet on more than equal terms his more orthodox brethren, with whom he was in continual dispute. He became a tutor at the Academy at Warrington, and afterwards acted as a minister in various places.

Priestley began his scientific experiments at Nantwich in 1758. An acquaintance in 1766 with Dr. Price, John Canton and Benjamin Franklin—a man, as Mirabeau says, " able to restrain both thunderbolts and tyrants "—turned Priestley's attention to the study of electricity. He became F.R.S. in 1766, and in 1767 he published his *History of Electricity*. In this work he anticipates Cavendish's proof of the inverse square law by a method suggested by Franklin, depending on the absence of force inside a hollow charged conductor. In 1767 Priestley removed to Leeds, where he began his experiments on gases as a result of living next to a public brewery, where he noticed that fixed air was evolved in the fermentation vats. He had previously attended some lectures on chemistry given at Warrington by a Dr. Turner of Liverpool, but at that time his knowledge of the subject was small, and he did not know that in the investigation of fixed air he had been anticipated by Black in 1755. His *Directions for Impregnating Water with Fixed Air*, London, 1772, announced his invention of soda-water, and attracted a good deal of attention. From Leeds Priestley went in 1773 as " literary companion " to Lord Shelburne (afterwards first Marquis of Lansdowne, and Prime Minister in 1782)

* J. T. Rutt, *Life and Correspondence of Joseph Priestley*, 2 vols., London, 1832, with list of works, ii, 535 f. ; *Scientific Correspondence of Joseph Priestley*, edited by H. C. Bolton, New York, 1892, with portrait, list of portraits, of contents of laboratory, etc. ; Hartog, *Dict. Nat. Biogr.*, 1896, xlvi, 357 ; *Annals of Science*, 1941, v, 1 ; Anne Holt, *A Life of Joseph Priestley*, London, 1931; Thorpe, *Essays*, 1902, 32; Tilden, *Famous Chemists*, 1921, 32 ; Partington, *Nature*, 1933, cxxxi, 348 ; McKie, *Science Progress*, 1933, xxviii, 17.

with a good salary, the use of a splendid library, provision for experiments, and the promise of an annual pension of £150 on retirement, and it was in Lord Shelburne's country mansion, Bowood, near Calne, in Wiltshire, that Priestley discovered oxygen. Very

FIG. 53.—JOSEPH PRIESTLEY (1733-1804).

soon after he travelled with Lord Shelburne on the Continent, and he met Lavoisier in Paris in the autumn of 1774.

After seven years Priestley and Lord Shelburne parted amicably, and in 1780 he settled in Birmingham, where he continued his experiments. He was indebted to Josiah Wedgwood, the potter, for many gifts of apparatus and funds for his work. In consequence of Priestley's freely expressed liberal opinions, he was the object of hatred from both the common people and persons in high places, and on 14th July, 1791, when a dinner was held in Birmingham to celebrate the fall of the Bastille, the mob was incited by authority to sack his house and laboratory, although he himself was not present at the dinner. The pillage was carried out with such hearty patriotism and gusto that for three days Birmingham was the scene of an orgy of political violence and bestiality which has rarely been equalled in England. "Persons in the habit of gentlemen" were seen rummaging among Priestley's papers in the vain hope of finding incriminating documents.

Priestley made a dignified protest, and several of the ringleaders in the mob were afterwards severely dealt with. He found it necessary to escape in disguise to Worcester and then went to London, where he lived happily for a time, although ostracized by the Royal Society: it is incorrect to say that he resigned his fellowship. At last, however, he was driven in April 1794 to follow his sons to America, where he settled on the banks of the Susquehanna, and there, after declining a professorship of chemistry at Philadelphia, he died after two years of painful illness in 1804. With many gifts from friends and his pension from Lord Shelburne, Priestley was well provided for and had ample means for his scientific researches.

In private life Priestley was a courteous gentleman of attractive and winning personality. His theological and political writings made him many enemies, though whether the fault lay with him or with his opponents has been disputed. On one point, that his sole object was the attainment of what he believed to be the truth, there can be no dispute. He was a man of prodigious industry: as he says himself, he often wrote till he could hardly hold the pen; in 1800 he began to learn Chinese, and in 1802 he was experimenting with the newly discovered voltaic battery. He published a great number of theological works. In science he was somewhat of an amateur, and his defective knowledge of chemical analysis exposed him to many mistakes. He was, however, a very clever and ingenious manipuator in his own field, the study of gases, and when he made mistakes in experiments he freely admitted them. Although heterodox in religion, he was orthodox in chemistry, and to the last he clung tenaciously to the theory of phlogiston, which his own experiments had done so much to overturn.

Priestley's Experiments on Air

Priestley's chemical experiments are mostly contained in his *Experiments and Observations on Different Kinds of Air* (3 vols., London, 1774-77) and their continuation as *Experiments and Observations relating to Various Branches of Natural Philosophy* (3 vols., vol. i London, vols. ii-iii Birmingham, 1779-86). A French translation appeared in 1777-80. A revised and abridged edition of the whole was published as *Experiments and Observations on Different Kinds of Air and other Branches of Natural Philosophy* (3 vols., Birmingham, 1790). He also published *Considerations on the Doctrine of Phlogiston and the Decomposition of Water* (Philadelphia, 1796), and *Doctrine of Phlogiston Established and that of the Composition of Water Refuted* (Northumberland, U.S.A., 1800; 2 edit., Northumberland and Philadelphia, 1803). Some of his papers in the *Philosophical Transactions* are referred to below.

In 1770 Priestley wrote from Leeds to his friend the Rev. T. Lindsey, with whom he corresponded to the end of his life, saying: "I am now taking up some of Dr. Hales's enquiries concerning air." * His first chemical paper was on charcoal.†

In 1772 he published a long and important paper, *Observations on Different Kinds of Air*.‡ In this he describes a pneumatic trough with a shelf for collecting gases over water (already to some extent anticipated by Brownrigg §), also the methods of collecting and manipulating gases and the preparation of several new gases. Nitrous air (nitric oxide, already obtained by Boyle, p. 72) was obtained from copper, etc., and dilute nitric acid; phlogisticated air (nitrogen); nitrous vapour (nitrogen dioxide) from copper or bismuth and concentrated nitric acid; "nitrous air diminished" by iron or a mixture of iron and sulphur (largely nitrous oxide); acid air (hydrochloric acid) which, as it was very soluble in water, was collected over mercury (see Cavendish, pp. 102-3); and a gas

* Rutt, i, 113.

† *Phil. Trans.*, 1770, 211.

‡ *Phil. Trans.*, 1772, 147-267, read March, 1772; see his letter to Price, November, 1771, in Rutt, i, 185.

§ *Phil. Trans.*, 1765, 218—the paper had been read 24 years before, and was known to Hales; Cavendish's trough, 1766, see p. 101, had no shelf.

afterwards shown to be carbon monoxide. Cavendish had already collected gases over water and mercury in 1766, and had discovered nitrogen before 1772. Priestley's apparatus is shown in Fig. 54, and Fig. 55 shows part of his laboratory.

In Fig. 54 we see the pneumatic trough *a* with the shelf *bb*, filled with water in which stand the jars, *cc*, containing gases, and another jar, *c*, is on the shelf receiving gas from the generating bottle, *e*.

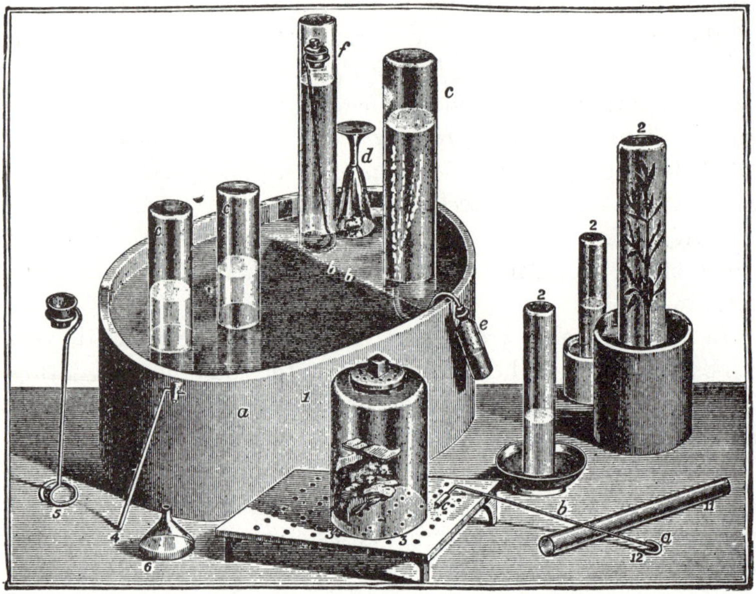

FIG. 54.—PRIESTLEY'S PNEUMATIC TROUGH AND OTHER APPARATUS.
(From his *Experiments and Observations on Air*, vol. i, 1774.)

In the inverted glass, *d*, is a mouse, these animals being kept in the arrangement 3. The jar, *f*, contains a cup supported on a wire stand, also shown in 5, in which a substance may be exposed to a gas. A plant growing in a gas is shown in 2, and other apparatus for the manipulation of gases is depicted. In Fig. 55 we see a substance being heated in a gun-barrel in the fire and the gas evolved collected in a suspended pneumatic trough over mercury. A substance in the tube 8*c* is heated by a candle and the gas, after passing through *d* to condense any liquid, is collected over mercury in 8*b*. In 10 a gas evolved in *c* is collected over water, passing through a bladder on the way so as to allow of the agitation of the bottle. (India-rubber tubing was then quite unknown.) Fig. 18 on the blackboard at the back is like a eudiometer but was filled with a liquid through which sparks were passed. Fig. 19 is the same as the apparatus used later by Cavendish (see our Fig. 62).

In 1772 Priestley measured the " goodness " of air by mixing a known volume of air with nitrous air (nitric oxide) over water and

FIG. 55.—PART OF PRIESTLEY'S LABORATORY.
(From his *Experiments and Observations on Air*, vol. i, 1774, very slightly modified).

noting the contraction (due to removal of oxygen and solution of the resulting higher oxides of nitrogen in the water). This method gives very irregular results unless carried out under standard conditions, as Cavendish (p.104) and Dalton (p.171) afterwards showed, so that Priestley was at first misled into thinking that the " goodness " of air varied rather considerably in different places at different times, whereas it is really practically constant.*

By using the mercury trough Priestley was able to collect several

* Priestley later found the variation " generally inconsiderable " ; *Experiments on Air*, 1790, ii, 259.

gases soluble in water. These he describes in his book, *Experiments and Observations on Different Kinds of Air*, viz. alkaline air (ammonia, 1773-1774); vitriolic acid air (sulphur dioxide, November 1774) the discovery of which was the result of an accident when mercury was drawn back from the trough into a bottle in which he was heating concentrated sulphuric acid, when his hand was badly burned; and fluor acid air (silicon fluoride, before November 1775; first obtained by Scheele in 1771, see p. 109). Oxygen was discovered in 1774 (see below), although he had obtained it in 1771, without recognizing its character, by heating nitre, the gas obtained supporting combustion.* An inflammable air, which was carbon monoxide, was obtained in 1772 (when he distinguished it by its flame from the inflammable air produced from metals and acids); in 1785 and in 1799 (in America) he obtained it by heating smithy scale (ferrosoferric oxide) with charcoal, but Priestley did not then differentiate it from hydrogen.

In 1771 † Priestley recognized that air which has been vitiated by putrefaction, by the breathing of animals or by the burning of candles, is restored by growing green plants (mint, spinach, or groundsel), and in the spring of 1778 he found oxygen in the bladders of seaweed. In June and the following months of 1778 he found that aquatic plants growing in water containing dissolved fixed air (carbon dioxide) give off oxygen. He says: "The injury which is continually done to the atmosphere by the respiration of such a large number of animals ... is, in part at least, repaired by the vegetable creation." Almost the same is said in a note written about 1773.‡ Priestley did not at first realize the important part played by light in the change, now called photosynthesis. He did not publish this until 1781, § and in 1779 he had been anticipated by Ingen-Housz, || who published the main facts, including the action of light. Senebier in 1782 showed that green plants, under the influence of light, convert fixed air into de-

* *Phil. Trans.*, 1772, p. 245.

† Rutt, i, 148 f.; published in *Phil. Trans.*, 1772.

‡ Rutt, i, 196.

§ *Experiments and Observations relating to various branches of Natural Philosophy*, Birmingham, 1781, ii, 16.

|| *Experiments upon Vegetables*, London, 1779.

phlogisticated air.* Priestley had not satisfied himself until September 1779 that the "green matter producing the pure air" was a living vegetable matter formed from invisible seeds in the atmosphere, and that all plants produce the effect when they are healthy, and he did not previously mention the specific action of light.

Priestley's Discovery of Oxygen

Priestley says in 1775† that although the elementary nature of air was a philosophical maxim, he was in the course of his experiments "soon satisfied that atmospherical air is not an unalterable thing". He obtained oxygen on 1st August, 1774, at Calne, by heating red oxide of mercury by a burning glass, and showed that (i) it was practically insoluble in water, (ii) it supported the combustion of a candle in a dazzling manner. His own description of the experiment is as follows.‡

Priestley had been presented with a large convex lens or burning glass, and by its aid he tried to extract "air" from a miscellaneous collection of chemicals given to him by his friend Warltire. Among these was red precipitate, or *mercurius calcinatus per se*, obtained by heating mercury in air, the nature of which had long been a puzzle to chemists (see p. 76). The substances were heated by focusing the sun's rays on them in small phials filled with, and inverted over, mercury.

"Having afterwards procured a lens of twelve inches diameter, and twenty inches focal distance, I proceeded with great alacrity to examine, by the help of it, what kind of air a great variety of substances, natural and factitious [i.e. *artificially* prepared] would yield. . . . With this apparatus, after a variety of other experiments, . . . on the 1st August, 1774, I endeavoured to extract air from *mercurius calcinatus per se*; and I presently found that, by means of this lens, air was expelled from it very readily. Having got about three or four times as much as the bulk of my materials, I admitted water to it, and found that it was not imbibed by it. But what surprized me more than I can well express,

* *Mémoires Physico-Chimiques*, 3 vols., Geneva, 1782.

† *Experiments and Observations on Air*, 1775, ii, 30.

‡ *Experiments and Observations on Different Kinds of Air*, 1775, ii, 29-103; *Alembic Club Reprint* No. 7.

was, that a candle burned in this air with a remarkably vigorous flame. . . . I was utterly at a loss how to account for it."

Early in 1775 Priestley found that a mouse lived twice as long in the new air as in the same confined volume of common air, and revived afterwards when taken out. He breathed it himself, and fancied his " breast felt peculiarly light and easy for some time afterwards "—hence he recommended its use in medicine (it is now used in the treatment of gas poisoning and pneumonia). " Who can tell," he says, " but that, in time, this pure air may become a fashionable article in luxury. Hitherto only two mice and myself have had the privilege of breathing it." He says, however, that " the air which Nature has provided for us is as good as we deserve".

Priestley had really obtained oxygen in 1771 by heating nitre, but he had confused it with air. In his paper on *Observations on different Kinds of Air,** he says: " In one quantity [of air] which I got from saltpetre a candle not only burned, but the flame was increased, and something was heard like a hissing, similar to the decrepitation of nitre in an open fire " ; and when he reprinted this in the first volume of his *Experiments and Observations on Different Kinds of Air* (1774, p. 155) he says: " Experiments, of which an account will be given in a second part of this work, make it probable, that a candle burned even *more than well* in this air . . . . At the time of this first publication, however, I had no idea of this being possible in nature."

Priestley assumed, from the teachings of Stahl, that a candle on burning gives out phlogiston, and is extinguished in a closed vessel after a time because the air becomes saturated with phlogiston. Ordinary air, therefore, supports combustion because it is only partially saturated with phlogiston, and can absorb more of it. Substances burn in air with only a moderate flame, whereas in the new air the flame is vivid. Priestley, therefore, concluded that the new gas must contain little or no phlogiston, and hence he called it *dephlogisticated air.* The gas left when bodies burnt out in ordinary air was named, for a similar reason, *phlogisticated air* :

Phlogisticated Air [Nitrogen] = Air + ϕ. (Scheele's Foul Air)
Dephlogisticated Air [Oxygen] = Air − ϕ. (Scheele's Fire Air.)

It has been suggested that Priestley in August 1774 considered

* *Phil. Trans.*, 1772, p. 245.

the new gas to be nitrous oxide, which he had previously shown to support combustion. If so, he must have overlooked two important differences between the two gases, viz. that nitrous oxide is appreciably soluble in water, whereas he says the new gas was not; and the difference between the dazzling combustion in oxygen and the peculiar flame of a candle burning in nitrous oxide, which he described rather minutely in 1774, when he says that treatment of nitrous air (nitric oxide) with iron "makes it not only to admit a candle to burn in it, but enables it to burn with an *enlarged flame*, by another flame (extending every where to an equal distance from that of the candle, and often plainly distinguishable from it) adhering to it", and says this property is removed by agitating the gas with water.* The third test distinguishing nitrous oxide from oxygen, viz. that the first gas is not acted upon by nitric oxide, whereas oxygen forms a red gas at once soluble in water, Priestley does not seem to have made until 1st March, 1775, after his return from Paris. He says himself that he was unaware of the real nature of the new gas until this date,† and wrote in April, 1775: "I have now discovered an air five or six times as good as common air," ‡ saying that he had "got it first from *mercurius calcinatus, per se*, red lead, &c.; and now, from many substances, as quick-lime (and others that contain little phlogiston), and spirit of nitre, and by a train of experiments demonstrate that the basis of our atmosphere is spirit of nitre. Nothing I ever did has surprised me more, or is more satisfactory."

Priestley's explanation of his results is much less acute than Scheele's description, since Scheele had proved before 1773 that *the whole* of the "pure air" is absorbed on the combustion of phosphorus, whilst Priestley's theory would require a residue of phlogisticated air. The latter, which Scheele had called "foul air", had been independently noticed in 1772 by Daniel Rutherford, afterwards professor of botany at Edinburgh. In his *Dissertatio inauguralis de Aere fixo dicto, aut mephitico*, § he describes experi-

* *Experiments and Observations on Different Kinds of Air*, 1774, vol. i, p. 216.

† *Experiments, etc., on Air*, 1775, ii, 40. ‡ Rutt, i, 268.

§ Translated by Crum Brown, *J. Chem. Education*, 1935, xii, 370; McKie, *Science Progress*, 1935, xxix, 650. The relevant passage is quoted from Rutherford's thesis (p. 17) by Thomson, *System of Chemistry*, 3rd

ments in which mice were allowed to breathe in a confined volume of air, and the fixed air (mephitic air) then removed by caustic potash. There was a residue of another air which is noxious to animals and extinguishes flame, but is distinguished from fixed air by giving no precipitate with lime water. Rutherford seems to have thought this was composed of air and the phlogiston given off from the lungs or combustible, whilst Scheele knew that it pre-existed as part of the air.

Nitrogen was also discovered by Cavendish (who communicated his results to Priestley) by passing air over red-hot charcoal and absorbing the fixed air in caustic potash solution.* By his nitric oxide method (p. 115) Priestley found that air contains one-fifth of its volume of dephlogisticated air (oxygen), which is an accurate result; Scheele (p. 106) had found from one-fourth to one-third, and Lavoisier, by various methods, found from one-sixth to one-fourth. Priestley was on the whole an accurate experimenter and his quantitative results are generally good.

In 1772 Priestley found that air in which tin or lead is heated by a burning mirror contracts by a maximum of one-fourth, and the residue does not affect lime water or nitric oxide. His work on the oxygen compounds of nitrogen is very good.† Besides discovering nitrous oxide (p. 119), he obtained nitrogen dioxide by the action of concentrated nitric acid on copper or bismuth and by heating lead nitrate, and noticed that it darkened in colour on heating. He noticed (1777, also observed by Wenzel in 1777) that iron becomes passive in concentrated nitric acid, that nitrous oxide is respirable by a mouse (1781, although later he said it was fatal to animal life), and that nitric oxide gives a dark-coloured solution in ferrous sulphate solution (1779). By heating iron with a burning-glass in nitric oxide he obtained half the volume of nitrogen (1786). He obtained crystalline nitrososulphuric acid by the action of nitrogen dioxide on concentrated sulphuric acid (1779).

ed., Edinburgh, 1807, i, 102: "Sed aer salubris et purus respiratione animali non modo ex parte fit mephiticus sed et aliam indolis suae mutationem inde patitur. Postquam enim omnis aer mephiticus [CO_2] ex eo, ope lixivii caustici secretus et abductus fuerit, qui tamen restat [N_2] nullo modo salubrior inde evadit; nam quamvis nullam ex aqua calcis praecipitationem faciat haud minus quam antea et flammam et vitam extinguit."

* Priestley, *Phil. Trans.*, 1772, 162; Wilson, *Life of Cavendish*, 1851, 28.

† Meldrum, *J. Chem. Soc.*, 1933, 902.

Priestley, in his memoir on respiration,* says the function of the blood is to discharge phlogiston from the system. He showed that air can act upon blood through the medium of a piece of bladder.

Although most of Priestley's experimental results were accurate, he was led astray in explaining them by his use of the phlogiston theory. In this respect his own candid self-criticism † is very true: "I have a tolerably good habit of circumspection with regard to *facts*; but as to conclusions from them, I am not apt to be very confident", but he is unjust to himself in attributing most of his discoveries to "chance". His *Considerations on the Doctrine of Phlogiston* (1796) and his *Doctrine of Phlogiston Established* (1800) represent the last stand against Lavoisier's new theory.‡

* *Phil. Trans.*, 1776, 226; read 25th January; cf. Rutt, i, 288.

† Bolton, *Correspondence of Priestley*, 1892, 50.

‡ See his letter to Lindsey, January 1800; Rutt, i, 426: "I feel perfectly confident of the ground I stand upon . . . though nearly alone, I am under no apprehension of defeat": also, July 1801, he says Watt and Kirwan, "as good chemists as any in Europe", approved of the phlogiston theory, and "truth will in time prevail over any error"; *ibid.*, ii, 468.

CHAPTER VII

LAVOISIER AND THE FOUNDATION OF MODERN CHEMISTRY

Lavoisier

ANTOINE LAURENT LAVOISIER,* born in 1743 in Paris, was the son of a wealthy advocate. He was educated at the Collège Mazarin. He followed the course on astronomy of la Caille, and on botany of Bernard de Jussieu, and studied geology under Guettard and chemistry under Rouelle. In 1768 he became a member of the Academy, and in the same year became assistant to the *Fermier Général*, Baudon; in 1780 he received this title and position himself. He was later appointed inspector of powder and saltpetre by Turgot. After the Revolution he was a member of the commission for the establishment of the Metric System, and in 1791 Secretary of the Treasury, in which capacity he wrote a treatise on the territorial riches of France with the object of reducing imports. In November 1793 all the *Fermiers Généraux* were arrested by decree of the Convention and Lavoisier was imprisoned. He was tried by jury, found guilty of conspiracy against the people of France by correspondence with the enemies of France, and guillotined on 8th May, 1794.

Lavoisier discovered no new substances, devised no really novel apparatus, and worked out no improved methods of preparation. He was essentially a theorist, and his great merit lay in his capacity of taking over experimental work carried out by others (whose claims he, unfortunately, did not always adequately recognize), and by a rigorous logical procedure, reinforced by his own quantitative experiments, of expounding the true explanation of the results.

It has been well said † that: " Lavoisier, though a great archi-

* E. Grimaux, *Lavoisier*, Paris, 1888; D. McKie, *Antoine Lavoisier*, 1935, 1953; *Notes and Records of the Royal Society*, 1949, vii, 1; *Œuvres de Lavoisier*, 6 vols., Paris, 1864-1893; Tilden, *Famous Chemists*, 1921, 63; J. A. Cochrane, *Lavoisier*, 1931; Hartog, *Annals of Science*, 1941, v, 1; Partington, *Nature*, 1943, clii, 207; Guerlac, *Isis*, 1954, xlv, 51; Duveen and Klickstein, *Bibliography of Lavoisier*, 1955.

† Brande, *Manual of Chemistry*, 6 ed., 1848, p. lxxxv.

tect in the science, labored little in the quarry; his materials were chiefly shaped to his hand, and his skill was displayed in their arrangement and combination." He completed the work of Black,

FIG. 56.—A. L. LAVOISIER, 1743–1794.

Priestley and Cavendish, and gave a correct explanation of their experiments. As compared with Priestley, Lavoisier had a very extensive knowledge of chemistry, and his historical summaries are usually very complete, although sometimes the most important earlier work, which his own experiments confirmed, is either not mentioned at all, or very briefly and inadequately, as though he had got to know of it only after his own experiments were completed.

In the case of his memoirs presented to the Academy, there is the further difficulty that they frequently appeared several years after the date prefixed to the volume of memoirs (as was customary), and in the interval had been revised and extended, or had been first presented in the interval. Thus memoirs published in 1784 in the volume dated 1781 may have taken shape only in 1783-84.

Lavoisier's Quantitative Method

Lavoisier's work is characterized by its systematically quantitative character: he made constant use of the balance, but it is a mistake to say that he was the *first* to apply quantitative methods to chemistry, since these had been systematically used before his time by van Helmont, Boyle and Black. The researches of Black were the model for Lavoisier, as the latter himself says. Such a method necessarily assumes the law of indestructibility of matter, or of conservation of mass, as valid. This was specifically stated by Lavoisier in the form: "... for nothing is created in the operations either of art or of nature, and it can be taken as an axiom that in every operation an equal quantity of matter exists both before and after the operation, that the quality and quantity of the principles remain the same and that only changes and modifications occur. The whole art of making experiments in chemistry is founded on this principle: we must always suppose an exact equality or equation between the principles of the body examined and those of the products of its analysis." *

This "axiom" was applied in Lavoisier's memoir *On the nature of water and on the experiments which have been supposed to demonstrate the possibility of its conversion into earth* (1770; published in 1773). Van Helmont had performed the first experiment of this kind (p. 51). After his time, many chemists had shown that distilled water always leaves a slight earthy residue when evaporated in a glass vessel, and this was supposed to prove that part of the water had been converted into earth. Lavoisier weighed a glass vessel called a pelican (Fig. 57), an alembic provided with tubes returning into the body the distillate condensing in the head, and put into it a weighed quantity of water which had been distilled eight times. The vessel was slowly heated and the stopper closing the head lifted from time to time. It was then

* Lavoisier, *Traité Élémentaire de Chimie*, Paris, 1789, 140.

sealed down and the whole weighed. The vessel was then kept at a temperature of 60°-70° Reaumur for 101 days. A white solid slowly separated in the water, which increased in amount as time went on. The apparatus was cooled and weighed. There was no change in total weight. The water was poured out and the pelican dried and weighed. It had lost $17\frac{4}{10}$ grains. The white powder, when dried, weighed only $4\frac{9}{10}$ grains. The water was then evaporated in a glass alembic and finally in a weighed dish, and left a residue of $15\frac{1}{2}$ grains. But $4\frac{9}{10}+15\frac{1}{2}=20\frac{4}{10}$ grains, which is 3 grains more than the loss in weight of the pelican. This excess was explained as due to water combined in the solid and to matter dissolved from the glass alembic used to evaporate the water. The conclusion reached was that the " earth " had been dissolved by the water from the glass vessel and had not come from the water itself. This was confirmed by Scheele * by qualitative analysis of the " earth ", which was consistent with its origin from the glass.

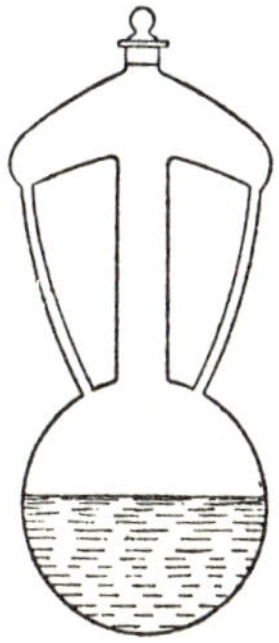

FIG. 57.—A PELICAN.

Lavoisier's Experiments on Combustion and Calcination

In November 1772 Lavoisier deposited with the Academy a sealed note which was opened in 1773. In it he says he had found that sulphur and phosphorus increase in weight when burnt in air. (Marggraf in 1740 had found this with phosphorus; Lavoisier never said how he obtained the result with sulphur.) Some air is absorbed (Hales in 1727 had found this with sulphur and phosphorus). Lavoisier suspected that " air " is absorbed in the calcination of metals, since on reducing litharge (lead oxide) with charcoal a large volume of " air " was evolved.

In his *Opuscules Physiques et Chimiques* (1774), completed in 1773, Lavoisier describes the burning of phosphorus in air over mercury in a bell-jar. Some air was absorbed, since the mercury rose in the jar. The excess of phosphorus could not be ignited in the residual air, which also extinguished a taper. The dry white

* *On Air and Fire,* 1777; *Collected Papers of Scheele,* transl. by Dobbin, London, 1931, 88.

powder (P_2O_5) formed was heavier than the phosphorus burnt: 1 grain of phosphorus gave about 2·7 grains of powder (should be 2·3 grains), and the increase in weight corresponded approximately with the weight of one-fifth of the volume of air used. He also describes an experiment in which 560 cu. in. of " elastic fluid " are evolved when 6 oz. of red lead is reduced by heating with charcoal in an iron retort. This was shown to be fixed air. Thus something is taken from the calx in forming the metal, and this must be " an air ".

In April 1774 Lavoisier presented a memoir to the Academy which was read in November and published in abstract in December.* In this he describes experiments on the calcination of tin and lead in sealed glass retorts, weighed before and after the calcination. There was no change in weight, and thus Boyle's theory of the fixation of ponderable fire particles (p. 75) was disproved. On opening the neck of the retort some air entered and there was a small increase in weight, which was assumed to be equal to the weight of the air which had combined with the metal.†

Boyle's experiments and conclusion had been criticized in 1745 by Lomonosov (1711-65),‡ professor in St. Petersburg ; he explained the increase in weight as caused by absorption of air, and in unpublished experiments in 1750 he found that in calcination in a sealed retort there was no increase in weight. Beccaria in 1759 had shown that the amount of tin or lead calcined in a closed vessel of air is greater the larger the vessel.

A much altered version of Lavoisier's work appeared in 1778,§ when it is said that air consists of two elastic fluids (Scheele had published this in 1777), one of which combines with the metal, and the other " mephitic " part is erroneously said to be " very complex " (*fort composée*). In December 1774 Lavoisier thought that air as a whole combines with metals on calcination.

But in October 1774 Priestley visited Paris with Lord Shelburne, and told Lavoisier at dinner of his discovery of dephlogisticated

* Rozier's *Observations sur la Physique*, 1774, v, 448.

† This is not strictly correct, since after the calcination the retort contains nitrogen, which is lighter than air.

‡ Ostwald's *Klassiker*, No. 178 ; Menshutkin, *Russia's Lomonosov*, Princeton, 1952 ; *Nature*, 1953, clxxi, 138.

§ *Œuvres*, ii, 105.

air, saying he "had gotten it from *precip. per se* and also *red lead*"; whereupon, he says, "all the company, and Mr. and Mrs. Lavoisier as much as any, expressed great surprise".*

Lavoisier also received a letter from Scheele,† dated 30th September, 1774, asking him to try heating silver carbonate with a burning-glass, absorbing the fixed air from the gas with lime water, and seeing if a candle would burn in it. Scheele then knew that oxygen is obtained in this way by ordinary heating.

In November 1774 Lavoisier tried Priestley's experiment of heating red precipitate with a burning-glass. It was repeated in February and March 1775, and a memoir was read on 26th April to the Academy. In May 1775,‡ he announced that the principle which combines with metals on calcination is not part of the air but "the entire air itself" (*l'air lui-même entier*). The air evolved on heating red precipitate has nearly the same density as common air and is diminished by nitric oxide to about the same extent (one-third) as common air. The air combined in nitre is common air deprived of its expansibility. The air which combines with metals and supports combustion and respiration is "even purer than the air in which we live".

A revised version of the memoir appeared in 1778,§ in which "the air itself" has become "the more salubrious and purer portion of the air" or "the eminently respirable part of the air", and some errors of fact are corrected. In his revised memoir on the calcination of metals, also published in 1778, Lavoisier says that "the whole of the air of the atmosphere is not in the respirable state; it is the salubrious part which combines with metals during their calcination and what remains is a species of *mofette*, incapable of sustaining the respiration of animals or the inflammation of bodies".

In 1776 || Lavoisier dissolved mercury in nitric acid, collected

* Priestley, *The Doctrine of Phlogiston Established*, 1800, 88.

† Scheele, *Collected Papers*, transl. Dobbin, 1931, 350; the original is in French.

‡ Rozier's *Observations sur la Physique*, 1775, v, 429.

§ *Œuvres*, ii, 122.

|| *Recueil de Memoires et d'Observations Sur la Formation & sur la fabrication du Salpêtre*, Paris, 1776, 601-17; Partington, *Annals of Science*, 1953, ix, 96 (correcting Hartog, *ibid.*, 1941, v, 1, and other writers); Lavoisier, *Mém. de l'Acad.*, 1776 (1779), 671; *Œuvres*, ii, 129.

the nitric oxide, evaporated the solution, and heated the mercuric nitrate, which was converted into red precipitate.

On heating this he obtained the original weight of mercury and " pure air " or " true air ", which in common air is mixed with three or four times its volume of " deleterious air " (*air nuisible*), which does not support combustion or serve for respiration. He refers to " pure air " as " the air which I obtained from calx or precipitate of mercury and M. Prisley [*sic*] obtained from a large number of substances by treating them with spirit of nitre [nitric acid] ". He should have said that Priestley had told him in October 1774 that he had obtained it by heating red precipitate.*

In the mercury experiment Lavoisier had decomposed nitric acid into nitrous air (nitric oxide) and pure air (oxygen), and by mixing about 2 vols. of nitrous air and 1 vol. of pure air over water he thought nitric acid was formed. He did not then know the composition of nitrous air, but this result would give NO_2 as the formula of nitric anhydride. Cavendish (1785) found $NO_{2 \cdot 33}$ (p. 138), the correct formula N_2O_5 being found by Berzelius (1814).

Lavoisier † confirmed his view of the composition of the atmosphere by a famous experiment described in a memoir presented in 1777 (when Scheele's work was published) and printed in 1780.

He heated 4 oz. of mercury in a retort which communicated with a measured volume of air in a bell-jar over mercury. The volume of air in the bell and in the retort was 50 cu. in. After a time he noticed the formation of red specks and scales of calx on the surface of the mercury. After twelve days the scales no longer increased; the fire was removed, and the experiment stopped. The air had contracted to 42 cu. in., and the gas left was " mephitic air ", which Lavoisier at first called *atmospheric mofette*. He afterwards named it *azote*, and Chaptal called it *nitrogen* in 1790. The scales of mercury calx (*mercurius calcinatus per se*) were collected and found to weigh 45 grains. They were transferred to a small retort and

* Pierre Bayen, in February 1775, had stated that red precipitate lost $\frac{1}{15}$ of its weight on heating and is converted into mercury without any addition of matter containing phlogiston, at the same time as the calx " lets escape a sufficiently large quantity of elastic fluid to which it owes its state " : *Opuscules chimiques*, 1797, i, 298; Rozier's *Observations sur la physique*, 1775, v, 154. Lavoisier never mentions Bayen.

† *Œuvres*, ii, 174; *Traité de Chimie*, 1789, 35.

heated; 8 cu. in. of dephlogisticated air, which was "an elastic fluid, much more capable of supporting respiration and combustion than ordinary air", were obtained, together with $41\frac{1}{2}$ grains of mercury. When this pure air was added to the atmospheric mofette, ordinary air was formed.

The apparatus used by Lavoisier in this experiment is shown in Fig. 58, which is reproduced from Lavoisier's *Traité de Chimie*, all the illustrations in that book being by Madame Lavoisier.

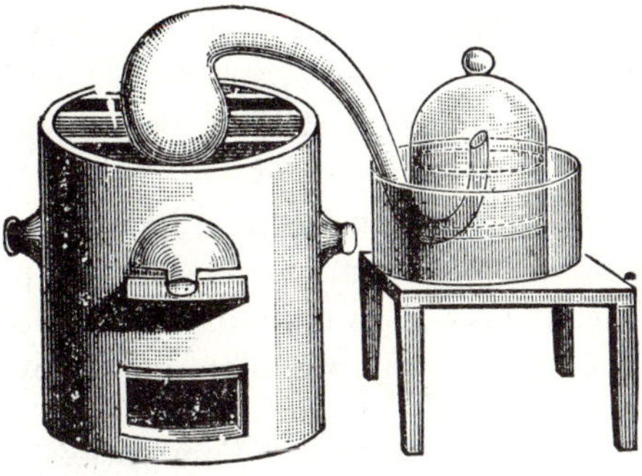

FIG. 58.—LAVOISIER'S APPARATUS FOR HEATING MERCURY IN A CONFINED VOLUME OF AIR.

Lavoisier * in describing oxygen speaks of "this air, which Mr. Priestley, Mr. Scheele and I discovered about the same time", but Priestley † says (as quoted above) that he told Lavoisier of his experiment of obtaining the gas by heating red precipitate, at the same time admitting that Scheele's work was independent of his own. Lavoisier has no claim to the discovery of oxygen.‡

Although Lavoisier did not discover oxygen, he was certainly the first to understand the consequences of its discovery, to realize its true nature as an element, and by ingenious quantitative experi-

* *Traité de Chimie*, 1789, p. 38.

† *Doctrine of Phlogiston Established*, 1800, p. 88; in his *Experiments on Air*, 1775, ii, 36, he says: "I frequently mentioned my surprize at this kind of air . . . to Mr. Lavoisier", in his visit to Paris in October, 1774.

‡ Berthelot, *Nature*, 1890, xliii, 1 ("it is certain that the discovery of oxygen is due to Priestley"); Thorpe, *Essays*, 1902, 149 f.; French, *J. Chem. Education*, 1950, xxvii, 83, suggests that Priestley was not sure he had used genuine red precipitate in August 1774; but Priestley specifically says he was assured by Warltire that it was genuine. See p. 119.

ments to establish the true chemistry of combustion and the calcination of metals which had been foreshadowed by Hooke and Mayow. He completely revolutionized chemistry—Berthelot speaks of his work as " La révolution chimique "—and although a book on chemistry written before his time would not be intelligible to a student unacquainted with the history of chemistry, Lavoisier's *Traité de Chimie* reads like a rather old edition of a modern textbook.

In making " modern " chemistry begin with Lavoisier, however, undue emphasis has perhaps been laid on the theory of combustion. There are other parts of the science, and Davy * had correctly said that if the theory of combustion had to be modified, " the great doctrines of chemistry, the doctrine of definite proportions, and the specific attractions of bodies must remain immutable ".

It is impossible even to mention here all the publications by Lavoisier which led to his new theory : the memoirs of the Academy in the period 1768-1787 contain over sixty papers by him.†

Lavoisier's Theory of Combustion

In 1777 (published in 1780) ‡ Lavoisier found that 1 grain of phosphorus burning in air absorbs 3 cu. in. or $2\frac{1}{2}$ grains, leaving about four-fifths or five-sixths of a *mofette atmosphérique*, which when mixed with " dephlogisticated or eminently respirable air " to make up the original volume gives a gas with all the properties of common air. He concluded that a quarter (*sic*) of the volume of the air unites with phosphorus to form " concrete phosphoric acid (i.e. P_2O_5 ; Lavoisier's " acids " are always the anhydrides). Sulphur on burning causes a contraction and concentrated vitriolic (sulphuric) acid is deposited, in weight two or three times that of the sulphur. (This error is repeated in the *Traité de Chimie*, 1789, p. 66.) Lavoisier felt able to conclude that phosphoric and sulphuric acids contain more than half their weight of eminently respirable air.

In 1777 (published in 1780) Lavoisier § determined the com-

* *Elements of Chemical Philosophy*, 1812, 488 ; *Works*, iv, 364.

† See the summary by Fourcroy, *Système des connaissances chimiques*, 1801, vol. i, p. 36 f.

‡ *Œuvres*, ii, 139.

§ *Œuvres*, ii, 194.

position of sulphuric acid by heating it with mercury, when sulphur dioxide was evolved, and then strongly heating the salt (mercuric sulphate), when it decomposed into mercury, sulphur dioxide and oxygen. He concluded that " volatile sulphurous acid is a vitriolic acid partially deprived of oxygen ".

In a memoir " On Combustion in General " (5th September, 1777, published in 1780) * Lavoisier presented his new theory of combustion:

(1) In every combustion there is disengagement of the matter of fire or of light.

(2) A body can burn only in pure air [oxygen gas].

(3) There is " destruction or decomposition of pure air " and the increase in weight of the body burnt is exactly equal to the weight of the air " destroyed or decomposed ".

(4) The body burnt changes into an acid by addition of the substance which increases its weight.

(5) Pure air is a compound of the matter of fire or of light with a base. In combustion the burning body removes the base, which it attracts more strongly than does the matter of heat, and sets free the combined matter of heat, which appears as flame, heat and light.

Lavoisier says the phlogiston theory explains the phenomena " in a very happy manner ", but it locates the matter of fire (phlogiston) in the combustible, whereas he proposed the alternative hypothesis that it is in the pure air.

Lavoisier proved that the matter of heat is weightless by an experiment published in 1786.† Phosphorus burnt in air in a closed flask, with no appreciable change in weight. Item (5) is an extension of Black's theory of latent heat (p. 95), which assumed that ice is changed to water, and water to vapour, by chemical combination with the matter of heat. In 1787 Guyton de Morveau named the imponderable matter of heat " calorique " (caloric).‡ Calcination, said Lavoisier, is only slow combustion, but instead of an acid there is produced a calx, a compound of the metal with the base of pure air.

In 1782 Lavoisier says Condorcet had proposed the name " vital

* *Œuvres*, ii, 225.

† *Œuvres*, ii, 618.

‡ *Méthode de Nomenclature Chimique*, 1787, 31 (see p. 136).

air " for pure air, but in a memoir * received in 1777, read in 1779, and published in 1781, entitled " General considerations on the nature of acids and on the principles composing them ", Lavoisier called the base of pure air the " acidifying principle " or " oxigine principle " (*principe oxigine*), which he later † changed to " oxygène ", derived " from the Greek ὀξὺς acid, and γείνομαι [a mistake for γεννάω] I produce", pure air being called " oxygen gas " (*gaz oxygène*). The name " gas " had often been used since van Helmont's time (p. 48) and Macquer had used it in France ; Lavoisier changed it to " gaz ".

Phosphoric, sulphuric, nitric, and probably all acids consist of different bases united with the oxygen principle. In a quantitative experiment sugar was oxidized by boiling with nitric acid, when nitrous air (NO) and " chalky acid " (*acide crayeux aériforme*, i.e. carbon dioxide) were evolved and oxalic acid produced, which " is formed by the combination of sugar with about a third of its weight of the acidifying principle ".

The composition of " chalky acid " was determined ‡ by burning charcoal in oxygen over mercury. After correcting for the water also formed (the composition of which he did not know until 1784), Lavoisier found that one part of chalky acid contains 0·234503 of carbon (*substance charbonneuse*) and 0·765497 of oxygen principle (Lavoisier always uses a superfluity of decimals). He later § adopted 72 and 28, the correct figures being 72·7 and 27·3, and this is one of the few good quantitative results which he obtained. In this memoir he calls chalky acid " carbonous acid " (*acide charbonneux*), which was changed in 1787 to " carbonic acid " (*acide carbonique*).||

Lavoisier and Laplace in 1783 ¶ made use of the ice calorimeter in measuring specific heats and the heat evolved in combustion and respiration. This work laid the foundations of thermochemistry, both on its experimental side and also on its theoretical side, since the authors stated the principle that as much heat is required to decompose a compound as is liberated on its formation from the

* *Œuvres*, ii, 248.

† *Traité de Chimie*, 1789, 54-5.

‡ *Œuvres*, ii, 403.

§ *Traité de Chimie*, 1789, 68.

|| *Méthode de Nomenclature Chimique*, 1787, 100.

¶ *Œuvres*, ii, 283 ; this paper is dated 18 June, 1783.

elements. They say that "respiration is a combustion, truly very slow but otherwise perfectly similar to that of charcoal; it occurs in the interior of the lungs" and "the heat developed in this combustion is communicated to the blood which traverses the lungs and is spread throughout the animal system". They measured the weight of ice melted, and the amount of carbon dioxide produced, by a guinea pig in oxygen in the ice calorimeter. The heat evolved in forming the carbon dioxide by burning charcoal in oxygen, determined in a separate experiment, melted 10·38 oz. of ice, that melted by the guinea pig was 13 oz.

FIG. 59.—LAVOISIER'S EXPERIMENTS ON RESPIRATION.

SEGUIN IS BREATHING AIR OR OXYGEN FROM A JAR STANDING IN A TROUGH THROUGH A TUBE FITTED IN A MASK. MADAME LAVOISIER IS SEATED AT THE DESK. (From a Drawing by Mme. Lavoisier.)

Lavoisier's Experiments on Respiration

Mayow, Scheele, Priestley and Lavoisier were all aware of the great similarity between combustion and respiration. Lavoisier * believed that the oxygen gas breathed into the lungs oxidizes the carbonaceous materials of the blood, producing carbon dioxide which is exhaled, and that *animal heat* is the result of this chemical

* *Œuvres*, ii, 676, 688; memoirs of 1785 and 1789.

process of oxidation. In memoirs with Seguin * (1789; printed 1793) Lavoisier describes experiments on respiration, in which he measured the oxygen consumed and the water and carbon dioxide produced (Fig. 59).

Lavoisier thought that oxidation, with production of animal heat, occurred in the lungs, but Mayow already knew that it occurred in the muscles (p. 83). Lavoisier and Seguin say the oxidation of carbon and hydrogen supplied by the food to produce heat takes place in the tubes of the lungs, in which a humour composed of carbon and hydrogen is secreted from the blood. Hassenfratz † says Lagrange supposed that oxidation occurs " not in the lungs alone, but in all parts of the body where the blood circulates ", but this view was not generally adopted until Gustav Magnus in 1837 showed that both arterial and venous blood gave off oxygen and carbon dioxide in a vacuum.‡

Chemical Elements

Lavoisier's definition of an " element or principle " as " the last point which analysis is capable of reaching " is essentially that of Boyle: it is tentative, for, " since we have not hitherto discovered the means of separating them, they act with regard to us as simple substances, and we ought never to suppose them compounded until experiment and observation have proved them to be so ".§ His table of elements is reproduced from his *Traité* on p. 135 (*acidifiables* in the third class should, obviously, read *salifiables*): the omission of the alkalis is noteworthy (see p. 183): by analogy with Berthollet's decomposition of ammonia (1785) he thought they might later be shown to be compounds.

The New Chemical Nomenclature

One of the first chemists to adopt Lavoisier's views must have been Black, who taught them before 1784.|| In 1785 Berthollet accepted the new theory, and was followed by Guyton de Morveau in 1786 and Fourcroy in 1786-7. Lavoisier and these three chemists

* *Œuvres*, ii, 688, 704. † *Ann. de Chimie*, 1791, ix, 261.

‡ *Ann. Physik*, 1837, xl, 583.

§ Traité, 1789, preface, and p. 192.

|| R. Lubbock. *Dissertatio de principio sorbile*, Edinburgh, 1784. p. 12.

	Noms nouveaux.	Noms anciens correſpondans.
Subſtances ſimples qui appartiennent aux trois règnes & qu'on peut regarder comme les élémens des corps.	Lumière	Lumière.
	Calorique	Chaleur. Principe de la chaleur. Fluide igné. Feu. Matière du feu & de la chaleur.
	Oxygène	Air déphlogiſtiqué. Air empiréal. Air vital. Baſe de l'air vital.
	Azote	Gaz phlogiſtiqué. Mofete. Baſe de la mofete.
	Hydrogène	Gaz inflammable. Baſe du gaz inflammable.
Subſtances ſimples non métalliques oxidables & acidifiables.	Soufre	Soufre.
	Phoſphore	Phoſphore.
	Carbone	Charbon pur.
	Radical muriatique.	Inconnu.
	Radical fluorique	Inconnu.
	Radical boracique	Inconnu.
Subſtances ſimples métalliques oxidables & acidifiables.	Antimoine	Antimoine.
	Argent	Argent.
	Arſenic	Arſenic.
	Biſmuth	Biſmuth.
	Cobolt	Cobolt.
	Cuivre	Cuivre.
	Etain	Etain.
	Fer	Fer.
	Manganèſe	Manganèſe.
	Mercure	Mercure.
	Molybdène	Molybdène.
	Nickel	Nickel.
	Or	Or.
	Platine	Platine.
	Plomb	Plomb.
	Tungſtène	Tungſtene.
	Zinc	Zinc.
Subſtances ſimples ſalifiables terreuſes.	Chaux	Terre calcaire, chaux.
	Magnéſie	Magnéſie, baſe du ſel d'Epſom.
	Baryte	Barote, terre peſante.
	Alumine	Argile, terre de l'alun, baſe de l'alun.
	Silice	Terre ſiliceuſe, terre vitrifiable.

LAVOISIER'S LIST OF THE ELEMENTS (1789).

published in 1787 a *Méthode de Nomenclature Chimique*, in which the names of chemical substances were changed so as to bring them into line with the new theory. This helped in its adoption, as also did Lavoisier's *Traité élémentaire de Chimie* (1789). Fourcroy (1755-1809) was a zealous propagator of the new theory in his lectures and books, his large *Systême des connaissances chimiques* (11 vols., 1801) being very detailed and clearly written. Chemists outside France followed suit, and only a few of the older conservative ones, like Priestley in England, Macquer in France, and Wiegleb in Germany refused conversion. The German chemists were just as ready to adopt the new views as any, the first text-book in German based on Lavoisier's views being C. Girtanner's *Anfangsgründe der antiphlogistischen Chemie* (1792).

The Composition of Water

Before Lavoisier's new theory could prove acceptable to other chemists a serious difficulty had to be removed. A metal like tin or zinc dissolves in an acid giving inflammable air, and on evaporating the solution a salt is left, which, on strong heating, parts with its acid and leaves the calx of the metal. The same salt is formed when the calx is dissolved in the acid, but no inflammable air is then evolved. Whence comes the inflammable air in the first experiment? This was an easy question for the phlogistonists. Inflammable air is phlogiston; the metal is (calx + phlogiston); and the salt is (calx + acid). In the first experiment:

$$(\text{calx} + \phi) + \text{acid} = (\text{calx} + \text{acid}) + \phi$$

This difficulty was serious and Lavoisier was unable to offer an explanation. The key was first supplied by the researches of Cavendish on the formation of water from inflammable air and dephlogisticated air.

In 1781 Priestley had exploded a mixture of inflammable air and dephlogisticated air in a bottle and noticed the deposition of dew. Priestley says this " random experiment " confirmed an opinion of Warltire's that " common air deposits its moisture when phlogisticated ". Warltire exploded the mixture of gases by an electric spark in a closed copper globe and found a small diminution in weight, which he attributed to the escape of heat. Priestley communicated these experiments to James Watt and to Cavendish, and the latter continued them, with Priestley's permission. Cavendish's experiments are described below (p. 138).

In March 1782 Priestley * investigated the reduction of calces of metals when heated in the inflammable air (hydrogen) discovered by Cavendish in 1766. This gas was confined over water or mercury in a bell-jar, the calx being supported in a vertical spoon and heated by the sun's rays focused on it by a burning-glass. By using red lead he obtained metallic lead and managed to cause the calx to absorb practically all the inflammable air. Since lead = calx + phlogiston, it would seem that inflammable air is identical with phlogiston, a view also held by Cavendish in 1766 and by Kirwan in 1782. Priestley thought the metal should have weighed more than the calx, but as this was never the case he concluded that the loss in weight was due to the sublimation of some of the calx. He says: †

"For seeing the metal to be actually revived, and that in considerable quantity, at the same time that the air was diminished, I could not doubt, but that the calx was actually imbibing something from the air; and from its effects in making the calx into metal, it could be no other than that to which chemists had unanimously given the name of *phlogiston*."

Priestley concluded that phlogiston is present in a combined state in metals, just as fixed air is present in chalk, both being equally capable of being expelled again by acids. He repeated the experiment with calx of mercury, the inflammable air being confined over mercury, and observed that "even though the inflammable air was previously well dried with fixed ammoniac" (calcium chloride), water was formed in "sufficient quantity". He thought this water was contained in the gas and is deposited when "this kind of air is decomposed".

Priestley also found that alkaline air (ammonia gas) reduced calx of lead heated in it, but left a large residue of phlogisticated air (nitrogen).

* Letter to Wedgwood, in Bolton, *Scientific Correspondence of Joseph Priestley*, 1892, 33, 35.

† *Experiments on Air*, 1790, vol. i, p. 248 f., 276.

Cavendish's Synthesis of Water

Cavendish's experiments on the formation of water from inflammable air (H_2) and dephlogisticated air (O_2) were begun as a continuation of the preliminary work of Priestley and Warltire (p. 136). His memoir is called " Experiments on Air ".* He begins by showing that when the two gases are exploded in a closed copper or glass vessel there is no loss of weight, as Warltire supposed. By using various mixtures he came to the conclusion that:

" When inflammable air and common air are exploded in proper proportions, almost all the inflammable air and near one-fifth of the common air lose their elasticity and are condensed into the dew which lines the glass."

" The better to examine the nature of the dew " he burnt the gases together, and " the burnt air " was passed through a glass tube 8 ft. long to condense the dew. " By this means upwards of 135 grains of water were condensed in the cylinder, which had no taste nor smell, and which left no sensible sediment when evaporated to dryness; neither did it yield any pungent smell during the evaporation; in short, it seemed pure water."

Cavendish then repeated Priestley's experiment, firing inflammable air and dephlogisticated air by means of an electric spark in " a glass globe, holding 8800 grain measures, furnished with a brass cock, and an apparatus for firing air by electricity. This globe was well exhausted by an air-pump and then filled with a mixture of inflammable and dephlogisticated air, by shutting the cock, fastening a bent glass tube to its mouth and letting up the end of it into a glass jar inverted over water, and containing a mixture of 19500 grain measures of dephlogisticated air, and 37000 of inflammable air; so that, on opening the cock, some of this mixed air rushed through the bent tube, and filled the globe." In a footnote he explains that the end of the bent tube was closed with a bit of wax, " which was rubbed off when raised above the surface of the water ". The gases were taken moist and no attempt was made to dry them, although Cavendish had previously used drying tubes (p. 102).

Cavendish gives only the above description, and no figure, of the

* *Phil. Trans.*, 1784, lxxiv, 119; *Alembic Club Reprint* No. 3; Partington, *The Composition of Water*, 1928.

glass globe used for exploding the gases, but it is believed that the original "Cavendish's eudiometer" is shown in Fig. 60. He carried out successive explosions in the globe, and the ratio of the volumes of inflammable and dephlogisticated air which appeared was 2·02 to 1, in good agreement with the experiments with air. About 30 grains of water had collected in the globe, but this was found to be "sensibly acid to the taste, and by saturation with fixed alkali, and evaporation, yielded near two grains of nitre; so that it consisted of water united to a small quantity of nitrous acid". He found that more acid was formed when the dephlogisticated air was used in excess, but when excess of inflammable air was used no acid was formed, and none was formed when inflammable air was exploded with common air. Cavendish was puzzled by this acid and started an investigation on the cause of its production, which delayed the publication of his paper. In the meantime he says:

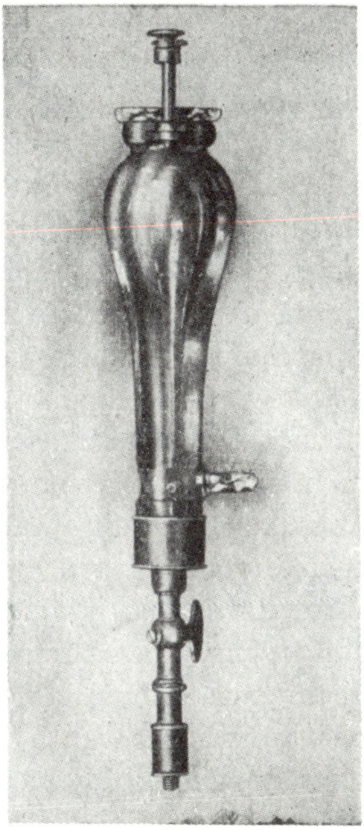

FIG. 60.—CAVENDISH'S FIRING GLOBE ("EUDIOMETER").

(From a Photograph of what is believed to be the original apparatus in the University of Manchester.)

"I think we must allow that dephlogisticated air is in reality nothing but dephlogisticated water, or water deprived of its phlogiston; or, in other words, that water consists of dephlogisticated air united to phlogiston; and that inflammable air is either pure phlogiston, as Dr. Priestley and Mr. Kirwan suppose, or else water united to phlogiston." It may be mentioned that Cavendish himself, in 1766, had supposed inflammable air to be pure phlogiston. Of the two alternatives, Cavendish thought the second more probable, i.e.:

$$\text{dephlogisticated air} = \text{water} - \phi,$$
$$\text{inflammable air} = \text{water} + \phi.$$

The water formed was thus supposed to pre-exist in the gases, and the cause of the explosion was a redistribution of phlogiston.

The reason for his preference is explained as follows. Common or dephlogisticated air will combine at the ordinary temperature with nitrous air (NO), which is a compound of nitric acid and phlogiston (p. 103), and since nitric acid is formed (in the presence of water), the dephlogisticated air can withdraw phlogiston from its combination with nitric acid. Inflammable air, however, does not react with dephlogisticated air until heated to redness, which is hard to explain if inflammable air is free phlogiston.

In the paper as printed in 1784, there is an addition by Cavendish saying that the results could also be explained by Lavoisier's theory, when "we must suppose, that water consists of inflammable air united to dephlogisticated air . . . and indeed . . . adding dephlogisticated air to a body comes to the same thing as depriving it of its phlogiston and adding water to it"; but the "commonly received principle of phlogiston explains all phaenomena, at least as well as Mr. Lavoisier's", and hence Cavendish adhered to it. He seems at times to be thinking on the lines of the new theory but with the old terminology, as when he says that red precipitate is "only quicksilver which has absorbed dephlogisticated air from the atmosphere during its preparation".*

Composition of Nitric Acid

Cavendish, in his second memoir on the subject, published in 1785,† had found an explanation of the cause of the acidity of the water. The nitric acid (which he calls "nitrous") is formed from nitrogen ("phlogisticated air") contained as an impurity in the dephlogisticated air. The acid is not formed in the explosion of inflammable air with common air because the flame is then not hot enough. Cavendish proved by experiment that nitric acid is formed when oxygen and nitrogen are sparked in presence of potash solution, when nitre is produced. (Priestley in 1774 knew that an acid is formed on sparking air over litmus solution ‡). A mixture of the two gases was enclosed by Cavendish in a bent glass tube inverted over mercury, and also containing some potash

* *Alembic Club Reprint* No. 3, p. 29. † *Phil. Trans.*, 1785, lxxv, 372.

‡ *Experiments on Air*, 1774, i, 183.

solution (" soap-lees ") (Fig. 62). The gases gradually disappeared on sparking and a solution of nitre was formed. Excess of oxygen

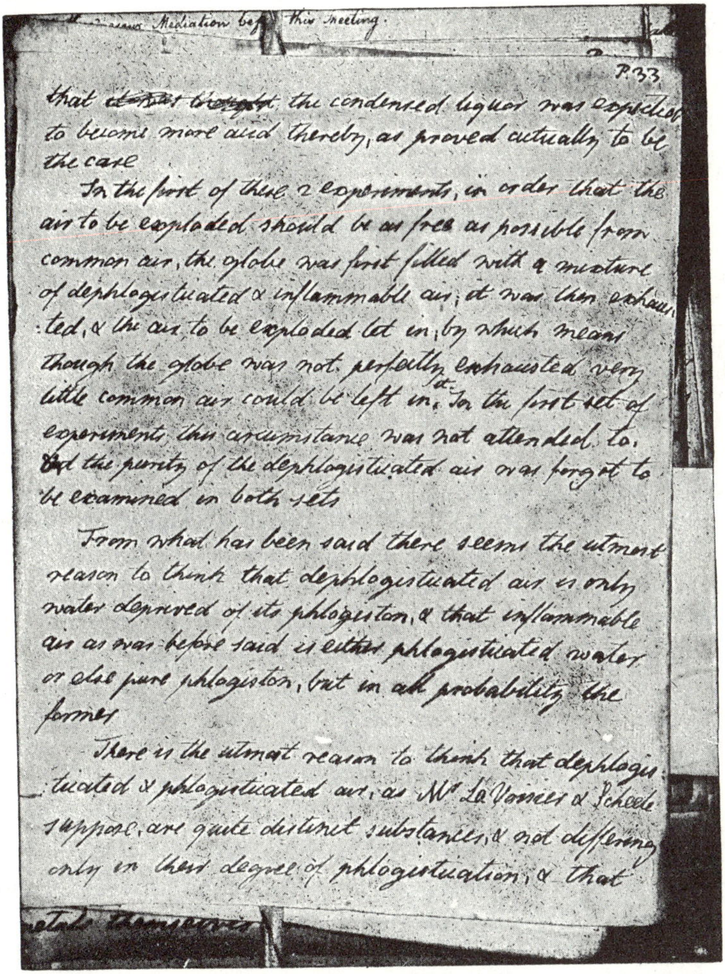

P. 33

that the condensed liquor was expected to become more acid thereby, as proved actually to be the case

In the first of these 2 experiments, in order that the air to be exploded should be as free as possible from common air, the globe was first filled with a mixture of dephlogisticated & inflammable air; it was then exhausted, & the air to be exploded let in, by which means though the globe was not perfectly exhausted very little common air could be left in. In the first set of experiments this circumstance was not attended to. And the purity of the dephlogisticated air was forgot to be examined in both sets

From what has been said there seems the utmost reason to think that dephlogisticated air is only water deprived of its phlogiston, & that inflammable air as was before said is either phlogisticated water or else pure phlogiston, but in all probability the former

There is the utmost reason to think that dephlogisticated & phlogisticated air, as Mr La Voisier & Scheele suppose, are quite distinct substances, & not differing only in their degree of phlogistication, & that

FIG. 61.—PART OF THE MS. OF CAVENDISH'S MEMOIR, IN THE ARCHIVES OF THE ROYAL SOCIETY.

was removed with a solution of liver of sulphur (potassium sulphide), and only a very small bubble of gas remained, " certainly not more than 1/120 of the bulk of the phlogisticated air let up into the tube ; so that if there is any part of the phlogisticated air of our atmosphere which differs from the rest, and cannot be reduced to nitrous acid, we may safely conclude that it is not more than 1/120 of the whole ".

This small residue of gas was long afterwards shown by Rayleigh and Ramsay to be argon, and its detection is a tribute to Cavendish's splendid experimental skill and to his absolute honesty in reporting his results. He says the experiments proved that " the phlogisticated air was enabled, by means of the electric spark, to unite to, or form a chemical combination with, the dephlogisticated air, and was thereby reduced to nitrous acid, which united with the soap-lees and formed a solution of nitre ; for in these experiments the two airs actually disappeared, and nitrous acid was actually formed in their room ".

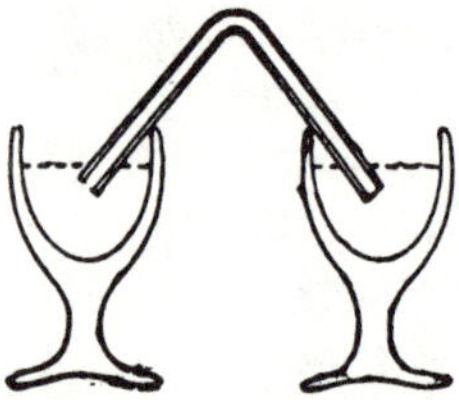

FIG. 62.—CAVENDISH'S APPARATUS FOR SPARKING AIR OVER POTASH SOLUTION.

Cavendish's great skill is also shown by his observation that, although the solution gave a precipitate with silver nitrate, this was not due to hydrochloric acid, since nitre, when strongly heated, parts with dephlogisticated air and leaves a residue which, on dissolving in water, precipitates silver nitrate (as silver nitrite). " This property of phlogisticated nitre is worth the attention of chemists, as otherwise they may sometimes be led into mistakes, in investigating the presence of marine acid by a solution of silver." Since some chemists expressed doubts as to the accuracy of Cavendish's results, a committee of the Royal Society repeated them in 1787-88, and fully confirmed them.

The Water Controversy

Whilst Cavendish had been carrying out his experiments, Priestley had not been idle. He now obtained his inflammable air by heating bits of charcoal in an earthenware retort, so that it would contain carbon monoxide, and his dephlogisticated air by heating nitre, so that it would contain some nitrogen. On exploding them together he obtained fixed air and nitric acid, as we should expect. The weight of water formed by the explosion was determined by mopping it off the globe with filtering paper and reweighing, and Priestley says that he was not very satisfied and " could wish for a nicer balance ", but " always found, as nearly as

I could judge, the weight of the decomposed air in the moisture acquired by the paper ".*

Two letters from James Watt, the first (dated 26th April, 1783) to Priestley, and the second (dated 26th November, 1783) to De Luc, gave his explanation of Priestley's experiments. Priestley submitted the letter of 26th April to Sir Joseph Banks, the President of the Royal Society, and it was seen by Blagden. It was to have been read, but Watt asked for this to be delayed, since he wished to have time to examine some new experiments, said by Priestley to contradict his theory. Cavendish's paper (" Experiments on Air ") was read on 15th January, 1784, whereupon Watt asked for his letters to be read, which was done, the first (dated 26th April, 1783) being read on 22nd April, and the second (dated 26th November, 1783) on 29th April, 1784. This was in consequence of some mischievous representations to Watt by De Luc, a foreign Fellow, that Cavendish was trying to steal his theory. In his letter to Priestley, Watt suggests that " water is composed of dephlogisticated air and phlogiston deprived of part of their latent or elementary heat ". Since Watt says that phlogiston is the same as hydrogen, it has been supposed that his claim to have first stated the composition of water has much to recommend it. In actual fact, Watt never seems to have fully understood what he meant. In a later letter of November 1783 to De Luc he says: " my assertion was simply that air was water deprived of its phlogiston and united to heat ".

It is probable that Watt's theory was the result of some letters from Priestley saying he had converted water into air by boiling it in (porous!) earthenware retorts. Watt wrote to Boulton on 10th December, 1782: " I have often said that if water could be heated red hot or something more, it would probably be converted into some kind of air, because steam would in that case have lost all its latent heat and that it would have been turned solely into sensible heat, and probably a total change of the nature of the fluid would ensue. Dr. Priestley has proved this by experiment." † Davy, in a letter of 1799, could say that Watt and Keir " are still phlogistians ",‡ and the claim of Watt—a man of the highest moral

* *Phil. Trans.*, 1783, 414.

† Bolton, *Correspondence of Priestley*, New York, 1892, 45.

‡ Paris, *Life of Davy*, 1831, i, 79.

character—seems to have been overestimated. The resulting " Water Controversy ", in which both eminent chemists and eminent lawyers took part, will be found in Wilson's *Life of Cavendish* (London, 1851) and in Muirhead's *Correspondence of the Late James Watt on his Discovery of the Theory of the Composition of Water* (London, 1846), which contain most of the original documents, and also deal with Lavoisier's claims.* Whatever verdict may be given for or against Watt, there can be little dispute as to Lavoisier. He has no claim to be regarded as having established the facts until he was informed of Cavendish's experiments, but he first gave the correct interpretation of them (see, however, Cavendish's statement, p. 140). The facts are as follows.

In May or June 1783 Blagden, who was Cavendish's assistant, visited Paris and told Lavoisier of Cavendish's discovery. Lavoisier had been searching for the acid which he thought should be formed on the combustion of inflammable air in oxygen, but without success. In his later memoir (see below) he mentions that Macquer in 1776-7 had noticed the deposition of drops of water on a cold porcelain saucer " licked " by a hydrogen flame, and that Bucquet had suspected that fixed air should be the product of combustion, whilst Lavoisier thought sulphurous or sulphuric acid should be formed. He describes an experiment made by Bucquet and himself in 1777 disproving both theories.† After meeting Blagden, Lavoisier on 24th June, 1783, made a crude experiment of burning hydrogen in oxygen and collecting the water formed. This was not quantitative in character, yet on the next day he sent a memoir to the Academy, in which it is asserted that the weight of water formed by the combination of hydrogen and oxygen is equal to the sum of the weights of the gases, and that water is, therefore, " composed, weight for weight of inflammable air and vital air ".

This result could not possibly have been established by Lavoisier's own experiments, and there is no doubt that he based it on what he heard of Cavendish's work from Blagden. Cavendish, however, is barely mentioned in the paper, which merely says that it was reported that Cavendish had burnt the two gases in a closed vessel and had " obtained a very sensible amount of water ". When the

* See also Thorpe, *Essays*, 1902, 98 f., 163 f.

† Partington, *The Composition of Water*, 1928, 34.

paper appeared, in 1784, in a volume of the *Mémoires* of the Academy dated 1781 (its publications were some years in arrear), it had also certainly been retouched since it was submitted in 1783, because it mentions Blagden as Secretary of the Royal Society. Blagden, however, did not become Secretary of the Royal Society until May 1784, so that the paper as finally submitted by Lavoisier was some months later than the reading of Cavendish's paper in January 1784. To add to the confusion, the reprints of Cavendish's paper are dated 1783 in mistake for 1784. Blagden, in a letter published in Crell's *Annalen* in 1786, i.e. during Lavoisier's life, bluntly charged the great French chemist with an attempt to plagiarize Cavendish, and this Lavoisier practically admitted in 1790 (*Ann. Chim.*, 1790, vii, 257). Arago,* who studied all the documents and the unpublished letters of Watt, although deciding for Watt's priority, did not think it necessary to consider a claim for Lavoisier.

In 1784 Lavoisier and Meusnier made some experiments, described in the memoirs of the Academy dated 1781,† in which an iron gun-barrel coated outside with clay was heated in a furnace in a slightly inclined position and water allowed to drip slowly into the hot tube from a tin funnel attached to the upper end. The lower end of the gun-barrel was attached to a spiral tube luted to a tubulated bottle to collect the liquid water coming over, and the hydrogen was then collected in a pneumatic trough. A modification of this experiment described in 1789,‡ in which steam is passed over heated iron in a glass tube, gave the composition of water by weight as 85 oxygen and 15 hydrogen.

Monge § also made experiments on exploding oxygen and hydrogen in a glass globe by an electric spark and obtained a large quantity of water (see Fig. 63). These experiments, published in 1786, were made at Mézières in 1783 and without knowledge of Cavendish's. The results are less accurate than Cavendish's.

* *Œuvres complètes*, 1854, i, 452; see also the re-examination of the matter by McKie, *Lavoisier*, 1935, 237 f., where the conclusions of Partington, *The Composition of Water*, 1928, 30 f., are confirmed; also Kahlbaum and Hoffmann, in Kahlbaum's *Monographien*, No. 1, 1897, 150 f.

† Lavoisier, *Œuvres*, ii, 360; Partington, *The Composition of Water*, 1928, 37, 42; Lowry, *Historical Introduction to Chemistry*, 1936, 118.

‡ *Traité élémentaire de Chimie*, 1789, 92.

§ *Mém. de l'Acad.*, 1783 (publ. 1786), 78; Partington, *The Composition of Water*, 1928, 49.

The name " hidrogène " for inflammable air was proposed by Guyton de Morveau ; * Lavoisier † called it " hydrogène " (from ὕδωρ water, γεννάω I produce).

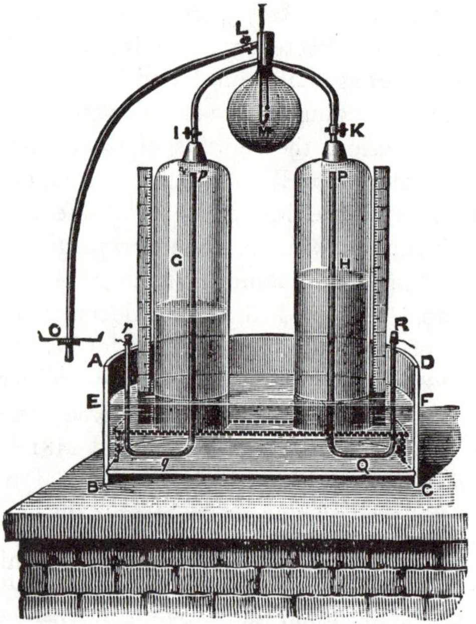

FIG. 63.—MONGE'S EXPERIMENT ON THE COMBINATION OF HYDROGEN AND OXYGEN GASES.

THE GASES WERE COLLECTED THROUGH SYPHON-TUBES, *pr* PR, IN THE CYLINDERS G AND H ; THEY PASSED THROUGH THE STOPCOCKS, I AND K TO THE GLOBE, M, PREVIOUSLY EXHAUSTED THROUGH THE TAP, L, LEADING TO AN AIR-PUMP AT O. THE MIXED GAS WAS THEN EXPLODED BY AN ELECTRIC SPARK AT M, AND THE PROCESS REPEATED.

For measuring the hydrogen and oxygen and supplying them to the combustion apparatus, Lavoisier had made at great expense what Arthur Young, who saw it in 1787, calls " a noble machine ".‡ The results obtained were inferior to those found by Cavendish and others (Bryan Higgins, Van Marum, etc.) with much simpler apparatus. Lavoisier in 1784 reported § that 12 vols. of oxygen

* *Méthode de Nomenclature Chimique*, 1787, 33.

† *Traité de Chimie*, 1789, 94.

‡ Young, *Travels in France*, Cambridge, 1929, 81 ; for the apparatus, see Lavoisier, *Œuvres*, ii, Plate IV, and *Traité*, 1789, Plate IV, Fig. 5 and Plate VIII.

§ *Œuvres*, ii, 334.

combine with 22·924345 (instead of 24) vols. of hydrogen, and with inaccurate densities calculated that 0·86866273 of oxygen and 0·13133727 of hydrogen, by weight, form one part of water. Results obtained with the apparatus by Fourcroy, Vauquelin, and Seguin * were no better.

It was now possible to explain the difficulty mentioned on p. 136. Lavoisier † says that Laplace in a letter of September 1783 had suggested that a metal such as zinc, dissolving in dilute acid, takes oxygen from water to form oxide of zinc, which dissolves in the acid to form a salt, and the hydrogen of the water is set free:

zinc + (hydrogen + oxygen) = zinc oxide + hydrogen,
zinc oxide + sulphuric acid = sulphate of zinc.

Lavoisier regarded the acid as an oxide; at present it is regarded as (oxide + water), $SO_3 + H_2O = H_2SO_4$, so that the hydrogen really comes from the acid.

In 1785‡ Lavoisier said that phlogiston is hypothetical and unnecessary, and in 1786§ he tried to show that the production of fire in combustion is better explained by his hypothesis that oxygen gas is a compound of a base and the matter of fire. His arguments involve some errors of fact, and although the chemical parts of his theory were soon accepted, his alternative to the phlogiston theory was regarded as unsatisfactory (see p. 89).||

Summary and Supplement

In this, and further summaries, the broad outlines of the work of each of the chemists dealt with in the text are summarized, and the opportunity is often taken of giving a brief mention of other chemists, or of work done by those dealt with in the text, in addition.

The Oxford Chemists

Robert Boyle (1627-91) defined element, compound, analysis; he believed in the atomic ("corpuscular") theory, which he used to explain chemical changes. He modified the air pump, first used

* *Mém. de l'Acad.*, 1790 (1797), 485.

† *Œuvres*, ii, 567.

‡ *Œuvres*, ii, 509.

§ "Réflexions sur le Phlogistique", *Œuvres*, ii, 623-55.

|| Partington and McKie, *Annals of Science*, 1938, iii, 337; 1939, iv, 113.

sealed thermometers in England, observed the effects of reduced pressure on boiling, made freezing mixtures, determined specific gravities of liquids and solids, prepared and collected hydrogen and showed that it was inflammable, discovered "Boyle's law" (1662), investigated phosphorus (1680-82) and wood spirit, examined combustion in relation to the air and the calcination of metals, and extended analytical tests, especially in solutions (acids and alkalis; iron salts; silver and copper salts). The tests for mineral waters, begun by Libavius (p. 56), were improved also by Friedrich Hoffmann (1660-1742), professor at Halle.

Robert Hooke (1635-1702) made further experiments on combustion, and suggested that air contains a substance fixed in nitre ("nitrous air") which acts as a solvent for combustibles.

John Mayow (1641-79), a physician, in his *Tractatus quinque* (1674) described very ingenious experiments with the object of proving that air and nitre contain a common constituent, "nitro-aerial particles", or "igneo-aerial particles", or the "nitro-aerial spirit". Combustion is due to collision of nitro-aerial particles with "sulphureous particles" in the combustible, a theory different from Hooke's. In the calcination of metals, nitro-aerial particles from the atmosphere are fixed in the metal, causing an increase in weight. Mayow, following Willis (1670), recognized that combustion and respiration are analogous processes. He states correctly that animal heat is developed in the muscles, and proved that arterial blood in a vacuum gives off a gas. He had very clear ideas on chemical affinity, including the effect of volatility.

The Phlogistic Period

Johann Joachim Becher (1635-82) introduced the idea of an inflammable earth (*terra pinguis*) to explain combustibility (1669), which was extended into the *Theory of Phlogiston* (the name phlogiston, given to Becher's *terra pinguis*, had been used before) by Georg Ernst Stahl (1660-1734), professor of medicine and chemistry at Halle. Stahl assumed that all combustible bodies and metals contained a common principle, phlogiston (ϕ), the same in all, which escaped on combustion or calcination, but could be transferred from one body to another, and restored to the metallic calces by heating with substances rich in phlogiston (charcoal, oil, etc.), when the metal was reproduced:

$$\text{metal} \rightleftharpoons \text{calx} + \phi.$$

Although it was well known as early as the sixteenth century, and had again been emphasized by Jean Rey in 1630, that metals increase in weight on calcination while phlogiston was supposed to have escaped, this fact was either disregarded or else phlogiston was later assumed to have negative weight.

Lomonsov (1745) explained the increase in weight as due to air fixed by the metal and (1750) disproved Boyle's theory that it was due to fixation of fire particles. Laurent Béraut (*Dissertation sur la Cause de l'Augmentation de Poids, que certaines Matières acquirèrent dans leur Calcination*, Bordeaux, 1747) concluded that the increase in weight comes from the addition of certain foreign particles mixed with the air and separated by the action of heat.

In the *old* Phlogiston Theory, phlogiston is equivalent to *minus* oxygen ; in the later theory it was sometimes assumed to be hydrogen (Cavendish, Kirwan, Priestley), or the matter of light (Macquer).

Although Hermann Boerhaave (1668-1738), professor in Leyden, does not mention the phlogiston theory in his famous text-book (*Elementa Chemiae*, 1732), he uses a similar theory in which a " food of fire " (*pabulum ignis*) replaces phlogiston : he thought this existed in a nearly pure state in alcohol, which produced water on combustion.

Nicolas Lemery (1645-1715), demonstrator at the Jardin du Roi, did not adopt the theory of phlogiston. His text-book, *Cours de Chymie* (1675), went through a large number of editions and translations.

Important chemists of the Phlogistic Period are :

German

Friedrich Hoffmann (1660-1742).
Caspar Neumann (1683-1737).
Johann Theodor Eller (1689-1760).
Johann Heinrich Pott (1692-1777).
Andreas Sigismund Marggraf (1709-1782).

French

Étienne François Geoffroy (1672-1731).
Wilhelm Homberg (1652-1715).*
Pierre Joseph Macquer (1718-1784).
Guillaume François Rouelle (1703-1770).

British and Irish

Joseph Black (1728-1799).
Henry Cavendish (1731-1810).
Joseph Priestley (1733-1804).
Richard Kirwan (1735-1812).
William Higgins (1763-1825).

Swedish

Torbern Bergman † (1735-1784).
Johann Gottlieb Gahn (1745-1818).
Carl Wilhelm Scheele (1742-1786).

Although the theory of phlogiston had the advantage of co-ordinating a large number of facts into a system, it retarded the progress of

* Homberg was born in Batavia of German parents but worked mostly in France.

† *Not* Bergmann.

chemistry, and prevented a number of the best investigators from seeing the correct explanation of the facts they brought to light.

Hoffmann clearly distinguished magnesia from lime. Eller determined the solubilities of salts. Pott, who was chemist to the Berlin porcelain factory, gave a clear description of zinc, and investigated the action of heat on a great number of substances from the point of view of the manufacture of porcelain. Marggraf, a pupil of Neumann, investigated phosphorus and microcosmic salt, discovered beet sugar (1747), distinguished alumina from lime (1754), and potash from soda (1757-9; already distinguished by Duhamel, a French chemist, in 1736). Macquer discovered potassium and sodium arsenates (1746-48) and made important researches on dyeing. He wrote a text-book and a dictionary of chemistry. Rouelle, demonstrator at the Jardin du Roi, clearly distinguished neutral, acid and basic salts (1744), preparing, for example, acid potassium sulphate ($KHSO_4$). He showed that sulphuretted hydrogen is inflammable. His brother, H. M. Rouelle, described urea in 1773. Geoffroy published affinity tables (1718). Homberg investigated quantitatively the combination of acids and bases (1699-1702). Bergman, professor at Uppsala, developed mineral analysis by the use of the blowpipe, the systematic use of reagents, and quantitative methods. His views on affinity (1775) and his " affinity tables " were very influential.

Development of the Chemistry of Gases

Stephen Hales (1677-1761) in his *Vegetable Staticks* (1727) describes the production of " air " from a variety of substances, and its collection and measurement, but he did not distinguish qualitatively the different gases (oxygen, hydrogen, nitric oxide, carbon dioxide, coal gas, etc.) which he must have obtained. Joseph Black (1728-99), professor in Glasgow and Edinburgh, by careful *quantitative* experiments, proved (1754) that " mild " alkalis, " mild " magnesia and limestone, are compounds of caustic alkalis, calcined magnesia and quicklime, respectively, with a peculiar gas which he called *fixed air*. This was carbon dioxide, already recognized and called *gas sylvestre* by van Helmont (p. 48). Black also discovered latent and specific heats. He adopted Lavoisier's oxygen theory in preference to the phlogistic hypothesis before 1784.

Henry Cavendish (1731-1810), who made important quantitative investigations in physics and chemistry, recognized and named *equivalents* (1766-88); investigated the properties of fixed air (CO_2) and inflammable air (H_2)—which he then identified with phlogiston, and collected gases over water and also over mercury (1766); determined the (nearly constant) composition of the atmosphere

(1781), and the volumetric composition of water (1784). He showed that nitric acid is formed by sparking nitrogen (" phlogisticated air ") with excess of oxygen (" dephlogisticated air ") over potash solution, which is converted into nitre, and on absorbing the excess of oxygen by liver of sulphur a small bubble of gas (later identified as argon) remains. Cavendish at that time remained faithful to the phlogistic hypothesis, although he admitted that his results could be explained by Lavoisier's oxygen theory.

Joseph Priestley (1733-1804), a theologian and chemist, very considerably improved the manipulation of gases (1770), using the mercury trough for collecting gases soluble in water. He discovered ammonia gas (" alkaline air "), hydrochloric acid gas (" acid air "), nitric oxide (" nitrous air "), nitrous oxide (" diminished nitrous air "), nitrogen dioxide (" nitrous vapours "), oxygen (" dephlogisticated air "), nitrogen (" phlogisticated air ")—also discovered by Daniel Rutherford in 1772—carbon monoxide (1772-99), and sulphur dioxide (" vitriolic acid air "). He recognized that green plants restore the goodness of air vitiated by the burning of candles or the respiration of animals (1771), and that there is here a production of oxygen, and that light is necessary (1781). He found that inflammable air (H_2) is rapidly absorbed by heated metallic oxides (calces) and reduces them, hence he (and Kirwan) identified it with phlogiston (1782-83). He noticed that water is formed even if the gas is dry. Priestley and Warltire (1781) noticed that water is formed in the explosion of inflammable air with common air or dephlogisticated air.

Carl Wilhelm Scheele (1742-86), a Swedish apothecary, was the first discoverer of oxygen (1772-4 ; first published in 1777), also chlorine, and characterized manganese and barium compounds (1774), with Gahn discovered phosphoric acid in bones (1770), obtained impure (1771) and then fairly pure (1786) hydrofluoric acid, and silicon fluoride (1771), investigated hydrogen sulphide and discovered hydrogen persulphide (1777), arsenic acid and arsenic hydride (1775) ; molybdic (1778) and tungstic (1781) acids ; distinguished between molybdenite and graphite (1779). Scheele discovered several organic compounds : tartaric (1770), oxalic (1776, 1784-5), lactic (1780), mucic and pyromucic (1780), uric (1780), prussic (1782-8), citric (1784), malic (1785), gallic and pyrogallic (1786) acids ; casein (1780) ; aldehyde (1782) and glycerol (1783-4). Scheele held the peculiar theory that phlogiston + fire air (oxygen) = heat.

Foundation of Modern Chemistry

Antoine Laurent Lavoisier (1743-94) had an excellent education in mathematics and physics, and in chemistry was a pupil of Rouelle. He paid particular attention to quantitative method and the use of the balance (following Black), and formally stated the " law of indestructibility of matter " (1789). He proved (1770) that water cannot be converted into earth, as was previously supposed. He showed that tin and lead increase in weight on calcination and an equal weight of air is absorbed (1774), and also found the same result for phosphorus on burning (1772). He proved that air consists of oxygen (so named in 1778) and " azote ", and that in combustion oxygen is absorbed, acids being formed by the combustion of non-metals (sulphur, phosphorus, carbon), and bases (" calces ") by the oxidation of metals. In 1783 he definitely replaced the phlogistic by the " antiphlogistic theory " (loss of phlogiston is absorption of oxygen and *vice versa*). Experiments on the composition of water (1783-4) were based on Cavendish's, but Lavoisier first clearly explained them. He analysed organic substances by combustion in oxygen, and collaborated with Guyton de Morveau (1737-1816), A. F. de Fourcroy (1755-1809) and C. L. Berthollet (1748-1822) in devising the new chemical nomenclature: *Méthode de nomenclature chimique* (1787). Lavoisier's *Traité élémentaire de Chimie* (1789) popularized the new theories.

After 1785 the theory of phlogiston rapidly disappeared except among a few very conservative chemists (Priestley, Macquer, Wiegleb, etc.).

CHAPTER VIII

LAWS OF COMBINING PROPORTIONS AND THE ATOMIC THEORY

PART I. CONSTANT PROPORTIONS

Proust

ISOLATED examples of the quantitative analysis or synthesis of compounds go back at least to the seventeenth century. Kunckel (d. 1703) says 12 parts of silver give 16 parts of chloride (white silver calx), which is very near the true weight of 15·9, and Bergman (d. 1784) made a large number of analyses of metallic salts for the purpose of determining metals, etc., in the form of insoluble compounds in quantitative analysis. The assumption that compounds were of definite composition seems, therefore, to have been tacitly recognized during the eighteenth century by all chemists who concerned themselves with quantitative investigation.

A long series of investigations on the composition of minerals' and artificial compounds of many metals was made by Joseph Louis Proust * (1754-1826), a French chemist, who was professor in Madrid from 1789 to 1808 (when his laboratory was pillaged during the siege by the French army). Proust showed that several metals can form more than one oxide (he discovered cuprous oxide in 1799) and sulphide, each of which has a definite composition. He also distinguished clearly between oxides and hydroxides or hydrated oxides.

Proust recognized the *law of constant proportions* in 1797, and stated it in 1799 and 1806 as follows :

" We must recognize an invisible hand which holds the balance in the formation of compounds. A compound is a substance to which Nature assigns fixed ratios, it is, in short, a being which

* *Ann. Chim.*, 1797, xxiii, 85; 1799, xxxii, 26; *J. de Physique*, 1801, liii, 89; 1802, liv, 89; 1802, lv, 324; 1804, lix, 260, 265, 321, 343, 350, 403; 1806, lxiii, 364, 421.

Nature never creates other than balance in hand, *pondere et mensurâ*."

In addition to his work on constant proportions, Proust isolated crystalline grape sugar (1802), and mannite (1806), showed that microcosmic salt contains soda (1775), and isolated leucine

FIG. 64.—J. L. PROUST, 1754-1826.

(" caseous oxide ") from the products of putrefaction of casein (1818).

Berthollet

Claude Louis Berthollet (1748-1822), born of French parents in Italy (Savoy), studied medicine in Turin. In 1772 he became physician to Mme. de Montesson, and studied chemistry in Paris, where Lavoisier was beginning his great researches on combustion. In 1794 he became professor at the École polytechnique, but he was not very successful as a teacher, his lectures being too difficult for his students, in whom (as Cuvier says) he mistakenly assumed the same degree of intelligence as his own.

In 1798 Berthollet accompanied the Napoleonic expedition to Egypt, and in 1799 he read at Cairo to a scientific society, founded by members of the expedition, a memoir on the action of mass, which was published in 1801.* The ideas in this work were ex-

FIG. 65.—C. L. BERTHOLLET, 1748-1822.

panded in 1803 into his *Statique Chimique* (2 vols.), in which he challenged the law of constant proportions.

Other chemical researches of Berthollet included the determination of the composition of ammonia (1785); researches on chlorine (oxymuriatic acid), hypochlorites and chlorates (1785-87), and on the application of chlorine to bleaching; the discovery that prussic acid is a compound of hydrogen, carbon and nitrogen free from

* *Recherches sur les lois de l'affinité*, Paris, An IX.

oxygen, and of cyanogen chloride (1787); and the discovery of the acidic properties of sulphuretted hydrogen and the investigation of hydrogen persulphide (1796; hydrogen persulphide was discovered by Scheele). As a result of the investigations of prussic acid and sulphuretted hydrogen, Berthollet concluded that acids need not necessarily contain oxygen, as Lavoisier had assumed (p. 131).

Controversy of Berthollet and Proust

The main evidence brought forward by Berthollet against constant proportions, and the corresponding reply by Proust, may be summarized as follows.

1. Metals such as copper, tin and lead, on heating in air can take up oxygen *continuously* in proportions increasing to a fixed limit, giving a continuous series of oxides, as shown in some cases (e.g. lead) by varying colour changes.

Proust showed that these oxides were mixtures of two, or a small number, of definite oxides, and he carefully distinguished between mixtures and solutions (*mélanges*; *dissolutions*) and chemical compounds (*combinaisons*). Thus the members of the supposed series of oxides of tin were mixtures of two definite oxides, the suboxide (tin 80, oxygen 20) and protoxide (tin 72, oxygen 28). Similar results were found for the sulphides.

2. A solution of an insoluble base, such as copper oxide, in an acid, such as sulphuric, on addition of an alkali is precipitated as basic salts containing amounts of acid which decrease continuously as more alkali is added.

Proust thought he had proved that these all consisted of the metallic hydroxide imperfectly freed from the metal salt by washing, but it is now known that they often contain basic salts of *definite* composition.

3. Mercury dissolves in nitric acid to take up oxygen and form a series of salts of continuously varying composition.

Proust showed that these are mixtures of two definite salts, mercurous and mercuric nitrates, the existence of which had also been recognized by Scheele before 1775.

4. Solutions, metallic alloys, amalgams and glasses are compounds formed in indefinite proportions.

This class gave Proust some trouble, but he refused to admit that solutions were chemical compounds, and wisely left the matter for further investigation. He says:

" Is the power which makes a metal dissolve in sulphur different from that which makes one sulphide dissolve in another? I shall be in no hurry to answer this question, legitimate though it be, for fear of losing myself in a region not sufficiently lighted up by the facts of science; but my distinction will, I hope, be appreciated all the same when I say: The attraction which causes sugar to dissolve in water may or may not be the same * as that which makes a fixed quantity of carbon and of hydrogen dissolve in another fixed quantity of oxygen to form the sugar of plants, but what we do clearly perceive is that these two kinds of attraction are so different in their results that it is impossible to confound them."

Berthollet was forced to recognize that in many cases substances of definite composition can be formed in chemical changes, but he regarded these as exceptional. They were formed as a result of the interference of extraneous physical forces. For example, certain proportions of the elements could produce a compound which was least soluble or of greatest density (influence of cohesion), or the most volatile (influence of elasticity) of all the possible compounds, and hence this compound was formed in preference.

Chemists seem to have decided in favour of Proust and definite proportions about 1805; the advent of Dalton's atomic theory in 1807 (see p. 173) finally put an end to the controversy, since it was incompatible with indefinite proportions. It was unfortunate that while rejecting Berthollet's law of indefinite proportions, chemists also did not develop his law of mass action, although Berzelius † emphasized that this is not incompatible with the atomic theory.

* Berthollet thought *all* forces of affinity were modified gravitational attraction.

† *Traité de Chimie*, Paris, 1831, vol. iv, pp. 513 f., 527 f., 586 f.: " Quoique les résultats des expériences de Berthollet aient paru d'abord si opposés à l'adoption du système général des proportions chimiques, nous trouvons maintenant qu'ils découlent, comme des conséquences nécessaires, des vues de la théorie corpusculaire." Berzelius stood well above his contemporaries in this as in many other matters.

PART II. MULTIPLE PROPORTIONS

The Law of Multiple Proportions

Although Proust and other chemists had recognized that two elements could combine in more than one proportion, they never

FIG. 66.—W. H. WOLLASTON, 1766-1828.

emphasized the ratio of the proportions of one element combining with identical weights of the other. In some cases the analyses were not sufficiently accurate to disclose a whole multiple relation. This was first realized by Dalton about 1803 as a consequence of his Atomic Theory, and he confirmed it by experiments with the oxides of nitrogen, and with marsh gas and olefiant gas, which are described below (p. 172). Dalton recognized clearly that when two elements combine to form more than one compound, the weights of one element which unite with identical weights of the other are in simple multiple proportions, but he did not state the result explicitly in this form.

This result was shown to be true also for compounds (acids and bases) in 1808 by Thomson, who directed attention to the two oxalates of strontium and of potassium, and showed that one salt (now called the normal salt) " contains just double the proportion of base contained in the second " (now called the acid salt). Wollaston in 1808 showed that a third oxalate of potash exists, the proportions of acid in the three oxalates being in the ratio 1 : 2 : 4 for identical amounts of base; that the amounts of carbonic acid (CO_2) combined with identical amounts of potash (K_2O) in the two carbonates were in the ratio 1 : 2; and that the same ratio holds for the proportions of sulphuric acid (SO_3) in the two sulphates of potash. Wollaston says: " the enquiry which I had designed appears to be superfluous, as all the facts that I had observed are but particular instances of the more general observation of Mr. Dalton." *

PART III. RECIPROCAL PROPORTIONS

Equivalents

The law of equivalents was first deduced from investigations on the combining proportions of acids and bases. Homberg in 1699 had determined the amounts of several acids required to neutralize identical weights of salt of tartar (K_2CO_3) and the weights of salt obtained on evaporation, but his results are not accurate.

Cavendish in 1766 found that identical weights of a given acid require different weights of different bases for neutralization, and he calls these weights of the bases *equivalent*. In 1788 he found that the weights of sulphuric and nitric acids which neutralized identical weights of potash also reacted with identical weights of marble, so that the ratio of the weights of the two acids was constant for the two bases. In unpublished notes, Cavendish in 1777

* Thomson and Wollaston's papers in *Alembic Club Reprint* No. 2: " Foundations of the Atomic Theory ".

records the making up of solutions of equivalent weights of salts and determining the electrical conductivities of the solutions by comparing the lengths of solution through which the discharge from a battery of Leyden jars gave equal shocks.*

Wenzel

The discovery of the general law of equivalent weights has persistently been attributed, owing to a slip on the part of Berzelius †, to Carl Friedrich Wenzel (1740-1793), who published in 1777 his *Theory of the Affinity of Bodies* (*Lehre von der Verwandtshaft der Körper*). An examination of the book, which is written in an involved and obscure style, lends no support to Berzelius's statement. Wenzel, in fact, so far from believing that pairs of salts exchange their components on double decomposition in such a way that no component is left in an uncombined state, asserts exactly the opposite. Thus, he found ‡ that when ½ oz. of silver chloride (horn silver) reacts with mercuric sulphide (cinnabar) to produce silver sulphide and corrosive sublimate, the analyses of the separate salts would indicate that " the acid of the horn-silver rises with the mercury out of 202½ grains of cinnabar as a corrosive sublimate; the silver, on the other hand, remains combined with only so much sulphur as is contained in 125½ grains of cinnabar." In other cases, Wenzel specifically refers to uncombined residues arising from double decompositions and suggests that they should be utilized by adding other substances, e.g. acids set free should be neutralized.

Bergman in 1783, in experiments made to determine the different quantities of phlogiston in metals, found the weights of various metals required to precipitate one another from solutions of the salts; e.g. to precipitate 100 parts of silver required 234 of lead and 31 of copper. These are really equivalent weights, although this was not recognized by Bergman.

* *Scientific Papers of Cavendish*, Cambridge, 1921, vol. i, p. 27.

† *Théorie des proportions chimiques*, Paris, 1819, p. 2; on p. 16 he correctly attributes the discovery to Richter; the examples given by Thomson, *History of Chemistry*, vol. ii, p. 279 f., as illustrating the law of equivalents, are not from Wenzel, and Thomson's account is often repeated by modern writers. The error was corrected by Hess in 1840.

‡ *Lehre von der Verwandtschaft*, p. 453.

Kirwan in 1783 determined the amounts of various metals, and of various bases, required to saturate 100 parts of each of the three mineral acids, and exhibited the results in tables, which are really tables of equivalents, although Kirwan does not draw this general conclusion.

Richter

Between 1780 and 1790 there was an accumulation of material which could lead some alert mind to the generalization now known

FIG. 67.—J. B. RICHTER, 1762-1807.

as the law of reciprocal proportions or the law of equivalents. This generalization is the work of Jeremias Benjamin Richter * (born in Silesia in 1762), who had studied under Kant at Königsberg and graduated in 1789 with a dissertation on " The Use of Mathematics

* Partington, *Annals of Science*, 1951, vii, 173 ; 1953, ix, 289.

in Chemistry ". Richter worked as an assayer in Breslau and from 1798 as second chemist in the porcelain factory in Berlin, where he died in 1807. During his short life Richter received little encouragement, and after his death his discoveries were mostly credited to others. He was permeated with the idea that chemistry is a branch of applied mathematics, and he busied himself in detecting regularities among combining proportions where Nature has not provided any.

In 1791 Richter found that if solutions of calcium acetate and potassium tartrate are mixed, when calcium tartrate is precipitated and potassium acetate remains in solution, the mixture remains neutral, and he remarks that " this is found by experience to hold for all decompositions by double affinity, in so far as the compounds used in the decomposition are also neutral ".* This " law of neutrality " has been incorrectly attributed to Wenzel. It was discovered by Guyton de Morveau (1787),† who showed that analyses of salts by Kirwan were incompatible with it. Richter drew from it the conclusion that if " the components of two neutral compounds are $A-a$, a and $B-b$, b, then the mass ratios of the new neutral compounds produced by double decomposition are $A-a:b$ and $B-b:a$ ". In his *Rudiments of Stoichiometry* ‡ in 1792-4 he generalized the result in the statement that " the elements must have among themselves a certain fixed proportion of mass ", and gave a large number of combining proportions and decomposition ratios. These are sometimes collected in tables; for each acid the combining proportions of various bases are stated, and so on.

In 1795 § Richter says: " If P is the mass of a determining element, where the masses of its elements determined are a, b, c, d, e, etc., Q the mass of another determining element, where α, β, γ, δ, ϵ, etc., are the masses of its elements determined, so that a and α, b and β, c and γ, d and δ, e and ϵ shall represent the same elements; and further if the neutral masses P + a and Q + β,

* *Ueber die neuern Gegenstände der Chymie*, Breslau and Hirschberg, 1791, i, 74.

† *Encyclopédie méthodique, Chymie*, 1787, i, 582, 595; *Annales de Chimie*, 1798, xxv, 292.

‡ *Anfängsgründe der Stöchyometrie*, 3 vols., Breslau and Hirschberg, 1792-1794: he defines the new name Stoichiometry as " the art of measuring the chemical elements ", i.e. their combining ratios by mass.

§ *Ueber die neuern Gegenstände der Chymie*, 1795, iv, 67.

P + a and Q + γ, P + c and Q + α, P + a and Q + γ, etc., are decomposed by double affinity, so that the resulting products are neutral, then the masses a, b, c, d, e, etc., have the same quantitative ratio among one another as the masses α, β, γ, δ, ϵ, etc., or conversely." This is the most general statement of the Law of Reciprocal Proportions.

Fischer's Table of Equivalents

In the German translation of Berthollet's *Recherches sur les lois de l'affinité* (see p. 155) by Ernst Gottfried Fischer (1754-1831) in 1802 *, a clear summary of Richter's views (which Fischer says were practically unknown even in Germany) is given, and the numerous tables of Richter are combined into a single table of equivalent weights of acids and bases referred to 1000 parts of sulphuric acid. Part of this is reproduced below. "The meaning of this table", says Fischer, "is the following. If a substance is taken from one of the two columns, say potash from the first, to which corresponds the number 1605, the numbers in the other column indicate the quantity of each acid necessary to neutralize 1605 parts of potash.... If a substance is taken from the second column, the numbers in the first column show how much of each of the substances in this column will be necessary for its neutralization."

Bases		*Acids*	
Alumina - -	525	Fluoric - -	427
Magnesia - -	615	Carbonic - -	577
Ammonia - -	672	Muriatic - -	712
Lime - - -	793	Oxalic - -	755
Soda - - -	859	Phosphoric - -	979
Strontia - -	1329	Sulphuric - -	1000
Potash - -	1605	Nitric - - -	1405
Baryta - -	2222	Acetic - -	1480

Fischer's table and part of his note were reproduced in a French translation in Berthollet's *Statique Chimique* in 1803 †, and thus became generally known. Richter himself published a similar table of 18 acids and 30 bases in 1803.

* *Claude Louis Berthollet über die Gesetze der Verwandtschaft*, Berlin, 1802, p. 229.

† *Statique Chimique*, Paris, An XI, 1803, vol. i, p. 134

PART IV. THE ATOMIC THEORY

Origins of the Atomic Theory

The theory of atoms goes back to the early Greek philosophers. Boyle in 1661 * could speak of "those Theories of former Philosophers, which are now with great applause revived, as discovered by these latter ages". Although Strabo † mentions a Phoenician Mochos, "a Sidonian who lived before the Trojan war", as the originator of "the ancient opinion about atoms", nothing more is known of him, and it is generally agreed that speculations about atoms began with the ancient Greeks. ‡

Anaxagoras of Klazomenai (near Smyrna) (498-428 B.C.) taught that bodies are divisible without limit and retain their characteristics. Gold is composed of little "seeds" of gold, flesh of little fleshes, etc., but the heart is not composed of little hearts but its flesh of little fleshes. Aristotle (384-322 B.C.), who reports this, called the "seeds" *homoiomeres*. They were not atoms, but a step towards them.

The Eleatic School, founded by Xenophanes of Kolophon (570-480 B.C.), had as chief master Zeno of Elea (b. 489 B.C.), who taught that matter is continuous and, since the universe is full of matter, there is no motion. He invented several ingenious paradoxes, such as that of "Achilles and the tortoise", to prove that things do not really move and we only think they do.

Aristotle says Leukippos (about 450 B.C. ?) was the real founder of the atomic theory. All his writings are lost. Aristotle says he argued against the Eleatic School somewhat as follows :

Eleatic : Without a vacuum there is no motion ; there is no vacuum, hence there is no motion.

Leukippos : Without a vacuum there is no motion ; there is motion, hence there is a vacuum.

Aristotle says : "Leukippos thought he had a theory which was in harmony with sense, 'for', said he, 'that which is strictly speaking real is an absolute *plenum* ; but the *plenum* is not one [undivided

* *Sceptical Chymist*, 1661, p. 23.

† *Geography*, XVI, ii, 2, p. 756 C.

‡ J. Burnet, *Early Greek Philosophy*, 3 edit., 1920 ; Ida Freund, *The Study of Chemical Composition*, Cambridge, 1904, 226 f. ; Partington, *Annals of Science*, 1939, iv, 245.

whole]. On the contrary, there is an infinite number of them, and they are invisible owing to the smallness of their bulk. They move in the void (for there is a void); and by their coming together they effect coming into being; by their separation, passing away.' "

The atomic theory was adopted and extended by Demokritos of Abdera (460-370 B.C.), one of the most famous and acute thinkers of antiquity, whose writings are unfortunately lost except a few fragments. Aristotle says he taught that atoms are hard, and have form and size and perhaps weight (this is a disputed point). They are invisible by reason of their very small size. They have no colour, taste or smell, since these are merely secondary or subjective properties. The atoms move spontaneously and ceaselessly in the vacuum; they come together by " necessity " (*ἀνάγκη*, not the same as " fate ", *τύχη*) and form aggregates by a sort of hook-and-eye mechanism, not by attractive forces. The motion of atoms is like that of dust particles seen in a sunbeam in still air in a room.

The atomic theory was adopted by Epikouros (341-270 B.C.), who definitely attributed weight to the atoms. They fall perpendicularly through the vacuum, but at indetermined times and in indetermined places they " swerve " and so enter into collision. This " swerve ", ridiculed by opponents of Epikouros, has some resemblance to the modern theory of indeterminancy, and the " necessity " of Demokritos is not unlike the statistical considerations of the kinetic theory of gases. The atomic theory was eloquently used by Lucretius about 57 B.C. in his long poem *De Rerum Natura* (" Of the Nature of Things ").

Lucretius, who calls the atoms " seeds ", says (in the translation of Thos. Creech, 1714):

> " But solid seeds exist, which fill their place;
> And make a diff'rence betwixt full and space.
> These, as I prov'd before, no active flame,
> No subtle cold can pierce, and break their frame,
> Tho' ev'ry compound yields: no pow'rful blow,
> No subtle wedge divide, or break in two.
> For nothing can be struck, no part destroy'd
> By pow'rful blows, or cleft without a void,
> And things that hold most void, when strokes do press,
> Or subtle wedges enter, yield with ease.

If seeds then solid are, they must endure
Eternally, from force, from stroke secure."

Asklepiades of Prusa, about 100 B.C., introduced the idea of small clusters of atoms (*onkoi*), corresponding with what we call molecules.

Revival of the Atomic Theory

Interest in the atomic theory was greatly revived by the writings of Gassendi (1592-1655), and during the seventeenth century it was very familiar. Boyle and Lemery made extensive use of it.* Boyle says†: "It seems not absurd to conceive that at the first Production of mixt Bodies, the Universal Matter whereof they among other Parts of the Universe consisted, was actually divided into little Particles of several sizes and shapes variously mov'd."

Newton was a thoroughgoing atomist. He also assumed ‡ that "particles attract one another by some force, which in immediate contact is exceeding strong, at small distances performs the chymical operations, and reaches not far from the particles with any sensible effect." By assuming that a gas is composed of particles repelling one another with a force varying inversely as the distance, Newton deduced Boyle's law.§ He rejected Descartes' idea that by their endless encounters some of the particles might have suffered attrition, and become "old worn particles" different from the rest.||

Higgins

In 1789 William Higgins published *A Comparative View of the Phlogistic and Antiphlogistic Theories*, to the advantage of the latter, in which he puts forward some very interesting speculations ¶ on chemical combination of "particles". These (which are partly based on earlier publications by Bryan Higgins, in 1775-86) certainly foreshadow the law of multiple proportions and valency bonds, but they cannot be regarded as anticipating Dalton's atomic theory. Dalton knew nothing of Higgins until his theory was completed.

* J. C. Gregory, *A Short History of Atomism*, London, 1931.

† *Sceptical Chymist*, 1661, p. 37. ‡ *Opticks*, 1730, Query 31, p. 364.

§ *Principia*, book 2, prop. 23, theorem 18; London, 1687, p. 301.

|| *Opticks*, 1730, p. 376.

¶ Partington, *Annals of Science*, 1939, iv, 245; *Nature*, 1955, clxxvi, 8.

In discussing the compounds of nitrogen and oxygen, Higgins uses diagrams (Fig 68).

Let the force between the ultimate particle of oxygen and the ultimate particle of nitrogen be 6. This is assumed to be divided

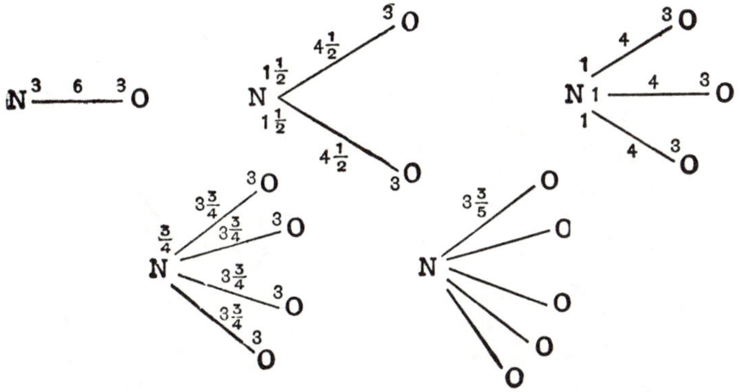

FIG. 68.—COMBINATION OF PARTICLES ACCORDING TO HIGGINS (1789).

equally between the two particles. If two particles of oxygen combine with one of nitrogen, the force 3 of the nitrogen atom is divided into two forces of $1\frac{1}{2}$, so that the force holding each oxygen particle to the nitrogen is now only $4\frac{1}{2}$. By adding 3, 4 and 5 particles to one of nitrogen the forces become, as is seen, 4, $3\frac{3}{4}$ and $3\frac{3}{5}$. The force which would arise by further addition of oxygen particles is too small to keep them attached, so that compounds beyond N + 5O do not exist.

Higgins did not attempt to find the relative *weights* of the particles, and it is not clear whether he thought they had the same or different weights.

Dalton

The atomic theory did not prove fruitful in chemistry until John Dalton (1766-1844) endowed the atoms of the chemical elements with fixed and different weights.

John Dalton * was born at Eaglesfield, a village near Cockermouth in Cumberland, his father being a weaver, and his mother, a woman of strong character, a member of a yeoman family. The parents were Quakers, to which Society John Dalton was attached

* Roscoe, *John Dalton*, 1895; Partington, *Scientia*, 1955, xc, 221.

throughout his life. In 1776 he entered the service of Elihu Robinson, a wealthy Quaker of scientific attainments, who taught him mathematics, and in 1778 he began to teach, single-handed, a village school for pupils of all ages. When this closed, Dalton worked in the fields. In 1781 he joined his brother in teaching in

FIG. 69.—JOHN DALTON, 1766-1844.

a school in Kendal, where he met John Gough, a blind philosopher, who encouraged him to study languages and mathematics, and on whose recommendation Dalton was, in 1793, appointed professor of mathematics and natural philosophy in New College, Manchester, in which city he spent the rest of his life. He resigned from New College in 1799, and thereafter he supported himself by giving private tuition, at the same time continuing his research.

From his note-books, found in the archives of the Manchester Literary and Philosophical Society by Roscoe and Harden in 1895, the origin and development of the atomic theory can be traced: it arose from the influence of Newton.

Dalton was not brilliant, and modestly attributed his success to "perseverance". His great independence led him to apply his powerful mind to problems which might have seemed to some almost trivial, but which led him to the great generalization of the Atomic Theory, which now reigns paramount in all departments of modern Chemistry and Physics.

Dalton's Atomic Theory, as Lothar Meyer said, is so simple that "at first sight it is not illuminating". It asserts that:

(1) The chemical elements are composed of very minute indivisible particles of matter, called atoms, which preserve their individuality in all chemical changes.

(2) All the atoms of the same element are identical in all respects, particularly in *weight*. Different elements have atoms differing in weight. Each element is characterized by the weight of its atom.

(3) Chemical combination occurs by the union of the atoms of the elements in simple numerical ratios, e.g. 1 atom $A + 1$ atom B; 1 atom $A + 2$ atoms B; 2 atoms $A + 1$ atom B; 2 atoms $A + 3$ atoms B, etc.

These statements are all implied in an entry in his note-book on 6th September, 1803 (his birthday), except the 2 : 3 ratio, which is given for nitrous anhydride (N_2O_3) in an entry in October 1803. In September Dalton calculated the atomic weights of oxygen, nitrogen, sulphur, and carbon from analyses of compounds by other chemists (which are very inaccurate), and represented the atoms of these elements and the particles of some of their compounds by circular symbols (p. 176). He took the atomic weight of hydrogen, the lightest element, as unity.

These assumptions explain the laws of constant, multiple, and equivalent proportions. Dalton's theory could not determine the relative weights of atoms from the combining proportions unless the number of atoms in the particle of the compound is known, and this problem remained for long years the main point of dispute in the theory. Dalton attempted to solve it by making use of the general rules:

(1) If only one compound of two elements is known, it is presumed to be binary (1 atom A + 1 atom B), unless some cause appear to to the contrary.

(2) When two compounds exist, they are assumed to be binary ($A+B$) and ternary ($A+2B$ or $2A+B$).

(3) When three compounds are observed, one is binary and the other two ternary.

(4) When four compounds are observed, one is binary, two are ternary, and one quaternary ($3A+B$ or $A+3B$).*

Thus water and ammonia were HO and NH. Dalton, who thought (with Newton) that atoms of the *same* element repel one another, regarded a binary compound as the most stable one.

Origin of Dalton's Atomic Theory

A careful study of the order of publication of Dalton's memoirs had led W. C. Henry † and Geo. Wilson ‡ to the conclusion, now regarded as practically certain, that Dalton arrived at his atomic theory on the basis of speculations on the physical properties of gases, and then made experiments on multiple proportions in order to confirm it, rather than the converse, which was long believed on the authority of Thomas Thomson (see below). The investigation of manuscript material led Roscoe and Harden § and A. N. Meldrum || to the following scheme of the origin of the theory.

In 1802 Dalton read before the Manchester Philosophical Society a paper on an "Experimental Enquiry into the Proportions of the Several Gases or Elastic Fluids constituting the Atmosphere", which was first published in 1805. In the paper as printed there is a description of some experiments on mixing nitric oxide ("nitrous gas") and air over water, both in narrow tubes (9 in. by $\frac{3}{10}$ in.) and in wide jars. In the tube, 100 measures of air were mixed with 36 of nitrous gas and left 80 measures of nitrogen; in the jar,

* *Alembic Club Reprint* No. 2: "Foundations of the Atomic Theory".

† *Life of Dalton*, 1854, 222, 230: "I am even inclined to suspect that the framing of the atomic hypothesis may have been the antecedent, and the discovery of multiple proportions the consequence, rather than the converse."

‡ *Religio Chemici*, Cambridge, 1862, 331 f.

§ *A New View of the Origin of Dalton's Atomic Theory*, 1896—fundamental.

|| *Chemical News*, 1910, vol. 102, p. 1; *Mem. Manchester Lit. and Phil. Soc.*, 1910, vol. 55, nos. 3 to 6.

100 measures of air mixed with 72 measures of nitrous gas left 80 measures of nitrogen. Hence Dalton concludes that: "the elements of oxygen * may combine with a certain portion of nitrous gas or with twice that portion, but with no intermediate quantity." † These experiments are mentioned in Dalton's note-books in the period October 10–November 13, 1803, and were added to the paper after it was read.

On 21st October, 1803, Dalton read to the Society a paper "On the Absorption of Gases by Water", which was also printed in 1805. In this he gives the first list of atomic weights. After referring to the discovery of William Henry, of Manchester, that the amount of gas absorbed by a liquid is proportional to the pressure (*Henry's law*) ‡ and generalizing it to apply to mixtures of gases, in accordance with his own *law of partial pressures*,§ Dalton says: "I am nearly persuaded that the circumstance depends upon the weight and number of the ultimate particles of the several gases; those whose particles are lightest and single being least absorbable, and the others more, according as they increase in weight and complexity."

In a footnote Dalton says: "Subsequent experience renders this conjecture less probable", and he proceeds in the text: "An enquiry into the relative weights of the ultimate particles of bodies is a subject, as far as I know, entirely new: I have lately been prosecuting this enquiry with remarkable success. The principle cannot be entered upon in this paper; but I shall just subjoin the results, as far as they appear to be ascertained by my experiments.'

* Oxygen and Caloric, see p. 131.

† The result depends on the fact that *half* the nitric oxide is *rapidly* oxidized (in the wide jar) to NO_2 and the absorption occurs as:

$$NO_2 + NO + H_2O = 2HNO_2,$$

whilst the oxidation of the second half, by the termolecular reaction: $2NO + O_2 = 2NO_2$, is slow (in the narrow tube with limited contact with water) and the absorption occurs as: $2NO_2 + H_2O = HNO_2 + HNO_3$. The entries in the note-books are later than the table of atomic weights on 6th September, but inconclusive experiments on these lines are recorded in March and April 1803, and on 4th August 1803 Dalton says "oxygen joins to nit. gas sometimes 1·7 to 1, and at other times 3·4 to 1", which is a case of the law of multiple proportions.

‡ *Phil. Trans.*, 1803, pp. 41, 274.

§ *Mem. Manchester Lit. and Phil. Soc.*, 1802, v, pt. II, p. 535.

On the next page, just filling it, is a " Table of the relative weights of the ultimate particles of gaseous and other bodies ", including :

Hydrogen	1	
Azot	4·2	
Carbone	4·3	
Ammonia	5·2	
Oxygen	5·5	
Water	6·5	
Nitrous gas	9·3	[9·7]
Gaseous oxide of carbone	9·8	
Nitrous oxide	13·7	[13·9]
Sulphur	14·4	
Nitric acid	15·2	
Sulphuretted hydrogen	15·4	
Carbonic acid	15·3	
Sulphureous acid	19·9	
Sulphuric acid	25·4	
Carburetted hydrogen from stag. water	6·3	
Olefiant gas	5·3	

A table of atomic weights, with symbols (see p. 176), also appears in Dalton's note-books in the autumn of 1803, and Roscoe and Harden consider that the experiments with nitrous gas (see above) were made *after* the theory. The values for carburetted hydrogen (marsh gas) and olefiant gas were added in 1804—the results are in Dalton's note-book for 1804—in August of which year Thomas Thomson visited Dalton in Manchester and received from him an account of the atomic theory.* Thomson says Dalton told him that " the atomic theory first occurred to him during his investigations of olefiant gas and carburetted hydrogen gases, at that time imperfectly understood, and the constitution of which was first fully developed by Mr. Dalton himself." Dalton found that " if we reckon the carbon in each the same, then carburetted hydrogen gas contains exactly twice as much hydrogen as olefiant gas does ".

This is the usual account of the origin of Dalton's atomic theory

* Thomson, *History of Chemistry*, ii, 289 ; see Partington, *Annals of Science*, 1949, vi, 115.

and it is obviously incorrect, since a table of atomic weights was written out in 1803, whilst the experiments in question were made in 1804.

Thomson enthusiastically adopted Dalton's atomic theory, and he gave the first published account of it in 1807 in the third edition

FIG. 70.—T. THOMSON, 1773-1852.

of his *System of Chemistry* *, the first part of the first volume of Dalton's *New System of Chemical Philosophy* being published in 1808.

In his *Life of Dalton* † W. C. Henry says his father (William Henry) and he were told by Dalton that the theory was deduced

* Edinburgh, 1807, iii, 425, and later sections.

† 1854, 84.

to explain Richter's table. In Dalton's note-books, however, Richter is not mentioned until 1807. Dalton may have seen the table in Berthollet's *Statique Chimique* (see p. 163), but as Thomson in his account of Dalton's theory in 1807 (see p. 173) says it explains Richter's law of equivalents, Dalton may have heard of Richter from him.

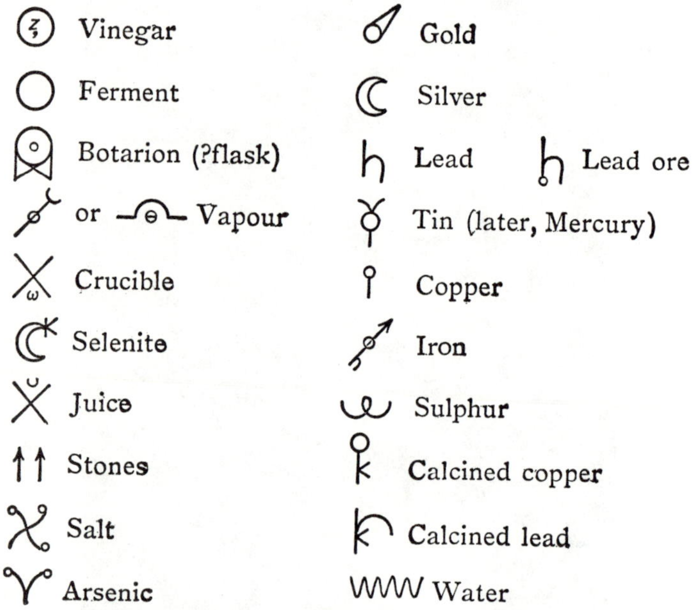

FIG. 71.—CHEMICAL SYMBOLS IN GREEK MSS.

In 1810 Dalton, in a lecture at the Royal Institution, attributed the origin of his theory to attempts to explain his law of partial pressures (1801-2). Gases diffuse into one another, and one gas does not press on another, because the particles are of different sizes; and he may later have had the idea that they have different weights. The experiments with nitric oxide and on the solubilities of gases (p. 171) may also have had some influence on his thoughts. Exactly how Dalton arrived at the statements which he wrote down in September 1803 is not known.

Chemical Names, Symbols and Formulae

One of the difficulties felt by a beginner in any science is in learning the technical terms, and sometimes symbols, of that science. This applies, naturally, to chemistry. The student may derive comfort from the fact that the present system of naming chemical substances, and of representing them by symbols, is very

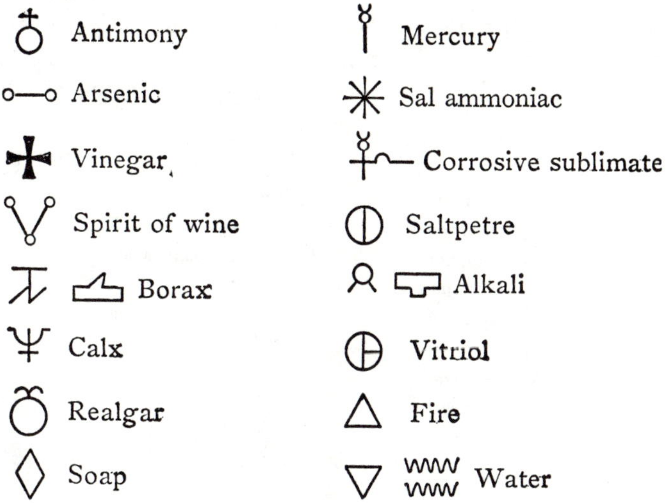

FIG. 72.—ALCHEMIST'S SYMBOLS; FROM OSWALD CROLL'S *Basilica Chymica*, 1609.

much simpler and more easily remembered than it was until about the beginning of the nineteenth century, when a new system of chemical nomenclature (naming of substances) was introduced by Lavoisier, etc., in 1787 (see p. 150) and, shortly afterwards, a new system of notation (representing names by symbols) by Berzelius in 1813 (see p. 197).

Let us first, in order to make this clear, look at the list of symbols in Fig. 71, which are taken from the early Greek chemical MSS. mentioned on p. 21. We notice that some are merely contractions of the Greek names for the materials (e.g. vinegar, ὄξος, crucible, χόανος, juice, χυμὸς, salt, ἅλς, arsenic, ἀρσενικόν); others, the symbols of the metals, are the symbols of the planets: Sun = gold; Moon = silver; Saturn = lead; Mars = iron; Venus = copper (see

p. 22). The alchemists also made much use of the planetary names and symbols, and in their books we must understand by *sol* (= Sun) gold, and so on. We still use the name *mercury* for the metal (after the planet) and speak of silver nitrate as *lunar caustic*. Some of the other symbols have obvious derivations, but there are some which have not.

The next list, in Fig. 72, from a text-book published in 1609, is similar. Some of the symbols are obvious modifications of those in the first list (e.g. arsenic), but there are several new symbols, e.g. for antimony, spirit of wine, borax, and soap (the shape of which is

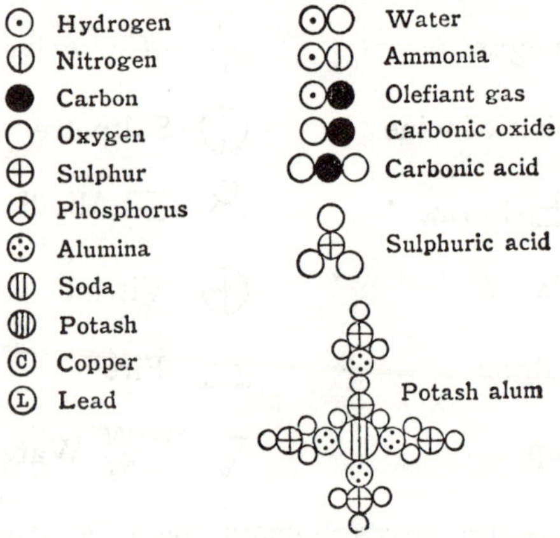

Fig. 73.—Dalton's Symbols and Formulae
From his *New System of Chemical Philosophy*, Vol. I, Part I, 1808; Part II, 1810.

that of a flake of " Lux "). The symbol for corrosive sublimate is made up of the symbols for " mercury " and " vapour ".

The early chemical names were purely empirical, and a name frequently had one meaning to the adept and quite another to the ordinary man. The same substance had a variety of names, depending on its mode of preparation. Names were often based on accidental resemblances. Thus *butter of antimony* was classed along with ordinary butter, and *oil of vitriol* (sulphuric acid) with olive oil. Such names as *liver of sulphur* (impure potassium sulphide) and *cream of tartar* (potassium hydrogen tartrate) arose in this way. Salts were often named after their discoverers, or the places where they were found (Glauber's salt, Epsom salt).

The symbols introduced by Dalton were, as we see from the few specimens reproduced in Fig. 73, not much better than the old ones, since there was no help to the memory in them. There was, however, one important step in advance, viz. that (besides the fact that each symbol represented one atom) the formula of a compound was made up of the symbols of its elements and also showed how many atoms of these were present in the molecule.

Berzelius's system of symbols is very simple, and is explained later (p. 197).

SUMMARY AND SUPPLEMENT

Claude Louis Berthollet (1748-1822), professor in Paris, visited Egypt with the Napoleonic Expedition (1798). He adopted Lavoisier's Theory in 1785, the first of the French chemists (Black had taught it before 1784), investigated the composition of ammonia (1785), prussic acid (1787) and sulphuretted hydrogen (1796), and recognized that the two latter are acids free from oxygen. He attempted to show (1785) that chlorine is oxymuriatic acid, discovered hypochlorites (1785) and chlorates (1787). Berthollet explained the part played by " mass " in chemical changes (1799-1803) and modified the old theory of " elective affinity " (Mayow, 1674; Geoffroy, 1718; Bergman, 1775): this led him (1803) to his theory of combination in " indefinite proportions " and his controversy with Joseph Louis Proust (1754-1826), who showed (1797-1809) that many metals form two or more definite oxides and sulphides, and that basic salts of definite composition (basic copper carbonate, etc.) exist. Proust also isolated grape sugar (1802), mannite (1806) and leucine (1818: " caseous oxide ").

Antoine François de Fourcroy (1755-1809) succeeded Macquer as professor at the Jardin des Plantes and later occupied other official positions. His work, largely in animal chemistry, was mostly done in collaboration with Vauquelin. Fourcroy was an excellent lecturer and writer, and did much to popularise Lavoisier's antiphlogistic theory, which he called " la théorie des chimistes français ". Lavoisier, however, said " elle est la mienne ".

Louis Nicolas Vauquelin (1763-1829), professor in Paris, discovered chromium compounds (1797) and chromium (1798), beryllium (" glucinum ") oxide and compounds (1798), and isolated several natural organic compounds: pectin (1790), allantoin (1800), asparagine (1805), and daphnin (1817).

Jean Antoine Chaptal (1756-1832), who mostly worked at technical problems, proposed the name " nitrogen " in 1790.

Martin Heinrich Klaproth (1743-1817), the first professor of chemistry in the University of Berlin, effected great improvements in mineral analysis, rediscovered strontium compounds (1793, independently of Crawford, 1790: see Partington, *Annals of Science*, 1942, v, 157; 1951, vii, 95), discovered zirconium oxide (1789), titanium compounds (1795, independently of Gregor, 1789), tellurium (1798), chromium (1798, independently of Vauquelin); uranium (1789), cerium oxide (1803, independently of Berzelius), and mellitic acid in honeystone (1799).

Carl Friedrich Wenzel (1740-1793) made experiments on the compositions of salts and on the rate of solution of metals in acids. His statement (1777) that the rate of solution of a metal is proportional to the concentration of the acid is an early recognition of the action of mass.

Jeremias Benjamin Richter (1762-1807) introduced the name " Stoichiometry " for the doctrine of the combining proportions of substances and defined and determined the equivalents of various substances (including acids and bases, and metals), especially by the method of double decomposition (from 1792). His results are obscured by his method of presentation. A clear table of equivalents was drawn up from Richter's results by E. G. Fischer in 1802.

Richard Kirwan (1733-1812) also investigated the combining proportions of acids and bases (1783). He suggested that hydrogen is phlogiston (1782), and discovered phosphine (1786) independently of Gengembre (1783, published in 1785).

William Hyde Wollaston (1766-1828) gave experimental proofs of the identity of frictional and voltaic electricity (1801), investigated emission and absorption spectra (1802), invented the total reflexion refractometer (1802), discovered some examples of " multiple proportions " (1808) in the cases of alkali oxalates, sulphates, and carbonates, devised a method of making platinum malleable and discovered palladium and rhodium (1803-4), improved the method of crystal measurement (" reflective goniometer ") (1809), and discovered cystin (1810). Wollaston invented the cryophorus (1813), proposed the use of " equivalents " instead of Dalton's " atoms ", and drew up tables of equivalents (and a " sliding scale ") (1814).

William Higgins in 1789 put forward some views of chemical combination which may be regarded as foreshadowing the law of multiple proportions and the theory of valency bonds, although he failed to develop the conception of atomic weight.

John Dalton (1766-1844), a teacher and private tutor in Manchester, discovered the law of expansion of gases by heat (1801), and the law of partial pressures (1801), measured vapour pressures (1801), and discovered the diffusion of gases (1801). He proved that liquids conduct heat (contradicting Rumford) (1799), and made extensive measurements in meteorology, inventing the dew-point hygrometer. On 6th September, 1803, he originated the chemical atomic theory, calculating some atomic weights and inventing chemical symbols; the atomic theory was first published by Thomas Thomson in 1807, after a conversation with Dalton, who published it in 1808 (*New System of Chemical Philosophy*, vol. I). Dalton established the law of multiple proportions by experiments with oxides of nitrogen (1803) and with marsh gas and olefiant gas (1804). He showed by exploding hydrocarbons with insufficient oxygen to burn them completely that the carbon burns to carbon monoxide preferentially to the hydrogen (1805), a most important result for the theory of combustion; he determined the composition of ether by exploding the vapour with oxygen, determined its vapour density by a method later (1826) used by Dumas, and gave the correct formula $C_4H_{10}O$ (1819); discovered propylene (1820; confirmed by Faraday, 1825); showed that the composition of the atmosphere is practically constant to a height of 15,000 ft. and correctly explained this by convection currents (1837); found that some salts do not increase the volume of water when they dissolve in it (1840); explained the nature of the aurora borealis and suggested that the atmosphere contains a magnetic constituent (1793; the paramagnetism of oxygen was discovered by Plücker in 1850, and Dalton's theory was independently proposed by Faraday in 1851); and gave the first detailed account of colour-blindness from observations of his own vision (1794). Dalton's quantitative experiments were usually accurate and he was an expert gas analyst.

CHAPTER IX

DAVY AND BERZELIUS AND THE ELECTRO-CHEMICAL OR DUALISTIC THEORY

Davy

Humphry Davy, the "Poet and Philosopher" * was born in Penzance in 1778, not far from the place now occupied by his statue. In 1795, soon after the death of his father, he became apprenticed to an apothecary and began to prepare himself for the medical profession. In 1797 Kerr's English translation of Lavoisier's *Elements of Chemistry* came into Davy's hands. He made experiments with simple apparatus from the surgery and in the following year proceeded to Clifton as superintendent of the laboratory established by Dr. Thomas Beddoes in the Pneumatic Institution, where the effects of the various newly discovered gases on the human body were to be investigated under the guidance of Beddoes, who was a qualified medical man.†

By experimental investigation Davy collected the materials for his publication, *Researches Chemical and Philosophical, chiefly concerning Nitrous Oxide or Dephlogisticated Nitrous Air and its Respiration*, 1800. In it he describes a number of analyses of oxygen compounds of nitrogen, the preparation of some substances, including potassium nitrososulphate, the preparation (already discovered by Berthollet in 1785) of nitrous oxide by heating ammonium nitrate, and the respiration of this gas. In April 1799 he breathed pure nitrous oxide and discovered its remarkable physiological action. This book made the reputation of Davy. After a period of convalescence from the effects of the various gases breathed, he

* Thorpe: *Humphry Davy*, 1901; *Works of Sir H. Davy*, edit. by J. Davy, 9 vols., 1839-40.

† T. Beddoes and James Watt, *Considerations on the Medicinal Use and on the Production of Factitious Airs*, Bristol, 1794, 3rd ed., 1796.

began in 1800 to work on the newly discovered Voltaic pile, still at the Pneumatic Institution, but in January 1801 we find him writing to his mother: "You have perhaps heard of the Royal Philosophical Institution, established by Count Rumford, and others of the aristocracy. It is a very splendid establishment, and

FIG. 74.—SIR HUMPHRY DAVY, 1778-1829.

wants only a combination of talents to render it eminently useful. Count Rumford has made proposals to me to settle myself there." He was appointed in the following month.

Benjamin Thompson, a Royalist American in the service of the Elector Palatine of Bavaria, by whom he was created Count Rumford of the Holy Roman Empire, is known chiefly for his important researches on heat, which showed that it was not material (caloric) but a form of motion of the small particles of bodies, and gave the first value of the mechanical equivalent of heat.* Rumford was instrumental in founding, in 1799, the Royal Institution in London, which is one of the most famous scientific associations in the world, its early reputation being mainly due to Davy and Faraday. Its objects were to be " diffusing the knowledge . . . of new and useful mechanical inventions and improvements ; and also for teaching, by regular courses of philosophical lectures and experiments, the applications of the new discoveries in science to the improvements of arts and manufactures ". The advent of Davy, and the inhalation of nitrous oxide, lent additional interest to the lectures, and science became fashionable. It was Davy who gradually moulded the character of the Royal Institution to that which it has ever since possessed. He resumed his work on electricity, and in 1806 read the Bakerian Lecture to the Royal Society : " On some Chemical Agencies of Electricity ",† which gained him the prize of 3000 francs founded by Napoleon, when First Consul, for the most important electrical research in the year. England and France were then at war. This work was soon put in the shade by the great Bakerian Lecture to the Royal Society in November 1807, entitled : " On some new phenomena of chemical changes produced by electricity, particularly the decomposition of the fixed alkalies, and the exhibition of the new substances which constitute their bases ; and on the general nature of alkaline bodies." ‡

Davy's Discovery of the Alkali Metals

In reading Lavoisier's book Davy must have been struck by the statement : " it appears that oxygen is the bond of union between metals and acids ; and from this we are led to suppose that oxygen is contained in all substances which have a strong affinity with acids. Hence it is very probable that the four eminently salifiable earths contain oxygen . . . they may very possibly be metallic oxyds § ". Kerr, the translator, suggested that " we may presume that potash

* *Phil. Trans.*, 1798; G. E. Ellis, *Memoirs of Sir Benjamin Thompson, Count Rumford*, Boston, 1871.

† *Phil. Trans.*, 1807, 1-56; *Works*, v, 1-56.

‡ *Phil. Trans.*, 1808, 1-44; *Works*, v, 57 f.; *Alembic Club Reprint No.* 6.

§ Lavoisier, *Elements of Chemistry*, Eng. tr., 3rd ed., 1796, p. 232.

is . . . a metallic substance, in some hitherto unknown state of combination ".* Lavoisier, although he suspected that potash and soda were not elements, did not regard them as oxides of metals, but (by analogy with ammonia) thought they might contain nitrogen. Davy at first considered that they might contain phosphorus or sulphur combined with nitrogen.† At the end of the memoir of 1806 Davy makes at least three prophetic suggestions, one of which is to the following effect: " If chemical union be of the nature which I have ventured to suppose‡, however strong the natural electrical energies of the elements of bodies may be, yet there is every probability of a limit to their strength; whereas the powers of our artificial instruments seem capable of indefinite increase." We may therefore " hope that the new mode of analysis may lead us to the discovery of the *true* elements of bodies ".

In 1807 Davy announced the decomposition of the alkalis by electricity. " Potash, perfectly dried by ignition, is a non-conductor, yet it is rendered a conductor, by a very slight addition of moisture, which does not perceptibly destroy its aggregation; and in this state it readily fuses and decomposes by strong electrical powers. A small piece of pure potash, which had been exposed for a few seconds to the atmosphere, so as to give conducting power to the surface, was placed upon an insulated disc of platina, connected with the negative side of a battery of 250 of the power of 6 and 4 [i.e. zinc and copper plates 6 in. by 4 in.], in a state of intense activity; and a platina wire, communicating with the positive side, was brought in contact with the upper surface of the alkali. The whole apparatus was in the open atmosphere. Under these circumstances a vivid action was soon observed to take place. The potash began to fuse at both its points of electrization. There was a violent effervescence at the upper surface; at the lower, or negative surface, there was no liberation of elastic fluid; but small globules having a high metallic lustre, and being precisely similar in visible characters to quicksilver, appeared, some of which burnt with explosion and bright flame, as soon as they were formed, and others remained, and were merely tarnished, and finally covered with a white film which formed on their surfaces. These globules, numerous experiments soon

* *ibid.*, p. 213. † *Works*, v, 103.

‡ On Davy's Electrochemical Theory, see p. 190.

shewed to be the substance I was in search of, and a peculiar inflammable principle the basis of potash."

This discovery was made on 6th October, 1807; sodium was discovered in a similar manner a few days later. We are told by John Davy * that when his brother "saw the minute globules of potassium burst through the crust of potash, and take fire as they entered the atmosphere, he could not contain his joy—he actually bounded about the room in ecstatic delight; and some little time was required for him to compose himself sufficiently to continue the experiment." An entry in Davy's notebook concludes with the statement: "Capital experiment, proving the decomposition of potash."

A paper read in 1808, "Electrochemical researches on the decomposition of the earths; with observations on the metals obtained from the alkaline earths, and on the amalgam procured from ammonia" † described the preparation of small quantities of magnesium, calcium, strontium and barium by heating the amalgams, obtained electrolytically. The earths were difficult to decompose, since they could not be fused and rendered conducting, but when "barytes, strontites, and lime, slightly moistened, were electrified by iron wires under naphtha . . . gas was copiously evolved, which was inflammable; and the earths where in contact with the negative metallic wires became dark coloured, and exhibited small points having a metallic lustre." It was then found that more promising results were obtained by the electrolysis of a mixture of baryta and oxide of mercury, when an amalgam of barium was formed. The work was suspended until, in May 1808, a new battery was ready. Davy then heard from Berzelius and Pontin in Stockholm that they had succeeded in reducing lime and baryta by negatively electrifying mercury in contact with them. Davy verified this and found that the soluble salts succeeded better than the earths.‡

Attempts to reduce alumina, zirconia and beryllia were far from successful, but those on the production of an amalgam from ammonia § were interesting. Berzelius and Pontin had informed

* *Works of Sir H. Davy*, 1839, vol. i, p. 109.

† *Phil. Trans.*, 1808, 333 f.; *Works*, v, 102 f.

‡ *Works*, v, 109, 176. § *Works*, v, 122.

Davy that mercury negatively electrified in contact with a solution of ammonia swelled up and became solid. Davy found that the most convenient way of obtaining "the amalgam from ammonia" was to connect a globule of mercury, resting in a cavity in a piece of moist sal ammoniac, with the negative wire, and the salt with the positive wire, or by the action of potassium amalgam on moist sal ammoniac. "It is scarcely possible to conceive that a substance which forms with mercury so perfect an amalgam, should not be metallic in its own nature; and on this idea . . . it may be conveniently termed ammonium." *

A long paper, read in 1808 †, "An account of some new analytical researches on the nature of certain bodies, particularly the alkalies, phosphorus, sulphur, carbonaceous matter, and the acids hitherto undecomposed; with some general observations on chemical theory," described the isolation of boron by the electrolysis of moist boracic acid between platinum surfaces and by heating boracic acid with potassium in a gold tube.

Davy's Researches on Chlorine

A paper read on 12th July, 1810, is entitled "Researches on the oxymuriatic acid, its nature and combinations; and on the elements of the muriatic acid: with some experiments on sulphur and phosphorus, made in the Laboratory of the Royal Institution." ‡ This is a short paper, but of an importance comparable with that on the alkali metals. Scheele in 1774 had obtained a new gas from "marine acid" (muriatic acid), itself obtained from salt and sulphuric acid, by the action of manganese dioxide. He regarded it as marine acid *minus* phlogiston, or since he considered phlogiston as the same as hydrogen, as muriatic acid *minus* hydrogen. § Lavoisier and the French school, on the other hand, regarded the gas as oxidised muriatic acid. Lavoisier and Fourcroy emphasized that the composition of muriatic acid from oxygen and an unknown radical, and of oxymuriatic acid from this radical and more oxygen, was hypo-

* *Works*, v, 131.

† *Phil. Trans.* 1809; *Works*, v, 140-204.

‡ *Works*, v, 284; *Alembic Club Reprint* No. 9.

§ *Collected Papers of C. W. Scheele*, tr. by Dobbin, 1931, p. 29; *Alembic Club Reprint* No. 13.

thetical, but Berthollet in 1785 had tried to bring forward experimental evidence, particularly that a solution of oxymuriatic acid in water on exposure to light evolves oxygen and leaves a solution of muriatic acid.

Gay-Lussac and Thenard in 1809-1811 were unable to detect the presence of oxygen in oxymuriatic acid by passing it over strongly heated charcoal. They found that dry muriates (e.g. of silver) were not decomposed with evolution of muriatic acid when heated with boric anhydride (B_2O_3) and carbon unless water was present. By passing muriatic acid gas over heated lead oxide they obtained lead muriate and water, and they concluded that dry muriatic acid gas contained a quarter of its weight of water. Gay-Lussac and Thenard in 1809 said that their experiments could be explained on the assumption that oxymuriatic acid was an element, but " it seems to us that they are still better explained by regarding oxymuriatic acid as a compound ".*

Attempts had been made by Davy to obtain muriatic acid free from water. Dry sulphate of iron, phosphoric glass, and boric anhydride, when heated strongly with dry muriate of lime in porcelain or iron tubes gave no gas, " though when a little moisture was added to the mixtures, muriatic acid was developed in such quantities, as almost to produce explosions ". By distilling corrosive sublimate with phosphorus a liquid was obtained, and when phosphorus was burnt in oxymuriatic acid gas in a retort " a white sublimate [PCl_5] collected in the top of the retort, and a fluid as limpid as water [PCl_3] trickled down the sides of the neck. The gas seemed to be entirely absorbed ".†

" One of the most singular facts that I have observed on this subject . . . is, that charcoal, even when ignited to whiteness in oxymuriatic or muriatic acid gases, by the Voltaic battery, effects no change in them ; if it has been previously freed from hydrogene and moisture by intense ignition in vacuo. This experiment, which I have several times repeated, led me to doubt of the existence of oxygene in that substance ". By the action of oxymuriatic acid on tin, Libavius's liquor [$SnCl_4$] is formed : " if this substance is a combination of muriatic acid and oxide of tin, oxide of tin ought to be separated

* *Mém. Soc. d'Arcueil,* 1809, ii, 358.

† Davy, *Phil. Trans.,* 1809, 39 ; *Works,* v, 194 f.

from it by the action of ammonia . . . a solid result was obtained, which was of a dull white colour ; some of it was heated, to ascertain if it contained oxide of tin ; but the whole volatilized, producing dense pungent fumes. . . . I made a considerable quantity of the solid compound of oxymuriatic acid and phosphorus by combustion, and saturated it with ammonia . . . on which it acted with great energy, producing much heat ; and they formed a white opaque powder." If this consisted of phosphate and muriate of ammonia, the latter should be driven off by heat, leaving phosphoric acid, but "to my great surprise . . . it was not at all volatile nor decomposable at this degree of heat, and . . . it gave off no gaseous matter." The powder was not attacked by common reagents : "The only processes by which it seemed susceptible of decomposition were by combustion, or the action of ignited hydrat of potash." *

"It is evident from this series of observations, that Scheele's view (though obscured by terms derived from a vague and unfounded general theory,) of the nature of the oxymuriatic and muriatic acids, may be considered as an expression of facts ; whilst the view adopted by the French school of chemistry, and which, till it is minutely examined, appears so beautiful and satisfactory, rests in the present state of our knowledge, upon hypothetical grounds."

"I have caused strong explosions from an electrical jar, to pass through oxymuriatic gas, by means of points of platina, for several hours in succession ; but it seemed not to undergo the slightest change." No oxygen was separated from its compounds with phosphorus, sulphur and tin when exposed to a current from 1000-2000 double plates. "Few substances, perhaps, have less claim to be considered as acid than oxymuriatic acid. . . . May it not in fact be a *peculiar* acidifying and dissolving principle, forming compounds with combustible bodies, analogous to acids containing oxygene, or oxides . . . ? On this idea muriatic acid may be considered as having hydrogene for its basis, and oxymuriatic acid for its acidifying principle." † From its colour, oxymuriatic acid was called *chlorine* by Davy. After some opposition, Davy's views were generally accepted, and chlorine now occupies a place in the list of elements.

* Davy, *Phil. Trans.*, 1810, 231 ; *Works*, v, 285 f.

† Davy, *Works*, v, 291 f.

In 1811, and 1815, respectively, Davy described two compounds of chlorine with oxygen, obtained by the action of hydrochloric (muriatic) acid and sulphuric acid, respectively, on potassium chlorate. The first, called *euchlorine*, was shown subsequently by Pebal to be a mixture of chlorine with the second, chlorine dioxide. The compounds of chlorine with phosphorus and sulphur were examined in 1810, and in 1813 Davy described experiments with the compound of chlorine and nitrogen, discovered by Dulong in 1812, and obtained by the action of excess of chlorine on ammonia. It is violently explosive. In the later stages of this work Davy was assisted for the first time by Faraday.

He published the first (and only) part of his *Elements of Chemical Philosophy* in 1812, and in April 1813 began his work on fluorine. Guided by a suggestion of Ampère, he was able to bring forward evidence that fluorine is an element similar to chlorine, and that "fluoric acid", obtained by the action of sulphuric acid on fluorspar *, is similar in composition to hydrochloric acid. He was unable to isolate fluorine, the element being first obtained by Moissan in 1886 by a process tried by Davy.

Davy's Researches on Iodine

Davy resigned the professorship at the Royal Institution in 1813, becoming honorary professor, and set out to travel on the Continent, taking Faraday as secretary and assistant. Special permission was obtained from Napoleon, who undoubtedly had a high opinion of Davy, to pass through France, then at war with England, on the way to Italy. Davy was warmly received in Paris and on 13th December, 1813, he was elected a Corresponding Member of the Institut.†

During November 1813 Ampère presented to Davy a small quantity of a "substance X" discovered in 1811 by Courtois, in the liquor from the lixiviation of kelp (seaweed ash). On the day of Davy's election to the Institut a letter from him to Cuvier was read in which he first apologizes for intruding in a field in which Gay-Lussac was known to be engaged, and then proceeds to point

* Partington, "The Early History of Hydrofluoric Acid", *Mem. Manchester Lit. and Phil. Soc.*, 1923, lxvii, No. 6, 73.

† See Faraday's Diary in Bence Jones, *Life and Letters of Faraday*, 1870.

out that the substance X is an element analogous to chlorine. He arrived at a true estimate of its nature, prepared many compounds of it (including potassium iodate and phosphonium iodide) by means of a small portable laboratory, and suggested for it the name *iodine.* (The French chemists called it *iode*, from the violet colour of the vapour.) Gay-Lussac's and Davy's researches on iodine were carried out simultaneously and it is now very difficult to decide priority. Their detailed publications appeared in 1814.* Gay-Lussac was annoyed at what he considered to be an intrusion into his own field, although he and Thenard in 1808 had taken up the study of the alkali metals immediately after Davy's publication.

From France, Davy went to Italy, and at Florence he and Faraday made use of the great burning glass of the Accademia del Cimento to burn the diamond in oxygen and show that it had the same composition as charcoal. The ancient colours from Rome and Pompeii were examined: the reds were minium, iron ochres and cinnabar; the blues were mainly mixtures of chalk with Egyptian blue (a frit of lime, sand and copper); the greens were mostly copper compounds, and the purples organic, probably similar to the so-called Tyrian purple of shellfish which is now known to be a dibromo-indigo.†

Davy's Researches on Flame

Davy arrived home in April 1815, and an invitation from the Chairman of a "Society for preventing Accidents in Coal Mines", which had been formed after disastrous explosions, set Davy to work in August on flame and explosions, and the invention of the safety lamp for mines was the result. He showed that inflammable gases have definite ignition points, that of methane, the chief constituent of the explosive gas (fire-damp) of coal mines, being relatively high. A flame can be extinguished by cooling, and cannot be propagated to an explosive mixture of gases through the mesh of wire gauze, which conducts away the heat of the flame and prevents

* Davy, *Phil. Trans.*, 1814-5; *Works*, v, 437, 457, 492, 510; Gay-Lussac, *Ann. Chim.*, 1814, xci, 5.

† Pliny, ix, 36-38; Partington, *Origins and Development of Applied Chemistry*, 1935, 136, 458.

the unburnt gas reaching its ignition point. Davy's safety lamp consists of an ordinary lamp encased in a cylinder of wire gauze. It was adopted early in 1816, and remains practically in its original form to-day. Davy's researches on flame * are very important.

The *electric arc* was discovered by Davy †: in 1812 he mentions the point of light between carbon points as "so vivid that even the sunlight compared with it appeared feeble", and in 1821 he obtained an arc 10 cm. long by the discharge between carbon poles from a battery of 2000 cells.

In 1820 Davy was elected President of the Royal Society, but he resigned in 1827, since his health was poor, and he went abroad to recuperate. His illness became acute and he died at Geneva in 1829; he was buried at his own request outside the town, where an inscription marks his tomb.‡

Davy's Electrochemical Theory

We have left for separate consideration the electrochemical theory of Davy. The work on the chemical effects of electricity was begun in 1800 §: the theory put forward was never very precisely formulated, and Faraday || said, rather bluntly, that "probably a dozen precise schemes of electrochemical action might be drawn up, differing essentially from each other, yet all agreeing with the statement there given" [i.e. by Davy], and he afterwards took the trouble to specify just twelve such schemes.

Davy showed that Volta's arrangement of two metals and one liquid could be replaced by one metal and two liquids. Of two metals, the one with the stronger affinity for oxygen forms the positive pole in the liquid.¶ Electrification modifies the "chemical powers" of matter, and in one passage of the 1806 memoir Davy suggests that electromotive force measurements can serve to deter-

* *Works*, vol. vi; Ostwald's *Klassiker* No. 247; Partington, *Annals of Science*, 1945, v, 229.

† *Elements of Chemical Philosophy*, 1812, 152; *Phil. Trans.*, 1821, cxi, 425.

‡ Reproduced by Guye, *J. Chim. Phys.*, 1906, iv, 263.

§ *Works*, ii, 139.

|| Faraday, *Experimental Researches*, 1849, series v, p. 136.

¶ *Elements of Chemical Philosophy*, 1812, p. 148; *Works*, iv, 107.

mine affinities, as we now know to be the case.* Electrification produced by the contact of substances is compared with the chemical activities of the substances, which lose their electrification of contact when chemical combination actually occurs.† Even apparently insoluble substances, such as glass, when moist, can be decomposed by the current, and acids may be transported through alkaline solutions provided they are not precipitated on the way.‡

Davy rejects the theory that electrification is primarily produced by chemical action §, on the ground that electrification is produced by mere contact of substances without appearance of chemical change. This was Volta's theory. Faraday thought chemical change was necessary, but recent experiment seems to have decided this rather delicate matter in favour of Volta and Davy.

Davy says chemical changes and electrical changes " are conceived . . . to be *distinct* phænomena; but produced by the *same power*, acting in one case [electrical] on masses, in the other case [chemical] on particles ". The origin of the electromotive force is contact, but when the metals are connected, the charges tend to become neutralized, and chemical change restores the electromotive force. " The same arrangements of matter, or the same attractive powers, which place bodies in the relations of positive and negative, i.e. which render them attractive of each other electrically . . . may likewise render their particles attractive, and enable them to combine, when they have full freedom of motion. . . . It has been supposed that the idea was entertained, that chemical changes were occasioned by electrical changes; than which nothing is further from the hypothesis which I have ventured to advance." || Davy adds that: " Some modern writers have asserted the existence of an electrical fluid with as much confidence as they would assert the existence of water . . . but it is impossible in sound philosophy to adopt such hasty generalizations ".

Berzelius and Hisinger

Whilst Davy was working at the Royal Institution in London, a memoir appeared in 1803,¶ bearing the names of Berzelius and

* *Works*, v., 42.
† *Chem. Phil.*, 1812, p. 158.
‡ *Chem. Phil.*, 1812, p. 160.
§ *Chem. Phil.*, 1812, p. 162.
|| *Chem. Phil.*, 1812, pp. 164, 171.
¶ *N. Allgem. J. Chem.*, 1803, i, 115.

Hisinger, describing experiments in Stockholm on electrolysis and putting forward the following conclusions:

1. Chemical compounds are decomposed by the electric current and their components collect at the poles.

2. Combustible bodies (hydrogen), alkalis, and earths and metals, go to the negative pole; oxygen, acids and oxidized compounds go to the positive pole.

3. The extent of decomposition is in compound proportion to the affinities and the surfaces of the poles. It is proportional to the quantity of electricity and to the electrical conductivity. (This statement was much modified as a result of the researches of Faraday.)

4. The chemical changes in decomposition depend firstly on the affinities of the components for the poles (metals), secondly on the affinities of the components for one another, and thirdly on the cohesion of the compounds formed. (The last statement refers to the law of mass action put forward by Berthollet and adopted, with qualifications, by Berzelius.)

Faraday

Michael Faraday * was born at Newington Butts, Surrey, in 1791. His parents were in poor circumstances and at the age of thirteen Michael began to work for a newsagent and bookseller. He read many scientific books which came to the shop for binding, and one of the customers gave him a ticket for Davy's lectures at the Royal Institution. He wrote out the lectures and sent them to Davy with a request for employment. Davy replied kindly, saw Faraday, and in 1813 engaged him as assistant.

Faraday's talents as an experimenter soon gained him promotion, and on Davy's recommendation he was made Director of the Laboratory of the Royal Institution in 1825. Brande had succeeded Davy as Professor of Chemistry in 1813, but in 1833 the Fullerian Professorship of Chemistry at the Royal Institution was created for Faraday, and he held this until his death in 1867.

* J. H. Gladstone, *Michael Faraday*, 1873; Bence Jones, *Life and Letters of Faraday*, 2 vols., 1870—a later edition had portions of the text omitted; Thorpe, *Essays in Historical Chemistry*, 1902, p. 185; S. P. Thompson, *Michael Faraday*, 1898.

Faraday's earlier work was an extension of Davy's and he gratefully acknowledged his indebtedness to the latter. Injustice has been done to Davy by magnifying some incidents in Faraday's career.

Much of Faraday's work lay in the domain of physics and was of fundamental importance. His discoveries of electromagnetic in-

FIG. 75.—MICHAEL FARADAY, 1791-1867.

duction and of specific inductive capacity are especially noteworthy. The papers on these subjects, and on electrolysis, are collected in his *Experimental Researches in Electricity* *; those on chemical subjects are collected in his *Researches in Chemistry and Physics* (1859), and include the discovery of "carbon perchloride" (C_2Cl_6) in 1821, of the liquefaction of gases (1823-45), of benzene

* 3 vols., 1839 (2 ed. 1849) -44-55.

(" bicarburet of hydrogen ", $\underline{C}_2H$, where $\underline{C}=6$) in 1825, of the isomeric sulphonic acids of naphthalene in 1826, and the preparation of colloidal gold (including a description of the so-called " Tyndall phenomenon ") in 1857. Faraday's most important work from our point of view was that on electrolysis (1832-3).

Faraday in 1832 showed that * the amount of decomposition was proportional to the current strength and the time, i.e., to the quantity of electricity passing (as stated but not satisfactorily proved by Berzelius and Hisinger), and in 1833 that the weights of substances deposited by the same current are in the proportion of their chemical equivalent weights. He used the water voltameter for measuring current strengths. The names electrolysis, electrolyte, anode, cathode, anion and cation were invented for Faraday by William Whewell. A theory of the mechanism of current conduction in solutions was proposed by von Grotthuss in 1805.†

Berzelius

Jöns Jacob Berzelius‡ was born in 1779 at Väfversunda in Sweden, and began his chemical studies at an early age. Berzelius's circumstances were poor but he was able to study in the University of Uppsala, and after adopting medicine as a career he was appointed in 1802 as assistant professor of medicine, pharmacy and botany at the University of Stockholm, becoming professor in 1807. In 1815 he became professor of chemistry in the new Chirugico-Medical Institute at Stockholm. He had merely to lecture, and in a small ill-equipped kitchen laboratory he carried out his own researches and instructed a few pupils. In 1818 he was made Permanent Secretary to the Stockholm Academy; after 1832, when Mosander succeeded him in the chair, he devoted himself to literary

* *Experimental Researches in Electricity*, 2nd ed., 1849, i, 107, 127, 195; partly reprinted in the *Everyman* Series.

† *Annales de Chimie*, 1806, lviii, 54; Ostwald's *Klassiker* No. 152.

‡ Thorpe, *Essays*, 1902, 298; Tilden, *Famous Chemists*, 1921, 131; Söderbaum, in Kahlbaum's *Monographien*, Nos. 3 and 7; Berzelius's letters to Liebig (ed. Carrière, 1893) and Wöhler (ed. von Braun and Wallach, 1901); *Bref*, 1912 f.; *Bibliographie*, ed. Holmberg, 1933 and suppls.; Söderbaum, *Berzelius Levnadsteckning*, 2 vols., Uppsala, 1929; Holmberg, *Berzelius-Porträtt*, Stockholm, 1939.

work; in 1835 he was made a Baron by King Charles XIV. He died on 7th August, 1848.

Berzelius's experimental work ranged over many fields. His great aim was the establishment of the atomic theory and his dualistic system in all branches of chemistry. He made many

FIG. 76.—J. J. BERZELIUS, 1779-1848.

accurate determinations of combining weights, improved the methods of chemical analysis, and showed that the laws of chemical proportions applied to organic substances and to minerals. He discovered ceria in 1803, selenium in 1817, thorium in 1828, and investigated vanadium and molybdenum compounds, ferrocyanides, the oxides of nitrogen, the halogen compounds of boron and silicon, carbon disulphide, and many organic compounds. He drew attention to the existence of what he called catalytic action * and to isomerism.

Berzelius's main characteristics were thoroughness and perseverance, exactness in observation and clearness in description, a capacity for systematization, and a conservative attitude which caused him to cling to what he thought well established with great tenacity. "Against anything violent, to his mind revolutionary, he fought with all his energy; he did not shun heated polemics when anything he regarded as sound was at stake" (E. von Meyer). His knowledge of all branches of chemistry was profound and he was a fine experimenter. Wöhler says Berzelius used to remark on seeing a hasty experiment: "Doctor, das war geschwind aber schlecht!" (Doctor, that was quick but bad.) By his large textbook, which went through several editions and was translated into French and German, and his annual report to the Swedish Academy on the progress in chemistry, etc. (translated from Swedish into German by Wöhler, as *Jahres-Bericht*), Berzelius exercised an important influence on chemical thought, and for many years he was the foremost authority in the science.

Perhaps the greatest influence on the development of chemistry was exerted by that part of Berzelius's work which dealt with his Electrochemical Theory, and this deserves the somewhat detailed explanation which follows.

Berzelius's Electrochemical Theory

The basis of Berzelius's Electrochemical Theory was Dualism. For Lavoisier, oxygen was the central element in the system: an *acid* was a compound of a radical with oxygen. Davy extended this by showing that a *base* was a compound of a metal with oxygen, and Berzelius completed the dualistic system by assuming that, in all cases, a *salt* was a compound of an acid (really an acid anhydride) with a base (really a basic oxide):

Salt

− Acid + Base

(+ Radical + − Oxygen) (+ Metal + − Oxygen)

Radical + Oxygen Metal + Oxygen

* *Ann. chim.*, 1836, lxi, 146.

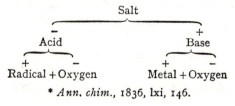

Berzelius noted * that a similar dualism exists in electricity, and since his work with Hisinger showed that salts in solution are decomposed by the electric current into bases, which go to the negative pole and therefore seem to carry a positive charge, and acids, which go to the positive pole and seem to carry a negative charge, the above conception became linked with electrical polarity. Bases were electropositive oxides, acids electronegative oxides, or, as Berzelius expressed it, they are electropositive and electronegative substances respectively. This polarity was extended to the elements. Oxygen was regarded as the most electronegative element: in an oxygen compound oxygen was always negative. All the elements could be arranged in a series with oxygen at one end and potassium (the most electropositive element) at the other.

The theory appeared first in De la Métherie's *Journal de Physique* for October 1811; in Berzelius's *System of Mineralogy* (English translation by John Black, 1814) it is stated that the theory was published in 1811 and a reference is made to the chemical symbols used by Berzelius. These (put forward in 1813) replaced Dalton's inconvenient circles by the initial letter, or letters, of what Berzelius took to be the Latin names of the elements. He remarks that he is following Thomas Thomson †, who used the symbols A, S, etc., for alumina, silica, etc., in mineral formulae. Berzelius indicated oxygen and sulphur atoms by dots and dashes over the symbols of the other elements; he later indicated two atoms (a "double atom") by drawing a bar through the symbol of the atom: $H^2O = \dot{\not{H}}$; $NaO, SO^3 = \dot{N}a\dddot{S}$. ‡ Symbols were little used in textbooks early in the century. They do not appear in Murray's textbooks nor in Thomson's *System* (5th ed., 1817). They are given in an appendix in the last edition of Henry's *Elements of Chemistry* (1829). Turner's *Elements of Chemistry* (4th ed., 1833) explains symbols and uses them, together with chemical equations, but with an apology in the preface. Liebig § uses equations but not liberally.

* *Traité de chimie*, French translation, Paris, 1831, vol. 4, p. 563; *Théorie des proportions chimiques*, Paris, 1819; 2nd ed., Paris, 1835, p. 43 f.

† *System of Chemistry*, 1802, vol. 3, p. 431.

‡ See Partington, *Chemistry and Industry*, 1936, vol. 55, p. 759.

§ *Traité de chimie organique*, 3 vols., Paris, 1840-44.

Symbols and equations are freely used in Gmelin's *Handbook* (English translation, 1848-72).

In the first edition of Berzelius's *Theory of Chemical Proportions*, published in Swedish in 1814 *, and in his *Text-Book* †, the electrochemical theory is explained in detail.

"Experience shows that heat is disengaged in every chemical combination when carried out in circumstances favourable to its perception, and that by the saturation of powerful affinities, the temperature often rises to the point of incandescence, whilst the satisfaction of the feeblest affinities is capable only of raising the temperature through a few degrees." There are some peculiar cases: hydrated oxides of chromium and zirconium, on heating, suddenly become incandescent and the oxide becomes insoluble in acids. Bodies may also be heated by the electric spark or current, and "the question arises whether the union of opposite electricities is not the cause of ignition in chemical combinations as well as in the electric discharge", especially in view of the close relation between chemical changes and electricity established by Davy. Rise of temperature increases the affinity between copper and sulphur; it also increases the intensity of the contact electrification until at a certain point incandescence occurs, combination ensues, and all traces of the electric charges vanish.

"In the present state of our knowledge the most probable explanation of combustion and of the ignition resulting from it is that in every chemical combination there is a neutralization of opposite electricities, and this neutralization produces fire (produit le feu) in the same way as it is produced in the discharges of the electric bottle [Leyden jar], the electric pile, and lightning, without being accompanied, in these latter phenomena, by chemical combination."

The hypothesis is put forward cautiously, perhaps as capable of replacement later by a better one. Some reactions, says Berzelius, are with difficulty explained by the theory. Hydrogen peroxide reacts with silver oxide, both compounds are decomposed and much heat is given out, whilst no combination seems to occur. Brodie (1850)‡ afterwards showed that there is chemical combination of two

* French in 1819; 2nd (revised) edition, 1835.

† French ed., 1831, vol. 4, p. 511 f., especially p. 554 f.

‡ *Phil. Trans.*, 1850, 759, 774.

atoms of oxygen, one from each substance, to form a molecule of gaseous oxygen, and this agrees with Berzelius's theory:

$$H_2O \; \boxed{\overset{+}{O} + \overset{-}{O}} \; Ag_2$$

Electrolysis, according to Berzelius, is the exact reverse of chemical combination: the electric charges lost on combination are restored to the two parts of the compound, and the latter appear in the free state. His theory differs from Davy's in that Davy assumed that substances acquire electric charges only on contact, and that the electric charges so acquired by the reacting particles are the cause of chemical action*, whilst Berzelius assumed that the particles always have electric charges, which they continue to have after combination.

A difficulty is pointed out by Berzelius, viz. that when two charged bodies exchange their charges the latter are neutralized and the bodies no longer attract each other, yet in compounds the parts are firmly held together. This shows that the atoms are still *polarized* in compounds. That combination is maintained by polarity and not by a superadded special chemical force of affinity is shown by the ready decomposition by electricity of compounds formed by the satisfaction of powerful affinities. The different electrochemical characters of elements are due to the preponderance of positive or negative electric charge in the same atom, which contains *both* charges (except oxygen, which is only negative); the atom is an electrical dipole.

Faraday's discovery that chemical equivalents carry equal charges was a great difficulty and Berzelius maintained that Faraday's results were inaccurate. Actually, Berzelius was thinking rather of intensity (potential) than quantity of electricity, and he does not distinguish the one from the other sufficiently clearly. This was a common mistake in his time.

Bodies may be divided into two groups, electropositive and electronegative, according to the charges they take up on contact, and the arrangement according to electroaffinities is the electrochemical

* See Davy's own words, p. 191.

series. (The arrangement does not differ much from modern tables of electrode potentials.)

O, S, N, F, Cl, Br, I, Se, P, As, Cr, Mo, W, B, C, Sb, Te, Ta, Ti, Si, H ; Au, Os, Ir, Pt, Rh, Pd, Hg, Ag, Cu, U, Bi, Sn, Pb, Cd, Co, Ni, Fe, Zn, Mn, Ce, Th, Zr, Al, Y, Be, Mg, Ca, Sr, Ba, Li, Na, K.

In this table, which may, says Berzelius, require modification with further experiments, the most electronegative elements are at the top and the most electropositive at the bottom. Hydrogen separates the two classes. The order may vary with temperature. The strength of an acid or base is greater the more pronounced the electrochemical character of the radical. A weak base may act as an acid towards a strong base (e.g. alumina and caustic soda), and a weak acid as a base towards a strong acid (e.g. boric acid and sulphuric acid). Salts may be feebly positive or negative as a whole and thus combine to form double salts (e.g. alum). This type of combination is contrasted with double decompositions between salts, due to the attractions between the electrical charges of the separate atoms.*

The difficult question of the distribution of electricity in bodies is discussed with care, and the results are stated to be tentative.† The modesty with which the whole theory is put forward, the candour with which its weak points are emphasized, the evident care to avoid dogmatism, and the resemblance between many parts of the theory and the modern views on chemical combination, are striking and come as a surprise to some readers.

Berzelius's Corpuscular Theory

In close relation to his electrochemical theory was the Atomic Weight system of Berzelius. A study of the little known work of Richter on equivalents (p. 163) led to the conclusion that careful analyses of a selected group of compounds would lead to the compositions of a number of others. The results of these analyses also confirmed Dalton's law of Multiple Proportions. These two empirical laws received a satisfactory explanation in Dalton's atomic theory. This, largely as a result of Berzelius's extensive and

* *Traité*, 1831, iv, 570 f.

† *Traité*, 1831, iv, 575 f.

accurate experimental work (published in 1811-12) on the composition of salts, etc.*, was now placed on a satisfactory foundation, although it was not accepted by some of the foremost chemists, e.g. not at first by Davy.

In adopting the theory Berzelius adhered to Dalton's plan of speaking of *compound atoms* and arranged these in orders of complexity. Those of the first order are composed of elementary atoms, as potash = potassium + oxygen; sulphuric acid = sulphur + oxygen. Compound atoms of the second order are composed of those of the first order: sulphate of potash = potash + sulphuric acid. Compound atoms of the third order are composed of those of the second order, as dry alum = sulphate of alumina + sulphate of potash. Compound atoms of the fourth order contain those of the third order, as crystalline alum = dry alum + water. The affinity between the components of compound atoms diminishes rapidly as their order increases: "the degrees of affinity which still exist in atoms of the third order are generally too feeble to be perceived in the hasty and disturbed operations of the laboratory . . . and ordinarily manifest themselves only in compounds formed when the earth passes slowly and tranquilly into the solid state, i.e. in minerals." †

The laws regulating the combinations of atoms Berzelius attempted to find empirically, after the manner of Dalton's rules. He calls this the *Corpuscular Theory*. In 1818 ‡ he says that if the amounts of oxygen absorbed by a radical are in the ratio 1 to $1\frac{1}{2}$, the oxides may be formulated as R + O and 2R + 3O, but as some chemists think sulphur and iron have unknown lower oxides, the known oxides are formulated as R + 2O and R + 3O. This makes the oxides of iron FeO^2 (ferrous) and FeO^3 (ferric). In 1826 he altered these to FeO and Fe^2O^3. (Berzelius objected to placing the figures below the symbols as in modern formulae, Fe_2O_3, etc., and quarrelled with Liebig because the latter printed one of Berzelius's papers in the *Annalen* with the figures below the symbols.) In 1826 Berzelius drew up series of combinations, such as the following:

1. A + B, 2B, 3B, . . . the limit probably being 12B, since 12 is the largest number of spheres which can surround one of equal size when in contact with it.

* Ostwald's *Klassiker* No. 35.

† *Traité*, 1831, iv, p. 539.

‡ *Essai*, 1819, p. 30.

2. 2A + 3B, 5B or 7B. Only a few such compounds are known. The question is raised whether 2A + 2B is, or is not, the same as A + B. This can be decided in some cases by experiment.*

3. The proportions in which *compound* atoms combine are:
   (*a*) A + B, 2B, 3B, . . . This is the commonest type.
   (*b*) 3A + 2B, or very rarely 3A + 4B.
   (*c*) 5A + 2B, 3B, 4B, 4½B, 6B.

Isomorphism

As a useful criterion in fixing the formulae of compounds Berzelius refers to Mitscherlich's Law of Isomorphism: "the same number of atoms, combined in the same manner, produces the same crystalline form; the crystalline form is independent of the chemical nature of the atoms and is determined solely by their number and mode of combination."

Eilhard Mitscherlich was born in 1794 in Oldenburg and died in Berlin in 1863. He was Klaproth's successor in Berlin. Mitscherlich began as a student of Oriental languages, then turned to medicine, and finally to chemistry, which he studied under Berzelius in Stockholm in 1819. His first paper on isomorphism (ἴσος, equal; μορφή, form) was read in 1819.† In 1812 Wollaston ‡ had found small but measurable differences between the angles of calcite, dolomite and spathic iron ore, which are isomorphous, and the approximate character of his law was later recognized by Mitscherlich. Mitscherlich also carried out (1833-5) important investigations on benzene, obtaining some of its derivatives.§

The first cases of isomorphism investigated by Mitscherlich were the pairs of salts now represented by the formulae: $Na_2HPO_4, 12H_2O$ and $Na_2HAsO_4, 12H_2O$; NaH_2PO_4, H_2O and NaH_2AsO_4, H_2O. "Every arsenate has its corresponding phosphate, composed according to the same proportions, combined with the same amount of water and endowed with the same physical properties; in fact the two series of salts differ in no respect, except that the radical of the acid in one series is phosphorus, whilst in the other it is arsenic."

* *Traité*, 1831, iv., 594, f.

† *Abhl. K. Akad. Berlin* (1818-19), 1820, 427 (summary of paper read 9 December, 1819); *Ann. Chim. Phys.*, 1820, xiv, 172; 1822 [mis-dated 1821], xix, 350; Ostwald's *Klassiker* No. 94.

‡ *Phil. Trans.*, 1812, 159.

§ Ostwald's *Klassiker* No. 98.

Isomerism

Berzelius then refers to Isomeric Compounds (ἰσομερής, composed of equal parts), which he defines as compounds having the same chemical composition " and the same capacity of saturation ", but with different properties. As examples he quotes * the phos-

FIG. 77.—E. MITSCHERLICH, 1794-1863.

phoric acids, fulminic acid and cyanic acid, the two kinds of stannic acid, and the two tartaric acids (tartaric and racemic). The case of fulminic and cyanic acids led to the recognition of isomerism by Berzelius. In 1823 Wöhler analysed cyanic acid and in 1824 Liebig analysed fulminic acid, both in their salts, and their results were identical. Berzelius at first thought some error had been made, but later work showed that compounds having the same compo-

* *Traité*, iv, 548.

sition could have different properties, and in 1827 Berzelius recognised the existence of the phenomenon and called it isomerism: "it would seem as if the simple atoms of which substances are composed may be united with each other in different ways." * This is clearly the beginning of the theory of structure, and the atomic theory had shown itself competent to explain the new facts.

In 1833 Berzelius distinguished between two cases of isomerism: *metamerism* when the two compounds contain the same number of atoms (fulminic and cyanic acids), and *polymerism* when the relative numbers of atoms are the same but the absolute numbers are different. As an example of the latter (to which he did not then give a special name) he referred in 1827 † to olefiant gas and a gas first noticed by Dalton in oil gas but more fully investigated by Faraday in 1825 (when he also discovered benzene in the liquid), which had the same composition as olefiant gas but twice the density (butylene, C_4H_8). Faraday says: "This is the first time that two gaseous compounds have been supposed to exist, differing from each other in nothing but density." Berzelius remarks that the "capacity of saturation" is different, from which we infer that by this term he understood the number of atoms in the molecule. In 1827 he states that Faraday's gas contains twice the number of atoms contained in olefiant gas, from which it is clear that at that time he regarded the density as a criterion of the relative weights of *compound* particles.

Berzelius's Theory of Volumes

In his discussion of the Theory of Volumes, based on the experiments of Gay-Lussac,‡ Berzelius states § that: "Experience shows that just as the elements combine in fixed and multiple proportions by weight, they also combine in fixed and multiple proportions by volume, so that one volume of an element combines with an equal volume, or 2, 3, 4, or more volumes of another element in the gaseous state. . . . The degrees of combination are absolutely the same in the two theories, and what is named *atom* in the one [Dalton's] is called *volume* in the other." It is clear that the theory is limited to

* *Jahres-Bericht*, 1832, p. 44; on *allotropy*, *ibid.*, 1841, xx, p. 13; Freund, *Study of Chemical Composition*, 1904, p. 549. † *Traité*, 1831, iv, 549.

‡ *Mém. d'Arcueil*, 1809, ii, 207 (read December 1808).

§ *Traité*, iv, 549.

elements. Berzelius evidently had in mind the objection of Dalton to the hypothesis that equal volumes of gases contain equal numbers of particles, which would seem to follow from Gay-Lussac's experiments, although it was not put forward by the latter. If 1 vol. of nitrogen contains *n* atoms, 1 vol. of oxygen contains *n* atoms, and since these unite to give 2 vols. of nitric oxide, 1 vol. of the latter can contain only $\frac{1}{2}n$ particles.* This difficulty had been explained as early as 1811 by Avogadro, as we shall see, but his work remained unnoticed. Berzelius refers to Avogadro's reasoning only in a later edition of his Treatise †, and without mentioning him by name.

By means of his Volume Theory, Berzelius deduced that the formulae of water, hydrochloric acid and ammonia, for example, are H^2O, HCl, and NH^3, since the combinations between the elementary gases occur in the ratios 2 : 1, 1 : 1, and 1 : 3, respectively. It should be noted that he does not make use of Avogadro's reasoning, and in particular that he does not assume that the free particles of hydrogen, oxygen, chlorine and nitrogen are H^2, O^2, Cl^2 and N^2.

Berzelius's Atomic Weight Tables

Berzelius remarked ‡ that " although our results on the atomic compositions of the majority of substances are probable, these results are, in the majority of cases, so uncertain that there exists but a very limited number of compounds of which one can say with certainty how many simple atoms form the compound atom." The one method which leads to certain results is the law of volumes, and its field of application is limited to gases. In this way the formulae H^2O, HCl, NH^3, for water, hydrochloric acid and ammonia, were derived.

When an element has several degrees of oxidation, if the ratios of oxygen for a given weight of the other element are 1 and 2, the compounds may be RO and RO^2; if the ratio is 2 to 3 the first compound may be RO and the second R^2O^3, or the two compounds may be RO^2 and RO^3. If the ratio is 3 to 4 the compounds may be RO^3 and RO^4, or R^2O^3 and RO^2. But if the ratio is 3 to 5 there is possible only

* Dalton, *New System*, 1808, I, i, 70, 188; Meldrum, *Avogadro and Dalton*, Aberdeen, 1904, and Edinburgh, 1906.

† *Traité*, 2nd French tr., 1845, vol. i, 64. ‡ *Traité*, 1831, iv. 591.

R^2O^3 and R^2O^5, or RO^3 and RO^5. When an electropositive oxide combines with an electronegative oxide, e.g. a base with an acid, the oxygen of the latter is a whole multiple of the oxygen of the former, and this number is generally the number of atoms of oxygen in the electronegative oxide. In this way Berzelius in 1826 found two typical series of oxygen compounds:

1. The Nitrogen Series: N^2O, NO, N^2O^3, N^2O^5.
2. The Sulphur Series: SO, SO^2, SO^3.

The preceding rule gives the formula CrO^3 to the anhydride of chromic acid, since in chromates the ratio of oxygen in the acidic oxide to oxygen in the basic oxide is 3 : 1. The ratio of oxygen in chromic anhydride to oxygen in chromic oxide is 2 to 1, hence the oxide is Cr^2O^3. This is isomorphous with ferric and aluminium oxides, the formulae of which are thus Fe^2O^3 and Al^2O^3. From the ratio of oxygen in ferrous and ferric oxides, the formula of the first must be FeO. This is isomorphous with the oxides of copper, cobalt, calcium, lead, etc., which thus have the formula RO. Analogies were skilfully used by Berzelius, and it will be seen that all the above formulae are those at present in use.

Since he thought that all strong bases had the formula RO, Berzelius wrote the formulae of potash, soda and silver oxide as KO, NaO and AgO. He had previously, in his tables of 1814 and 1818, written these as KO^2, NaO^2 and AgO^2. After the enunciation in 1819 of Dulong and Petit's law of atomic heats for solid elements: "the atoms of all simple bodies have exactly the same capacity for heat", or atomic weight × specific heat = atomic heat = constant *, Berzelius halved a number of atomic weights to bring them into line with the law, but he still wrote the oxides of potassium, sodium and silver as KO, NaO and AgO, although these gave atomic weights double those deduced from the specific heats. With this exception, his table of atomic weights published in 1826 is practically the same as the modern table. A selection of Berzelius's atomic weights is given in the table on p. 207.

Before 1830, therefore, Berzelius had drawn up a table of atomic weights identical in practically every respect (apart from small inaccuracies in the experimental determinations) with that now in use.

* *Ann. Chim. Phys.*, 1819, x, 395.

BERZELIUS'S ATOMIC WEIGHT TABLES

1	2	3	4	5	6	7	8
	1814		1818		1826		
Element							
O		16		16		16·03	
S	SO^2, SO^3	32·16	SO^2, SO^3	32·19	SO^2, SO^3	32·24	SO_2, SO_3
P	P^2O^3, P^2O^5	26·80	PO^3, PO^5	62·77	P^2O^3, P^2O^5	31·43	P_2O_3, P_2O_5
Cl		(35·16)		(35·41)	Cl^2O^5	35·47	
C	CO, CO^2	11·986	CO, CO^2	12·05	CO, CO^2	12·25	CO, CO_2
N		(14·36)		(14·05)	N^2O, NO	14·19	N_2O, NO
H	H^2O	1·062	H^2O	0·9948	H^2O	1	H_2O
As	AsO^3, AsO^5	134·38	AsO^3, AsO^5	150·52	As^2O^3, As^2O^5	75·33	As_2O_3, As_2O_5
Cr	CrO^3, CrO^6	113·29	CrO^3, CrO^6	112·58	Cr^2O^3, CrO^3	56·38	Cr_2O_3, CrO_3
Si	SiO^3	48·696	SiO^3	47·43	SiO^3	44·44	SiO_2
Hg	HgO, HgO^2	405·06	HgO, HgO^2	405·06	Hg^2O, HgO	202·86	Hg_2O, HgO
Ag	AgO^2	430·107	AgO^2	432·51	AgO	216·6	Ag_2O
Cu	CuO, CuO^2	129·03	CuO, CuO^2	126·62	Cu^2O, CuO	63·42	Cu_2O, CuO
Bi	BiO^2	283·84	BiO^2	283·81	Bi^2O^3	213·22	Bi_2O_3
Pb	PbO^2, PbO^3	415·58	PbO^2, PbO^3	414·24	PbO, Pb^2O^3	207·46	PbO, Pb_2O_3
Sn	SnO^2, SnO^4	235·29	SnO^2, SnO^4	235·3	SnO, SnO^2	117·84	SnO, SnO_2
Fe	FeO^2, FeO^3	110·98	FeO^2, FeO^3	108·55	FeO, Fe^2O^3	54·36	FeO, Fe_2O_3
Zn	ZnO^2	129·03	ZnO^2	129·03	ZnO	64·62	ZnO
Mn	MnO^2, MnO^3	113·85	MnO^2, MnO^3	113·85	MnO, Mn^2O^3	55·43	MnO, Mn_2O_3
Al	AlO^3	54·88	AlO^3	54·77	Al^2O^3	27·43	Al_2O_3
Mg	MgO^2	50·47	MgO^2	50·68	MgO	25·38	MgO
Ca	CaO^2	81·63	CaO^2	81·93	CaO	41·03	CaO
Na	NaO^2	92·69	NaO^2	93·09	NaO	46·62	Na_2O
K	KO^2	156·48	KO^2	156·77	KO	78·51	K_2O

In this table, columns 2, 4 and 6 give the formulae of the oxides assumed by Berzelius, column 8 the modern formulae of the oxides; columns 3 and 5 give the atomic weights recalculated from Berzelius's values, referred to oxygen = 100, to oxygen = 16; column 7 gives Berzelius's values on his alternative scale of hydrogen = 1.

By 1840 these atomic weights were nearly all abandoned by other chemists, and the so-called " equivalents " used instead.

Avogadro's Hypothesis

In 1811 the difficulties attending the application of the law of volumes had been cleared up by an Italian physicist, Amedeo Avogadro. He was born in Turin in 1776 and died there in 1856.

Fig. 78.—Amedeo Avogadro (1776-1856).

In 1796 he became Doctor of Law, and in 1806 he taught physics in a college in Turin, where he afterwards became professor of mathematical physics. He was removed from his chair in the revolutionary movement of 1822, but restored to it in 1835. Avogadro was little known in Italy and still less abroad. His important memoir * appeared in French in 1811, and another in 1814. His work passed unnoticed and a similar theory was advanced in 1814 by Ampère.†

* *J. de Physique*, 1811, lxxiii, 58; *Alembic Club Reprint* No. 4; Partington, *Nature*, 1956, clxxviii, 8.

† *Ann Chim. Phys.*, 1814, xc, 45; Ostwald's *Klassiker* No. 8.

Avogadro refers to Gay-Lussac's results and says that " it must be admitted that very simple relations also exist between the volumes of gaseous substances and the numbers of simple or compound molecules which form them. The first hypothesis to present itself in this connection, and apparently even the only admissible one, is the supposition that the number of integral molecules in all gases is always the same for equal volumes . . . ; the ratios of the masses of the molecules are then the same as those of the densities of different gases at equal temperature and pressure."

The difficulty raised by Dalton no longer exists if it is assumed that the smallest particles of gases are not necessarily simple atoms, " but are made up of a certain number of these atoms * united by attraction to form a single molecule ". In cases where molecules interact they may be divided, and the case cited by Dalton is easily explained in this way :

N_2	+	O_2	=	$2NO$
1 vol.		1 vol.		2 vols.
n molecules		*n* molecules		2*n* molecules
1 molecule		1 molecule		2 molecules

Avogadro considers molecules containing only even numbers of atoms ; he does not say why odd numbers are left out. In a similar way Avogadro showed, e.g., that the hydrogen and chlorine molecules contain 2 atoms :

H_2	+	Cl_2	=	$2HCl$
1 vol.		1 vol.		2 vols.

Berzelius, as mentioned previously, refers to this theory : " One may represent the atoms [molecules] by groups [of atoms], by admitting that in combination the atoms of the molecules [' molécules '] of one simple body exchange with the atoms of the molecules of the other, in such a way that in the gaseous compound obtained the number of molecules remains the same for a given volume ". He rejects the hypothesis because of difficulties introduced by the results of Dumas on vapour densities of mercury, sulphur and phosphorus, which gave results differing from the atomic

* Avogadro uses the name " molecule " for both atom and molecule ; for clearness the word " atom " has been used above whenever Avogadro speaks of " elementary molecule ".

weights which Berzelius, rightly, considered accurate, and the difficulty caused Berzelius to restrict his law of volumes to *permanent* gases, and to reject Avogadro's hypothesis.

In 1826 Dumas assumed that "gases in similar circumstances are composed of molecules or atoms placed at the same distance, which is the same as saying that they contain the same number in the same volume". The argument used by Avogadro also shows that "it is necessary to admit that the atoms of gaseous bodies are susceptible of division when they enter into combinations".* In 1832 Dumas found that the vapour densities of sulphur, phosphorus, arsenic and mercury gave anomalous results (he did not know that the molecules were peculiar, S_6, P_4, As_4, Hg), and he then stated that "gases, even when they are simple, do not contain, in equal volumes, the same number of atoms, or at least of chemical atoms".

Gaudin † in 1833 gave a clear account of the application of Avogadro's hypothesis to the determination of atomic weights, making use of volume diagrams :

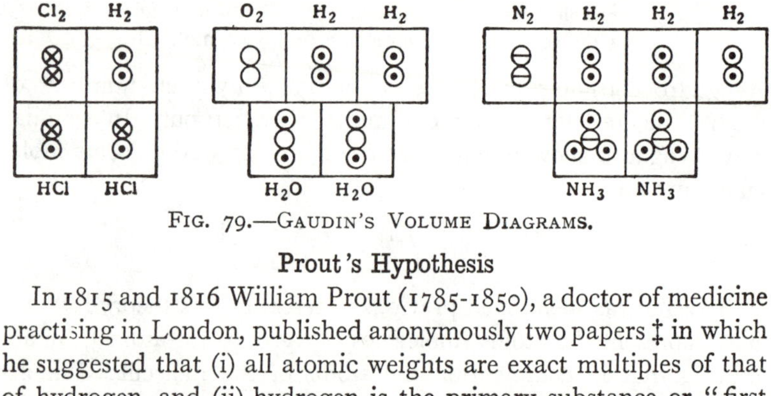

FIG. 79.—GAUDIN'S VOLUME DIAGRAMS.

Prout's Hypothesis

In 1815 and 1816 William Prout (1785-1850), a doctor of medicine practising in London, published anonymously two papers ‡ in which he suggested that (i) all atomic weights are exact multiples of that of hydrogen, and (ii) hydrogen is the primary substance or "first matter", respectively. Thomas Thomson was greatly attracted by this hypothesis and in 1825 he published *An Attempt to establish the First Principles of Chemistry by Experiment* (2 vols.). According

* According to Graebe, *Geschichte der organischen Chemie*, p. 225, Dumas was the only writer until 1850 to mention Avogadro by name in connection with the molecule.

† *Ann. Chim. Phys.*, 1833, lii, 113.

‡ *Annals of Philosophy*, 1815, vi, 321 ; 1816, vii, 111 ; *Alembic Club Reprint* No. 20. Similar views were proposed by Meinecke, *Ann. Physik*, 1816, xxiv, 159 ; *J. Chem. Phys.* (Schweigger), 1818, xxii, 137 ; 1819, xxvii, 39.

to Berzelius the title of the book was unhappily chosen, since "the experiments described in it appear to have been performed at the writing desk rather than in the laboratory". This was unjust, since Thomson was a conscientious man, but his enthusiasm for the hypothesis undoubtedly led him astray. Thus, he says,* after remarking that Berzelius had found the atomic weight of glucina to be 3·2522 (oxygen = 1): "I consider myself entitled to conclude from it, that the atomic weight of glucina is 3·25. I leave out the last two decimal places because they would destroy the law pointed out by Dr. Prout." Thomson repeated Berzelius's experiment and found † exactly 3·25.

Use of Equivalents

In spite of the general correctness of Berzelius's atomic weight table, chemists felt that there was no general method, apart from analogies and the combination of results from various fields of research, by which the atomic weight of an element could be found without doubt or uncertainty. The work of Dumas had also shaken the confidence of chemists in the law of volumes, which had seemed the most promising method, and Avogadro's hypothesis was entirely neglected. Between 1830 and 1840 it seemed as if the atomic theory was beginning to come to the end of its usefulness, and many chemists, who did not care much for theoretical discussions, used what were called "equivalents" directly ascertainable by experiment, rather than atomic weights. This attitude was strengthened by the adoption of equivalent weights by Leopold Gmelin in his large *Handbook of Chemistry* (English transl., 19 vols., 1848-72): H = 1, O = 8, C = 6, S = 16, etc., the formulae of water, potash, nitric acid, etc., being written ‡ HO, KO, NO_5. There was, however, the difficulty that several elements, e.g. iron, had more than one equivalent. The smallest number was often selected.

Summary and Supplement

Davy and Berzelius. The Electrochemical or Dualistic Theory

Sir Humphry (not "Humphrey") Davy (1778-1829), professor in the Royal Institution, London, investigated the properties of nitrous oxide (1799), and worked on electrolysis, putting forward an electro-

* *First Principles*, i, 321. † *ibid.*, 325.

‡ Gmelin, *Handbook*, vol. i, p. 42 f.

chemical theory (1806). He isolated the metals of the alkalis (1807) and alkaline earths (1808), and boron (1807), demonstrated the elementary nature of chlorine (1809-10) and worked on the oxides of chlorine (1811-15) and on iodine (1813-14), attempted to isolate fluorine (1813) and investigated the nature of flame (1815), inventing the safety lamp for mines.

Michael Faraday (1791-1867), at first an assistant to Davy in the Royal Institution, and later Fullerian professor of chemistry there, discovered two chlorides of carbon (1821), liquefied chlorine (1823) and other gases (1823-45), discovered butylene and benzene (1825), and the laws of electrolysis (1832-3).

Joseph Louis Gay-Lussac (1778-1850), professor in Paris, discovered the law of expansion of gases by heat (1802 ; independently of Dalton, 1801) and the law of gaseous volumes (1808). He investigated iodine (1813-14), and cyanogen compounds (1815), isolating cyanogen and making clear its nature as an organic radical; he distinguished the oxides of nitrogen N_2O_3 and NO_2 (1816); he applied volumetric analysis to acidimetry, chlorimetry and silver titrations (1824-32), and proposed improvements in the manufacture of sulphuric (1827) and oxalic (1829) acids. The Gay-Lussac tower was first used in 1842.

Louis Jacques Thenard (not "Thénard") (1777-1857) worked with Gay-Lussac on chlorine (1809-11) and the alkali metals (1808-11), the two discovering an improved method for the preparation of these metals. They also discovered sodium and potassium peroxides and potassamide, and showed that caustic potash and soda contain hydrogen as well as oxygen. Thenard discovered hydrogen peroxide (1818), and made investigations on the oxides of metals (1805) and phosphorus (1812 f.), and in organic chemistry, e.g. on sebacic acid (1801), bile (1805-7), ethers (1807), etc.

Jöns Jacob Berzelius (1779-1848), professor of chemistry in Stockholm and Secretary to the Swedish Academy of Sciences, by accurate quantitative researches showed that the laws of combination and the atomic theory applied in both inorganic and organic chemistry (1810 f.). He introduced a classification of minerals based on chemical composition (1814), devised the modern chemical symbols (1813), and developed (from 1811) an electrochemical ("dualistic") theory, independently of Davy's, from which it differs in points of detail. He drew up tables of atomic weights (1814, 1818, 1826), the later forms being practically identical with the modern values, except for the alkali metals and silver (double the modern values). He discovered ceria (1803, published in 1804), selenium (1817), and thorium (1828, published in 1829) ; isolated silicon (1810), titanium

(1824), and zirconium (1825), and made detailed investigations of the compounds of tellurium (1834) and of the rarer metals (V, Mo, W, U, etc.) and sulpho-salts (1825). Lithium compounds were discovered in his laboratory by Arfvedson in 1817, the metal being first obtained by Davy. Berzelius greatly improved analytical methods (use of rubber-tubing, water bath, desiccator, wash bottle, filter paper, blow-pipe analysis) and organic combustions (1814). He also discovered sarcolactic (1806) and pyruvic (1835) acids, and other organic compounds, and (also showing in 1830 that tartaric and racemic acids had the same composition) recognized and named isomerism (1827). He proposed the name catalysis for the supposed cause of a group of reactions (1835). His large textbook and annual reports were authoritative, and he had a great influence on the development of contemporary chemistry.

Eilhard Mitscherlich (1794-1863), professor in Berlin, discovered isomorphism (1819), polymorphism (1821), monoclinic sulphur (1823), selenic acid (1827), benzene sulphonic acid and nitrobenzene (1833), and obtained benzene from benzoic acid (1834). He put forward the "contact" theory of etherification (1834) and fermentation (1836), so recognizing the existence of catalysis, which he called "contact action".

Pierre Louis Dulong (1785-1838), director of the École polytechnique in Paris, discovered nitrogen trichloride (1813), worked on the oxides of nitrogen (1816), the lower oxy-acids of phosphorus (1816), and the gravimetric composition of water (with Berzelius) (1819). In conjunction with Alexis Thérèse Petit (1819) he put forward the law of the constancy of atomic heats, which was used by Berzelius (and later by Cannizzaro) in deciding atomic weights. Dulong measured the refractive indices (1825) and specific heats (1828) of gases. He put forward (1815) a "hydrogen theory" of acids similar to one proposed by Davy.

William Prout (1785-1850) suggested (1815-16) that all atomic weights are multiples of that of hydrogen, and that hydrogen is the fundamental element (protyle) from which other elements are formed by condensation. Prout also discovered uric acid in boa-constrictor excrement (1815), and murexide (1818), investigated and worked out methods of urine analysis, and organic combustion analysis (1815-1827) with the use of oxygen.

Amedeo Avogadro (1776-1856), professor of physics in Turin, put forward the hypothesis (1811) that equal volumes of all gases, at the same temperature and pressure, contain identical numbers of particles (molecules). The molecules of elementary gases, e.g. oxygen, hydrogen and chlorine, consist of two (or perhaps a multiple of (two) atoms, and not (as Dalton thought) of single atoms. Avogadro showed that this hypothesis would reconcile Dalton's atomic theory with Gay-Lussac's law of combining volumes, and overcome a difficulty pointed out by Dalton. Ampère (1814), who mentions Avogadro, suggested that the molecules of elements all contain four atoms. Avogadro's hypothesis was not adopted, and chemists long continued to base atomic weights on arbitrary rules, or to use equivalents instead of atomic weights.

Thomas Thomson (1773-1852), professor in Glasgow from 1818, first published an account of Dalton's atomic theory (*System of Chemistry*, 1807). He observed multiple proportions in the oxalates (1808), and discovered sulphur chloride (1803) and chromyl chloride (1827). He determined many atomic weights, which he thought supported Prout's hypothesis (1825), and so came into conflict with Berzelius, whose own values did not agree with it.

Surprisingly good early atomic weight determinations were made in 1833 by Edward Turner (1796-1837), the first professor of chemistry in University College, London, and by F. Penny in 1839.

Earlier workers in analytical chemistry, who extended the methods of Berzelius and founded the " group tables " for qualitative analysis, were Heinrich Rose (1795-1864), a pupil of Berzelius, professor in Berlin, and author of a *Handbuch der analytischen Chemie* (1829, and later editions) ; and Carl Remigius Fresenius (1818-1897), a pupil of Liebig, professor in Wiesbaden, and author of treatises on qualitative (1841) and quantitative (1846) analysis which are still in use in revised editions.

Antoine Jérome Balard (1802-1876), demonstrator in Montpellier, where he discovered bromine in 1826, then professor in the Sorbonne, Paris, also discovered hypochlorous acid and chlorine monoxide (1834). The discovery of bromine clinched the elementary nature of the halogens chlorine and iodine. (The name *halogen* was introduced by Berzelius in 1825.)

Leopold Gmelin (1788-1853), professor in Heidelberg, discovered potassium ferricyanide (1822), taurine (1824), croconic and rhodizonic acids (1825), haematin and pancreatin (with Tiedemann, 1826), introduced the names ester and ketone (1848) and wrote a large *Handbook of Chemistry*.

CHAPTER X

THE BEGINNINGS OF ORGANIC CHEMISTRY

Early Knowledge of Organic Chemistry

In the early part of the nineteenth century the chemistry of carbon compounds was much less advanced than that of the metals and commoner elements, such as sulphur, phosphorus and nitrogen. It was divided into the two groups of Vegetable Chemistry and Animal Chemistry, and in the textbooks of Thomson and Berzelius, for example, we find the various constituents of plants and animals described without indication of the real relations existing between their chemical compositions. In Vegetable Chemistry, sugar, acids, gum, indigo, bitter principle, extractive principle, tannin, camphor and indiarubber call for separate description; and in a similar way in Animal Chemistry gelatin, albumin, fibrin, urea, blood, saliva, urine, and the like, are described mostly from the point of view of the medical man. It was realized that all these materials have carbon and hydrogen, sometimes oxygen, nitrogen and sulphur, as essential constituents, but the isolation and characterization of definite substances from them was but little advanced. Several so-called " proximate principles " of vegetable and animal matter, e.g. those such as sugar, gum, camphor, etc., which can be extracted by solvents or by other simple methods, were examined by Fourcroy and Vauquelin (see p. 177), who also discovered some new substances. Many substances could not be crystallized, their purification presented great difficulties and their analysis was in a very primitive condition.

We can therefore understand Wöhler, as late as 1835, writing to Berzelius that " organic chemistry appears to me like a primeval forest of the tropics, full of the most remarkable things ". The entry into the dark forest, and the clearing of the undergrowth, was a task to which chemists first seriously applied themselves in the period at which we have now arrived, and the discoveries were destined to alter the whole aspect of chemistry in a most interesting fashion.

Isolated investigations of carbon compounds go back to much earlier times. Alcohol is described in a manuscript of the twelfth century; ether was described in a work of Valerius Cordus (*d.* 1544),

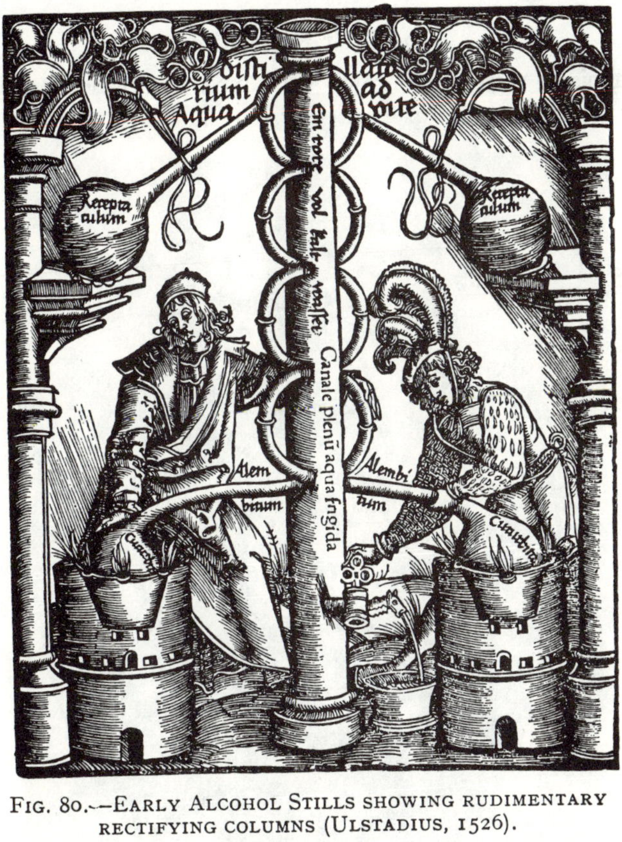

FIG. 80.—EARLY ALCOHOL STILLS SHOWING RUDIMENTARY RECTIFYING COLUMNS (ULSTADIUS, 1526).

and Boyle, in the *Sceptical Chymist* (1661), describes the separation of the distillate from boxwood into an acid and an " adiaphorous spirit " not kept back by distilling over salt and tartar. This would contain methyl alcohol and acetone. Benzoic acid was described by Blaise de Vigenère (1522-1596) in a work published in 1618.* The investigations of Scheele (1742-1786) resulted in the discovery of

* Patterson, *Annals of Science*, 1939, iv, 61.

many organic acids in plants and fruits, glycerin (which he showed is related to sugar both by its sweet taste and by giving oxalic acid when oxidized by nitric acid), hydrocyanic acid, and esters by distilling acids with alcohol. He showed that esters were " saponified " by boiling alkali.

The Older Radical Theory

Lavoisier regarded organic substances as formed on the same lines as inorganic: in particular, substances containing oxygen

FIG. 81.—L. N. VAUQUELIN, 1763-1829.

had the latter united with a *radical* (a name first used in the modern sense by de Morveau in 1787). This radical in the simpler compounds consisted of carbon and hydrogen (" hydrocarbon radical") and might be united with different proportions of oxygen: sugar was the lower and oxalic acid the higher oxide of a hydrocarbon radical. The oils, which were considered free from oxygen, might be free radicals.*

The existence of an organic radical, cyanogen, was perhaps first clearly demonstrated in the researches of Gay-Lussac on cyanogen compounds in 1815 (see p. 224).

In 1817 Berzelius † extended the dualistic theory to organic compounds. "All organic substances are oxides of compound radicals. The radicals of vegetable substances consist generally of carbon and hydrogen, those of animal substances of carbon, hydrogen and nitrogen."

Organic Analysis

In 1784 Lavoisier devised a method for the analysis of organic substances ‡ by burning them in a small lamp floating on mercury in a bell-jar containing air to which oxygen was added, the carbon dioxide being absorbed by potash solution. He also knew that acetic acid, first prepared in the pure state ("glacial acetic acid") by Stahl §, was formed by the oxidation of alcohol, and the relation between the two substances appeared. Dalton in 1804 made analyses of hydrocarbon gases by exploding them with oxygen in a eudiometer, and so determined the formulae of marsh gas and olefiant gas.

The first satisfactory method of organic analysis was worked out by Gay-Lussac and Thenard in 1810 ‖. They mixed the substance with potassium chlorate, formed the solid into pastilles, and dropped these through a large stopcock into a heated tube set vertically. The gas evolved was analysed. Hydrogen was calculated by difference. It was found that in substances such as sugar and starch the hydrogen and oxygen were combined in the proportions to form water, and such compounds were called "carbohydrates" (carbon + water) by Schmidt in 1844. Acids contained an excess of oxygen

* *Traité de chimie*, 1789, vol. i, p. 209.

† *Traité*, 1849, v, 29; the original statement is in the second Swedish edition, published in 1817, vol. i, p. 544; *Annalen*, 1839, xxxi, 5.

‡ *Œuvres de Lavoisier*, tome 2, p. 586.

§ *Specimen Becherianum*, 1703.

‖ *Recherches*, 1811, tome 2, p. 265; see Lowry, *Historical Introduction to Chemistry*, 1936, p. 391.

and oils an excess of hydrogen. The method was also applied to organic compounds containing nitrogen, which was measured as gas, only enough chlorate being used to oxidize the carbon and hydrogen.

The unexpected existence of organic bases, containing nitrogen, was first demonstrated by Sertürner's isolation of morphine from opium in 1805 and his recognition that it was a base in 1816; the alkaloids strychnine, brucine, quinine and cinchonine had all been isolated by Pelletier and Caventou by 1820.

The question as to whether the organic substances followed the laws of chemical combination had been considered by Dalton * in 1803, who had given the following formulae (re-written in modern notation) to some of them: olefiant gas CH, ether C_2O, alcohol CH_2O, sugar CHO_2.

Berzelius †, in an important research in 1814-15, made with the object of deciding whether the laws of chemical combination held in the organic world, improved the method of analysis. He heated the mixture of the substance and potassium chlorate, to which common salt was added to moderate the reaction, in a glass tube wrapped with tin-plate and heated by charcoal in a sloping position. This tube was connected with a receiver and a calcium chloride tube to absorb the water formed, and with a bell-jar standing over mercury, on the surface of which a small weighed flask of solid caustic potash floated, to absorb the carbon dioxide. The acids were mostly analysed in the form of their lead salts, and the formulae are therefore those of the anhydrides. The following formulae were obtained:

Substance	*Berzelius's Formula*	*Modern Formula*
Citric acid -	CHO	$C_6H_8O_7 - H_2O = (CHO)_6$
Tartaric acid	$C_4H_5O_5$	$C_4H_6O_6 - H_2O = C_4H_4O_5$
Oxalic acid -	$C_{12}HO_{18}$	$6(C_2H_2O_4 - H_2O) = C_{12}O_{18}$
Succinic acid	$C_4H_4O_3$	$C_4H_6O_4 - H_2O = C_4H_4O_3$
Sugar - -	$C_{12}H_{21}O_{10}$	$C_{12}H_{22}O_{11} - H_2O = C_{12}H_{20}O_{10}$
Starch - -	$C_7H_{13}O_6$	$C_6H_{10}O_5$

* Roscoe and Harden, *New View of Dalton's Atomic Theory*, 1896, 45.

† *Annals of Philosophy*, 1814, iv, 323, 401; 1815, v, 93, 174, 260; Lowry, p. 392.

The hydrogen was somewhat overestimated, owing to the difficulty of drying the substance, otherwise the results are very accurate.

Copper oxide was substituted for chlorate by Gay-Lussac in 1815, and by Döbereiner in 1816; the method was used and improved by Liebig *, and remains essentially the same to-day.† The combustion method of estimating nitrogen in organic compounds was devised by Dumas in 1830.

Berzelius's analyses showed that the laws of chemical combination and the atomic theory could be applied to organic compounds, although the latter often have complicated formulae. The results of inorganic chemistry could serve as a guide in studying "the mode of combination of the elements" in organic compounds ‡, and in the *Traité* compounds are classified as acids, bases and neutral substances. He emphasized § that "the electrical phenomena we have described manifest themselves principally in inorganic nature; in organic nature the state of things is different." In the organic world the existence of a mysterious "Vital Force" was assumed ‖ to intervene in the formation of compounds, and their preparation in the laboratory could hardly be expected. This theory received a severe blow when Wöhler in 1828 showed that urea, a typical organic compound, could be produced by merely heating ammonium cyanate, the two substances being isomeric, but it persisted long after this time.

Chevreul

Important investigations on organic compounds were carried out by Michel Eugène Chevreul, who was born in 1786, and lived in vigour to the age of 103.¶ He occupied technical positions in Paris, including the directorship of the famous Gobelin tapestry works. He began to work on saponification in 1810, taking the work of Scheele as a starting point. In 1816 he established the fact that soap is formed by the combination of alkali with an acid constituent

* *Annalen der Physik*, 1831, xxi, 1.

† History of organic analysis in H. L. Fisher's *Laboratory Manual of Organic Chemistry*, New York, 1924, p. 217 f.

‡ *Traité*, 1849, v, 28; *Annalen*, 1833, vi, 173.

§ *Traité*, 1831, iv, 582. ‖ *Traité*, 1831, v, 2.

¶ Hofmann, *Berichte*, 1889, xxii, 1163.

of the fat (a fact asserted by Tachenius in 1666), the other constituent (glycerol) being set free. In 1817, in conjunction with Braconnot, he distinguished between stearin and liquid olein and prepared stearic acid. He often used the melting point as a method of identification. In 1818-23 he obtained butyric, valeric

FIG. 82.—M. E. CHEVREUL, 1786-1889.

and caproic acids, and impure oleic acid. Chevreul's classical *Recherches chimiques sur les corps gras d'origine animale* appeared in 1823, and his *Considérations générales sur l'analyse organique et sur ses applications* in 1824. One of his later researches was on wool fat in 1853.

Chevreul also investigated various dyes, obtaining haematoxylin and brazilin from logwood and Brazil wood, also quercitrin, morin and luteolin. In 1812 he obtained indigo white by reducing indigo, in 1815 he showed that diabetic sugar is identical with grape sugar, and in 1832 he isolated creatine from meat extract.

As a result of the researches of Vauquelin, Gay-Lussac and Chevreul, Paris became famous as a centre of work in the new science of organic chemistry, and we shall later find that Liebig went to Paris in 1822 for the purpose of studying organic chemistry. Liebig's influence was later very important, and in some ways it represents an extension of that of the French school and a deviation from that of Berzelius.

Gay-Lussac and Thenard

Joseph Louis Gay-Lussac was born at St. Léonard in 1778 and died in Paris in 1850.* He was assistant to Berthollet and Fourcroy and later a professor in the École Polytechnique and the Jardin des Plantes. His lectures were published in 1828.† In many of his researches he was in collaboration with Louis Jacques Thenard, born in 1777 at Louptière, the son of a peasant. Thenard began as Vauquelin's laboratory boy, succeeded him as professor in the Collège de France, became wealthy as the discoverer of Thenard's blue, famous as the discoverer of hydrogen peroxide (1818), and died in 1857 Chancellor of the University of Paris and a Peer of France.

Gay-Lussac, like Davy, was a versatile man. In 1802 he published the law of expansion of gases by heat, and after preliminary work with Humboldt in 1805 he put forward the law of combining volumes of gases in 1808. Gay-Lussac and Thenard in 1808 investigated the action of ammonia gas on potassium, and obtained potassamide, KNH_2. They worked out a method for obtaining the alkali metals in quantity by the action of the fused alkalis on red-hot iron, and in this work they observed the presence of hydrogen in the alkalis, which Davy thought were oxides. They discovered peroxides of sodium and potassium. In 1809 they obtained nearly anhydrous hydrofluoric acid and boron fluoride, in 1810 barium peroxide, and in 1811 they observed the absorption of hydrogen by heated sodium and potassium. Gay-Lussac discovered hydriodic

* Arago, *Œuvres complètes*, iii, 1; Partington, *Nature*, 1950, clxv, 708.

† *Cours de chimie*, 2 vols., Paris, 1828; see Findlay, *Nature*, 1937, cxl, 22.

acid and potassium iodate in 1813 and chloric acid in 1814. His investigation on iodine * is a model of experimental research.

FIG. 83.—J. L. GAY-LUSSAC, 1778-1850.

In 1815 Gay-Lussac published an important paper † on cyanogen compounds. The composition of Prussian blue had been a mystery. By digesting the blue with water and mercuric oxide a salt (mercuric cyanide) was obtained by Scheele. Gay-Lussac found that this on heating gave off an inflammable gas composed of carbon and nitrogen. This gas, which Gay-Lussac called *cyanogen*, proved to be the basis of a series of compounds, the so-called cyanogen compounds. With hydrogen it formed prussic acid, it combined with chlorine and iodine, and its compounds with metals were the

* *Annales de chimie*, 1814, xci, 5.

† *Annales de chimie*. 1815, xcv, 136-230.

prussiates, now called cyanides. Hydrocyanic acid was an example of an acid free from oxygen, and cyanogen, which appeared in all

FIG. 84.—L. J. THENARD, 1777-1857.

the compounds, was a free organic radical, similar to chlorine, as Berzelius pointed out in 1839:

Cyanogen radical, CN or Cy	Chlorine atom, Cl
Cyanogen gas, C_2N_2 or Cy_2	Chlorine gas, Cl_2
Prussic acid, HCN or HCy	Muriatic acid, HCl
Potassium cyanide, KCN or KCy	Potassium chloride, KCl
Cyanogen iodide, ICN or ICy	Iodine chloride, ICl

In discussing cyanogen chloride, Gay-Lussac remarks that chlorine replaces hydrogen in hydrocyanic acid: "It is very remarkable that two bodies of which the properties are so different yet play the same rôle in combining with cyanogen " ; thus he foreshadows the theory of substitution. In 1814 he remarked * that acetic acid and " matière ligneuse " (cellulose) had the same composition, and drew the conclusion that " the arrangement of the molecules [atoms] in a compound has the greatest influence on the neutral, acid or alkaline character of the compound "—an early recognition of isomerism. Another anticipation of later work was his observation that potassium alum crystals grow in a solution of ammonium alum—a case of isomorphism. Gay-Lussac also introduced volumetric methods of acidimetry and alkalimetry and the titration of silver, and carried out work of technical importance: the Gay-Lussac tower was invented in 1827, in 1821 he showed that wood was rendered non-inflammable by treatment with borax, and in 1829 he obtained oxalic acid by fusing sawdust with caustic alkali.

Dumas

Jean Baptiste André Dumas was born in Alais in 1800 and died in 1884.† He was apprenticed to an apothecary, but wishing to improve his knowledge he set out on foot to Geneva, where he entered the pharmaceutical laboratory of Le Royer. Geneva was a centre of academic life, and Dumas came under the notice of Nicolas Théodore de Saussure and of De Candolle, the botanists. He also studied chemistry and carried out work which attracted the attention of De la Rive, who was then professor of chemistry in Geneva. In 1818 Dumas and Coindet introduced the use of iodine compounds as a remedy for goitre and, with the physiologist Prevost, Dumas carried out investigations on various important physiological problems. He then had the good fortune to meet Alexander von Humboldt, and as a result he determined to go to Paris, where he rapidly made himself a place in the world of science. At first he had to equip a laboratory at his own expense.

In studying the physical properties of liquids Dumas became interested in the esters, and his paper with Boullay in 1827 was the result. In 1831 he isolated anthracene from coal tar, in 1832 he studied ethereal oils and obtained formulae for camphor,

* *Annales de chimie*, 1814, xci, 149.

† Thorpe, *Essays in Historical Chemistry*, 1902, 318; Hofmann, *Berichte* 1884, xvii, 630.

borneol and artificial camphor, and in 1834 he and Péligot obtained cinnamic aldehyde and cinnamic acid, and methyl nitrate. His later work was mainly chemical, but after 1848 his interests were largely

FIG. 85.—J. B. A. DUMAS, 1800-1884.

political; he became Minister of Education and of Agriculture and Commerce, and played a prominent part in public affairs.

The Etherin Theory

In 1815 Gay-Lussac introduced a method of determining the vapour densities of liquids, afterwards improved by Hofmann, and determined the densities of alcohol and ether vapours. These were simply related to the densities of steam and olefiant gas as follows :

wt. of 1 vol. of alcohol vapour = wt. of 1 vol. of steam + wt. of 1 vol. of olefiant gas ;

wt. of 1 vol. of ether vapour = wt. of 1 vol. of steam + wt. of 2 vols. of olefiant gas.

In 1816 Robiquet and Colin showed that, just as alcohol and ether could be regarded as compounds of water and olefiant gas, so muriatic ether (ethyl chloride), obtained by distilling alcohol with hydrochloric acid, could be regarded as a compound of olefiant gas and muriatic acid ($C_2H_5Cl = C_2H_4 + HCl$). These preliminary observations were extended by Dumas and Boullay in 1827*, and the " ethers " (esters) were compared with ammonia salts as follows (modern formulae are used for simplicity) :

	Ammonia	*Olefiant Gas*
Hydrochlorate -	NH_3, HCl	C_2H_4, HCl
Nitrite - -	NH_3, HNO_2	C_2H_4, HNO_2
Acid Sulphate -	NH_3, H_2SO_4	C_2H_4, H_2SO_4

The research on esters was noteworthy in that the vapour densities of the compounds were determined. The substances could be represented as compounds of olefiant gas with acids, just as salts of ammonia could be considered as compounds of this substance with acids. Dumas and Boullay adopted the atomic weights $\underline{C} = 6$, $H = 1$, $O = 16$, and wrote olefiant gas C_4H_4 or $\underline{C}_2H_2$. The formulae of alcohol and ether were then written as follows :

$$\text{Alcohol, } 4H_2\underline{C}_2\ 2HOH\ ; \quad \text{Ether, } 4H_2\underline{C}_2\ HOH.$$

Berzelius at first opposed the theory, on the ground that olefiant gas has no alkaline properties †, but he explained it in his text-book and in 1832 called C_2H_4 (or $H_2\underline{C}_2$) the *aetherin* radical. Dumas did not call etherin a radical.

Liebig

Justus von Liebig ‡ was born at Darmstadt in 1803, and at first acted as assistant to his father, who was a druggist. He studied in

* *Annales de chimie*, 1828, xxxvii, 15.

† *Annales de chimie*, 1832, l, 240, 282.

‡ Shenstone, *Justus von Liebig*, 1901 ; Hofmann, *J. Chem. Soc*, 1875.

Bonn and Erlangen, but there was no practical instruction in chemistry, so that he went in 1822 to Paris. He there carried out in Gay-Lussac's laboratory some investigations on the fulminates, which in 1824 obtained for him the chair at Giessen, where he remained for twenty-eight years, until his migration to Munich in 1852. During this period his school became famous through-

FIG. 86.—J. VON LIEBIG, 1803-1873.

out the world, and the laboratory at Giessen was almost the first in Germany where practical instruction in chemistry was systematically given, although Thomson had a teaching laboratory in Edinburgh before 1807 and one in Glasgow from 1819.

Liebig was one of the outstanding chemists of the first half of the nineteenth century. He carried out a prodigious amount of experimental work in the field of organic chemistry, and made a great number of accurate analyses of organic compounds. He was also a very clear thinker, as distinguished in theory as in the laboratory. He was rather hot-tempered and sometimes wrote rather aggressive articles, but his letters to Berzelius (published by Carrière) show that Liebig was a man of very high moral character and inspired always by a desire to reach the truth, even if it meant disagreement with his friends. He died in 1873 and in his later years did not do very much experimental work. According to E. von Meyer *, Liebig "emphasized with all the force at his command that the true centre point of chemical study lay not in lectures but in practical work". When Baeyer succeeded Liebig at Munich, however, he found no laboratory of any kind.†

Liebig's improved method of organic analysis by combustion with copper oxide enabled him and his students to determine the formulae of numerous compounds, and in this way the relations of groups of compounds became clear. Besides his purely scientific work, Liebig also rendered important service in agricultural chemistry by introducing the use of mineral fertilizers, and his work in physiological chemistry, on the production of fat, the nature of blood and bile, and on extract of meat, although sometimes incorrect in points of detail, was useful and stimulating. Liebig regarded fermentation as brought about by the transmission of vibration from the particles of the ferment, which he assumed to be in violent motion, to the particles of the fermenting material, and he opposed both Berzelius's theory that it was a special case of catalysis, and Pasteur's theory ‡ that it was the result of living organisms. It is now known that fermentation is actually the result of the catalytic action of enzymes secreted by the living organisms and can be carried out in the absence of life.

Wöhler

Closely associated with Liebig in friendship and scientific collaboration was Friedrich Wöhler, born in 1800 at Eschersheim, near Frankfort on the Main. He studied under Leopold Gmelin, and in 1823 for a year with Berzelius in Stockholm. In 1825 he was a

* *History of Chemistry*, 1906, 644. † Perkin, *J. Chem. Soc.*, 1923, p. 1520.
‡ *Ann. Chim. Phys.*, 1860, lviii, 323.

teacher in the Berlin Technical School, in 1831 he was professor in the Technical High School at Cassel, and in 1836 in the University of Göttingen. He died in Göttingen in 1882 after raising the chemical school there to a high rank.* He differed entirely in temperament from Liebig, being cool and disliking controversy, and with a capacity for seeing the humorous aspects of things.

FIG. 87.—F. WÖHLER, 1800-1882.

Wöhler carried out important researches in organic chemistry, both in collaboration with Liebig and alone (e.g. on quinone, hydroquinone and quinhydrone), and also in inorganic chemistry. He carried out a surprising amount of experimental work, particularly on boron and silicon and their compounds, discovered silicon

* Thorpe, *Essays*, 1902, 294; *Ber.*, 1875, viii, 838; 1882, xv, 3127.

hydride, analysed a large number of minerals, and prepared compounds of the rarer metals. In 1824 he showed that oxalic acid is formed by the action of cyanogen on water, and in 1828 he obtained urea by evaporating a solution of the isomeric ammonium cyanate.

This was a synthesis of urea, since in 1782 Scheele obtained cyanide from ammonia, graphite, and alkali, and in 1781 Priestley reduced nitric acid to ammonia. Croconic acid, a benzene derivative, was obtained in 1825 by Gmelin from the residue in the preparation of potassium from potassium carbonate and carbon, and acetylene was prepared from the same residue by John Davy in 1836.

The Benzoyl Radical

In 1832 appeared that memorable research by Wöhler and Liebig on the benzoyl radical, which opened the way into the dark forest of organic chemistry.* The two chemists had become acquainted by reason of the work on cyanic and fulminic acids (p. 203). In 1831 Wöhler's wife died, and in the hope of providing some distraction Liebig invited him to undertake a joint research on oil of bitter almonds. The work, which was completed in a month, showed that oil of bitter almonds could be converted into a series of compounds containing the radical C_7H_5O, or as they wrote it, $C_{14}H_{10}O_2$, the first radical containing *three* elements, which they called *benzoyl*.†

Oil of bitter almonds. Benzoyl hydride	$C_7H_5O.H$
Benzoic acid. Benzoyl hydroxide -	$C_7H_5O.OH$
Benzoyl chloride - - - -	$C_7H_5O.Cl$
Benzoyl cyanide - - - - -	$C_7H_5O.CN$
Benzamide - - - - - -	$C_7H_5O.NH_2$

Berzelius, in a note at the end of the paper, hailed the work as "the beginning of a new day in vegetable chemistry", and proposed to call the benzoyl radical (which he formulated as Bz) *proïn* (πρωί, break of day), or *orthrin* (ὀρθρὸς, dawn of morning), the name *amide* being given to the NH_2 compound, the radical NH_2 being formulated Ad. On further consideration, however, it appeared that a radical containing oxygen was hardly admissible in the dualistic system, and Berzelius wished to regard the benzoyl compounds as containing a radical free from oxygen.

* *Annalen*, 1832, iii, 249; Ostwald's *Klassiker*, No. 22.

† The combustion analyses agree closely with the formulae, but they are based on a somewhat incorrect atomic weight of carbon.

The Ethyl Radical

In 1834 Liebig recognised that alcohol, ether and muriatic ether could be regarded as compounds of the radical C_2H_5, which he called *ethyl* (afterwards aethyl).* This, he says, "is more than an opinion, it is an incontrovertible fact." The same view had been proposed in 1833 by Robert Kane, of Dublin †, but he says it was "a subject of amusement and ridicule among the chemical circles" of that city. Their sense of humour was evidently better than their sense of judgment. Liebig, who doubled all the formulae, wrote:

$$C_4H_8 = \text{aetherin} = \text{Ae}\ ;\quad C_4H_{10} = \text{ethyl} = \text{E}.$$
$$\text{ether } C_4H_{10}O = \text{Ae} + H_2O = \text{EO}.$$
$$\text{alcohol } C_4H_{12}O_2 = \text{Ae} + 2H_2O = \text{EO} + H_2O.$$
$$\text{muriatic ether } C_4H_{10}Cl_2 = \text{Ae} + H_2Cl_2 = \text{E} + Cl_2.$$

These formulae (the aetherin formulae are due to Berzelius) were, of course, incompatible with the vapour densities and Avogadro's hypothesis, since alcohol vapour is lighter than ether vapour, but Liebig was in good company in ignoring such matters. Berzelius regarded alcohol not as the hydrate of ether but as the oxide of a radical C^2H^6, ether being the oxide of ethyl, $C^4H^{10}O$, and esters of oxy-acids compounds of ether with the acid anhydrides, e.g. acetic ester, $C^4H^6O^3 + C^4H^{10}O$.‡

The Methyl Radical

In 1834 Dumas and Péligot § announced experiments on the methyl radical. Boyle ‖ had separated the distillate of wood into an acid and an "adiaphorous" part by standing over marble or by slow fractional distillation. Taylor ¶ had prepared "pyroligneous ether" from it, but the French chemists proved that it is an alcohol, to which they gave the name methyl alcohol (μέθυ, wine; ὕλη, wood, or timber; "c'est à dire vin ou liqueur spiriteuse de bois"). They prepared several of its esters, and gaseous methyl ether. They

* *Annalen*, 1834, ix, 1.
† *Phil. Mag.*, 1839, xiv, 163.
‡ *Traité*, 1832, vi, 510 f.
§ *Ann. Chim. Phys.*, 1835, lviii, 5; 1836, lxi, 193 (read Oct.-Dec., 1834).
‖ *Works*, ed. Birch, 1744, vol. i, p. 390.
¶ *Phil. Mag.*, 1822, lx, 315; the work was carried out in 1812.

pointed out that these could be regarded as compounds of a radical CH_3 ($\underline{C}_4H_6$; $\underline{C}=6$) which Berzelius called methyl. This was a very important research, since it made known the existence of a group of related compounds, the alcohols.

In 1835 Regnault *, whose later work lay in the domain of physics, by treating ethylene bromide (obtained by the direct union of olefiant gas with bromine) with concentrated alcoholic potash, obtained the compound C_2H_3Br by removal of HBr. The radical C_2H_3 was called *aldehydene.* Ethylene bromide he wrote $C_2H_3Br + HBr$ or $C_4H_6Br_2 + H_2Br_2$; aldehyde and acetic acid, the two successive oxidation products of alcohol, were $C_4H_6O + H_2O$ and $C_4H_6O_3 + H_2O$. This work was prompted by Liebig, who called the radical C_4H_6 *acetyl.*† Ethyl was $C_2H_3.H_2$ and etherin $C_2H_3.H$, and the aetherin and ethyl theories were therefore combined.

In 1831 Liebig ‡ by the action of chlorine on alcohol obtained chloral, and with caustic alkali this gave chloroform, which he thought had the formula C_2Cl_5. Soubeiran § in the same year obtained chloroform by distilling alcohol with bleaching powder and gave it the formula CH_2Cl_2. Its true formula, $CHCl_3$, was obtained by Dumas on the basis of his theory of substitution (p. 241).

In a joint memoir presented by Dumas and Liebig to the French Academy on 23rd October, 1837, it was asserted that " in inorganic chemistry the radicals are simple ; in organic chemistry they are compounds—that is the sole difference. The laws of combination, the laws of reaction, are the same in the two branches of chemistry." In 1838 a definition of an organic radical was given by Liebig ‖ : (i) it is an unvarying constituent of a series of compounds, (ii) it can be replaced in these by elements, (iii) in its compounds with an element the latter can be separated or replaced by equivalents of other elements. At least two of these conditions must be fulfilled before an atomic complex can be called a radical.

* *Annales de chimie,* 1835, lix, 358.

† *Annalen,* 1839, xxx, 139 ; the modern acetyl radical is C_2H_3O.

‡ *Annalen,* 1832, i, 189.

§ *Annales de chimie,* 1831, xlviii, 131.

‖ *Annalen,* 1838, xxv, 3.

Bunsen

The Radical Theory, based on the views of Lavoisier and Berzelius, and extended by Liebig, was supported by the classical researches of Bunsen on the cacodyl radical. Robert Wilhelm

FIG. 88.—R. W. BUNSEN, 1811-1899.

Bunsen *, born in Göttingen in 1811, became assistant professor in the University there, succeeded Wöhler in Cassel, and was then appointed to the chair in Marburg in 1838. He was called in 1851 to Heidelberg, where he continued to work until his resignation in 1889. He died in 1899.

Bunsen's work as an investigator was of the first rank; his

* Roscoe, *J. Chem. Soc.*, 1900, lxxvii, 513; Bolton, *Chem. News*, 1899, lxxx, 283.

research on organic arsenic compounds, from 1837 onwards, was his only work in this field. From 1846 onwards he devised exact methods of gas analysis, with Kirchhoff he discovered spectrum analysis in 1859 and immediately brought to light by its aid the two new elements caesium (1860) and rubidium (1861). His work on mineral analysis was also important. As a teacher Bunsen was very successful: his inclinations were practical, and he took no part in the discussions on theory which raged during the first half of the nineteenth century. In 1840 he invented the Bunsen battery, in 1844 the grease-spot photometer, in 1853 the Bunsen burner, in 1855 the absorptiometer, in 1856 the actinometer (with Roscoe), in 1857 the effusion apparatus, in 1868 the filter pump and in 1870 the ice calorimeter. His other work included the production of cyanides by heating alkalis with carbon in a current of nitrogen (with Playfair in 1846), the preparation by electrolysis of magnesium (1852) and cerium (1858), and the joint investigation with Roscoe on the chemical action of light, begun in 1855.

The Cacodyl Radical

In 1760 Cadet, by distilling a mixture of equal weights of potassium acetate and arsenious oxide, obtained a heavy, brown, strongly fuming liquid of fearful odour. This repulsive liquid was fully investigated by Bunsen *, who isolated from it cacodyl oxide, $C_4H_{12}As_2O$, the oxide of the radical $C_4H_{12}As_2$, to which the name cacodyl (from κακώδης, stinking) was given by Berzelius. The chloride, iodide, cyanide and fluoride of cacodyl were obtained by the action of the acids on the oxide; oxidation of the latter gave cacodylic acid, $C_4H_{12}As_2O_3 + H_2O$. The free radical was obtained by heating the chloride with zinc in an atmosphere of carbon dioxide, dissolving out the zinc chloride with water, and drying the oily liquid, which easily inflames in the air. In this work Bunsen had to breathe through long glass tubes, and owing to an explosion he partly lost the sight of one eye. He found that if a drop of cacodyl oxide fell on a heated part of the distilling apparatus, "the apparatus was demolished by an explosion, and an arsenic

* *Annalen*, 1837, xxiv, 271; 1841, xxxvii, 1; 1842, xlii, 14; 1843, xlvi, 1; Ostwald's *Klassiker* No. 27; *Phil. Mag.*, 1842,(3), xx, 343, 382, 395.

flame several feet high arose, covering the surrounding objects with a black layer of foul-smelling arsenic."

Bunsen determined the vapour densities of all the volatile cacodyl compounds, and so obtained the correct formulae. In connection with the cacodyl compounds he says: "By a glance at this group of substances we recognise in it an unchangeable member, the composition of which is represented by the formula $C_4H_{12}As_2$. . . . The constituent elements of this member, combined with each other by a powerful affinity, take part only as a whole in the decompositions which characterise these bodies. They form in their compounds one of those higher units which we call organic atoms or radicals." The actual isolation of the cacodyl radical was a great triumph for the radical theory, which, as Bunsen remarks, became thereby to all intents and purposes a fact.

Summary and Supplement

The Beginnings of Organic Chemistry

The Radical Theory

Michel Eugène Chevreul (1786-1889) investigated the composition of oils and fats (from 1813; 1826), and vegetable colours (1808-64). He explained clearly the reaction of saponification, worked on the analysis of organic compounds, and must be regarded as one of the founders of modern organic chemistry.

Justus von Liebig (1803-73), professor in Giessen (1824) and Munich (1852), studied the fulminates in Gay-Lussac's laboratory (1821), discovered hippuric acid (1829), prepared chloroform and chloral (1831, published in 1832), worked with Wöhler on the benzoyl compounds and recognized the benzoyl radical (1832), and adopted the ethyl radical, simultaneously suggested by Kane and Berzelius (1833). Liebig's theory of polybasic acids (1838) extended Graham's researches to a number of organic acids, and revived Davy's and Dulong's "hydrogen theory" of acids. Liebig put forward a "vibratory" theory of fermentation and decay (1839), investigated uric acid derivatives (1834-1837), alkaloids (1839), amino-acids and amides from animal products (1846-52), and creatine and creatinine (1847); he developed theories of botanical and physiological interest, and introduced mineral manures, and "extract of meat". Liebig edited the *Annalen der Pharmacie* (1832-39), continued as *Annalen der Chemie und Pharmacie* from 1840, and as *Annalen der Chemie* from 1874. He discovered the cyanide process for separating nickel and cobalt (1848-53), and devised a method for silvering mirrors (1856).

Friedich Wöhler (1800-1882), a pupil of Berzelius, professor in Göttingen (from 1836), obtained urea from ammonium cyanate (1828), isolated aluminium (1827), beryllium (1828), and crystalline silicon and boron (1856-7), prepared silicon nitride (1857), and (with Buff) discovered silicon hydride and silicon chloroform (1857). Wöhler's other work in inorganic chemistry included the preparation of phosphorus by the modern method (1829), on metallic suboxides and peroxides, the discovery of calcium carbide and the preparation of acetylene from it (1862), and the recognition of the analogy of silicon and carbon compounds (1863). In organic chemistry, Wöhler collaborated with Liebig (see above); they investigated mellitic acid (1830) and discovered amygdalin (1837) and parabanic acid (1838); Wöhler discovered hydroquinone and quinhydrone (1843-4), and tellurium methyl (1855). Wöhler translated several editions of Berzelius's text-book, and the nearly complete run of his *Jahres-Bericht* (vols. 4-27), into German. Urea had been obtained synthetically by John Davy (*Phil. Trans.*, 1812, 144) by the action of ammonia gas on carbonyl chloride, which he obtained by the action of light on a mixture of chlorine and carbon monoxide and called phosgene. He did not know that urea was formed in the reaction.

Robert Wilhelm Bunsen (1811-99), professor in Marburg (1838) and Heidelberg (1858), investigated cacodyl compounds (1837-43), recognized that they contained the cacodyl radical, and isolated free cacodyl $As_2(CH_3)_4$. He observed the formation of cyanides from alkalis, carbon and nitrogen (1845), worked out analytical methods, e.g. for iodine titration (1853), studied "actinometry", i.e. the chemical action of light (with Roscoe) (1857-59), applied the spectroscope to chemistry, and found that each element has a characteristic emission spectrum (1859), thereby discovering caesium (1860) and rubidium (1861). He improved gas analyses and invented chemical apparatus, e.g. the Bunsen cell, Bunsen burner, filter pump, etc. Bunsen was very successful as a teacher. He had little interest in theory and was very practical. After his work on cacodyl compounds, Bunsen lost interest in organic chemistry: none of his assistants in Heidelberg was permitted to use the university laboratories for researches in that subject. (Bunsen's *Gesammelte Abhandlungen*, edited by Bodenstein and Ostwald, were published in 3 vols. in 1904.)

CHAPTER XI

SUBSTITUTION. THE UNITARY AND TYPE THEORIES

The Theory of Substitution

Just about the time when the radical theory seemed to have been firmly established, a new way of looking at the reactions of organic chemistry was introduced by the two French chemists, Dumas and Laurent. This led in the end to the downfall of the dualistic theory of Berzelius, and since the molecule as a whole was regarded as a structure which, by modification of its parts, could give rise to series of related molecules, this new point of view came to be known as the Unitary Theory. The basis of the theory was the fact of substitution.

Gay-Lussac *, in speaking of the bleaching action of chlorine, remarked that chlorine takes part of the hydrogen of oils to form hydrochloric acid, "and at the same time part of the chlorine combines with the oil and takes the place (prend la place) of the hydrogen removed." In an investigation of the cause of the irritating fumes given off by candles used in the Tuileries, Dumas found that the wax had been bleached with chlorine, and the fumes were hydrochloric acid.† He was thus led to investigate the action of chlorine on alcohol ‡ and obtained chloroform. He showed that this contains hydrogen, and a determination of its vapour density gave the formula $\underline{C}_2HCl_3$ ($\underline{C} = 6$; Dumas formulates chlorine as Ch): he remarks that if Liebig had determined the vapour density of chloroform he would have seen that his formula, C_2Cl_5, could not possibly be correct. By the action of chlorine on alcohol, $\underline{C}_8H_{12}O_2$ ($\underline{C} = 6$), chloral, $\underline{C}_8H_2Cl_6O_2$, is formed, so that 10 "volumes" of

* *Cours de chimie*, 1828, Leçon 28, pp. 11 and 22.

† Hofmann, *Berichte*, 1884, 630 f.

‡ *Annales de chimie*, 1834, lvi, 113.

hydrogen are turned out whilst only 6 of chlorine take their place. The correct formula of chloroform was deduced from its production by the action of alkali on chloral. Dumas also studied the action of chlorine on turpentine and found that for every atom of hydrogen removed as hydrochloric acid, one of chlorine entered the molecule.

As a result of these experiments Dumas formulated what he called " a theory or law of substitutions ", or of " metalepsy ", from μετάληψις, " which expresses well enough that the body on which one acts has taken one element in place of another, chlorine in place of hydrogen, for example." * He summarized the new theory as follows † :

" (i) When a substance containing hydrogen is submitted to the dehydrogenating action of chlorine, of bromine, of iodine, of oxygen, etc., for each atom of hydrogen which it loses it gains an atom of chlorine, of bromine, of iodine, or half an atom of oxygen." ($O=16$; $H=1$.)

" (ii) When the compound contains oxygen the same rule holds good without modification."

" (iii) When the hydrogenized body contains water, this loses its hydrogen without replacement, and then, if a further quantity of hydrogen is removed, it is replaced as in the former cases."

As an example of the first law Dumas was able in 1838 ‡ to bring forward the action of chlorine on acetic acid. The latter, $\underline{C}_4H_4O_2$ ($\underline{C}=6$; $O=16$), is converted into trichloracetic acid, $\underline{C}_4HCl_3O_2$, and $3HCl$ are evolved. The third rule was introduced to explain the action of chlorine on alcohol. This latter, $\underline{C}_4H_6O$ or $\underline{C}_8H_{12}O_2$, was regarded as a hydrate of etherin, $\underline{C}_8H_8+2H_2O$. The first action of chlorine was to remove hydrogen without replacement, forming aldehyde (" ether acétique "), $\underline{C}_8H_8O_2$, which by further action was substituted to form chloral :

$$(\underline{C}_8H_8+2H_2O)+Cl_4=\underline{C}_8H_8O_2+H_4Cl_4 ;$$
$$\underline{C}_8H_8O_2+Cl_{12}=\underline{C}_8H_2Cl_6O_2+H_6Cl_6.$$

* *Mém. de l'Institut,* 1838, xv, 548 ; read in 1835.

† *Traité de chimie,* 1835, v, p. 99, where he gives eleven examples ; *Journal de Pharmacie,* May 1834, vol. v, p. 285.

‡ *Comptes rendus,* 1838, vii, 474 ; 1839, viii, 609 ; on the subject of this section see Schorlemmer, *Rise and Development of Organic Chemistry,* 2 ed., 1894.

In this way Dumas thought he could distinguish between hydrogen united with carbon and that united with oxygen.

The oxidation of alcohol to acetic acid is an example of the substitution of half an atom of oxygen for one of hydrogen :

$$\frac{C_8H_8}{H_4O_2} + O_4 = H_4O_2 + \frac{C_8H_4O_2}{H_4O_2}$$

Criticism of the Electrochemical Theory

Trichloracetic acid is so like acetic acid in its properties that Berzelius's electrochemical theory was obviously challenged : electropositive hydrogen was replaced by electronegative chlorine and yet no fundamental change in properties occurred. Berzelius * thereupon announced that : " an element so eminently electronegative as chlorine can never enter an organic radical : this idea is contrary to the first principles of chemistry ; its electronegative nature and its powerful affinities would prevent it from entering except as an element in a combination peculiar to itself." In a further paper † he emphasized that the laws of inorganic chemistry, i.e. his own dualistic theory, must be the guiding principles in the new science of organic chemistry, and in this way he arrived at the formula $C^{14}H^{10}Cl^6 + 2C^{14}H^{10}O^3$ for benzoyl chloride. In a letter to Wöhler ‡ he said : " This representation [by Dumas] necessarily involves the overthrow of the whole structure of chemistry in its present form, and the revolution is based on the decomposition of acetic acid by chlorine! "

By this time Liebig found himself unable to agree with Berzelius, and in a footnote to the above paper in his *Annalen* he says : " I do not share the views of Berzelius, since they rest upon a mass of hypothetical assumptions of which proof of any kind is lacking ", and in connection with a further letter of Berzelius §, Liebig says : " I do not share the views by which he explains the composition of the compounds discovered by Malaguti. I believe, on the contrary, that these materials are produced by simple substitu-

* *Annales de chimie*, 1838, lxvii, 309.
† *Annalen*, 1839, xxxi, 1.
‡ *Annalen*, 1839, xxxi, 113 ; cf. *Traité*, French tr., 1849, v, p. 26 f.
§ *Annalen*. 1839, xxxii, 72 ; cf. Crum Brown, *B. A. Report*, 1874, 45.

tions." Liebig and Dumas now joined forces and issued a programme of research which was to be undertaken under their joint direction with the aid of the British Association, Berzelius being disregarded. The two men were really incompatible, and, as Berzelius predicted, they soon drifted apart, although they always regarded one another with respect. They were leaders of thought in chemistry for many years.

As to the analogy with inorganic chemistry, which was fundamental for Berzelius, Liebig says: "Up to a point, therefore, we follow the principles of inorganic chemistry, but beyond the point where they leave us in the lurch (wo sie uns verlassen) we require new principles." All this was keenly felt by Berzelius, who saw his former great authority slipping away. A further blow came from Dumas, who said: "These electrochemical conceptions, this special polarity which has been assigned to the elementary atoms, do they really rest on such evident facts that they may be accepted as articles of faith? Or, if we regard them only as hypotheses, do they possess the property of adapting themselves to facts, are they capable of explaining them, can we assume them with such complete certainty that in chemical investigations they appear as useful guides? We must admit that such is not the case."

The investigations of Bunsen on cacodyl were certainly a support for Berzelius. He writes to Wöhler: "This is a triumphal chariot which has run over and smashed the rickety theoretical barricades of Dumas."

Berzelius on Substitution

We must now show how Berzelius tried to explain the facts of substitution by means of the radical theory. In 1837 Malaguti showed that chlorine can be substituted for hydrogen in ether. The simplest representation is ($\underline{C}=6$; $O=16$; $H=1$):

$$(\underline{C}_8H_8+H_2O)+Cl_8=(\underline{C}_8H_4Cl_4+H_2O)+H_4Cl_4$$

in accordance with Dumas' scheme, but as this involves substitution of chlorine in the etherin radical, $\underline{C}_8H_8$, Berzelius preferred to regard "gechlortes Aether" as $\underline{C}^4H^6O+\underline{C}^4Cl^4$, i.e. a dualistic compound of methyl oxide and carbon chloride. This, however, no longer contains the ethyl radical of ether, the parent substance, but has a totally different structure. Dumas' trichloracetic acid Berzelius wrote as $H^2O + C^2O^3 + C^2Cl^6$ ($C = 12$; $O = 16$), i.e. oxalic acid $H^2O + C^2O^3$ combined with carbon chloride, whereas the parent acetic acid was the hydrate of the trioxide of the acetyl radical, $C^4H^6 + 3O + H^2O$.

In making up these strange formulae Berzelius relied chiefly on instinct, i.e. his extensive knowledge of the behaviour of chemical compounds. In a letter to Wöhler he says: "When I see an incorrect theoretical representation I feel, even when the correct one is unknown to me and when I cannot make it clear to myself, that it is wrong, just as when the ear apprehends a false note in music: it is something of the same kind as when one recognizes bad form (insozialer Takt) by feeling rather than by reason."

These views, however, were definitely disproved by an experiment made by Melsens, the assistant of Dumas, who found * that trichloracetic acid, which resembles acetic acid in its properties and in those of its salts and esters, could readily be converted into acetic acid by the action of potassium amalgam, or zinc and dilute sulphuric acid, when the chlorine was simply replaced by hydrogen. Acetic and chloracetic acids must, therefore, have similar formulae. Berzelius admitted this, and rewrote the formulae as ($C = 12$; $O = 16$):

$H^2O.C^2O^3 + C^2H^6$	$H^2O.C^2O^3 + C^2Cl^6$
oxalic acid + methyl	oxalic acid + carbon chloride
Acetic acid	Trichloracetic acid

Trichloracetic acid he calls "Chlorkohlenoxalsäure", and he speaks of compounds formed on this plan as Copulated ("Paarlinge oder gepaarte Verbindungen"). "The substances which form copulae," he says †, "are of diverse kinds: compound radicals or binary non-oxidized compounds, e.g. hydrogen carbides, amides, chlorides, oxides and chlorides of compound radicals, and even compounds which we can suppose to be copulated compounds themselves: all can be copulae for active oxides." Further, "Copulated compounds belong not only to the bodies which can play the part of acid or base, they may also occur with bodies which may be considered entirely

* *Comptes rendus*, 1842, xiv, 114.

† *Traité*, 1849, tome v, p. 43; the name is due to Gerhardt.

neutral. . . . There exist compound radicals, and their compounds with oxygen, sulphur, and halogens have a great tendency to produce copulated compounds, in which one of the terms (composé actif) conserves its property of uniting with other bodies, whilst the other term, which we call the copula, has lost all tendency to combination, with certain exceptions."

Laurent * says bitingly: "A word let fall from the pen of Gerhardt was thus transformed into a luminous idea for dualism. From this time everything was copulated. . . . So that to make acetanilide, for example, they no longer employed acetic acid and aniline, but they recopulated a copulated oxalic acid with a copulated ammonia. . . . What then is a copula? A copula is an imaginary body, the presence of which disguises all the chemical properties of the compounds with which it is united. . . . The dishonesty is flagrant."

Berzelius, in the new formula for chloracetic acid, had practically admitted substitution of chlorine for hydrogen in the copula, but hardly anyone was prepared to accept the new views which he put forward. It was reserved for Kolbe, long afterwards, to show that these formulae contained an important germ of truth.

Another attack on Berzelius's theory is contained in Liebig's long and wordy, but important, memoir on organic acids.† Previous to this the theory of Berzelius on the constitution of salts had been modified by the researches of Graham on the phosphoric acids.‡

Graham

Thomas Graham, born in Glasgow in 1805, studied under Thomson and became professor in Anderson's College in 1830. He succeeded Turner as professor in University College, London, in 1837, a position which he resigned in 1855 on his appointment as Master

* *Chemical Method*, 1855, 204.

† *Annalen*, 1838, xxvi, 113-189; it concludes with the words: "Durch die Nacht fährt unser Weg zum Lichte." Ostwald's *Klassiker* No. 26. Berzelius's severe criticism of this paper led to an estrangement between him and Liebig.

‡ *Phil. Trans.*, 1833, 253; *Alembic Club Reprint* No. 10; *Chemical and Physical Researches by Thomas Graham*, privately printed, Edinburgh, 1876, 321 f.; on Graham, see Thorpe, *Essays*, 1902, p. 206.

of the Mint. He died in 1869. Graham's valuable physico-chemical researches dealt with the adsorption of salts by charcoal (1830), on the diffusion of gases (" Graham's Law "; 1828-33), on the transpiration of gases through tubes (1846), on the colloidal state of matter (1849), on osmose (1854), and on the absorption of

FIG. 89.—THOMAS GRAHAM, 1805-1869.

hydrogen by metals (1866-9). His investigation on the arsenates and phosphates showed that the ortho-, pyro- and meta-phosphates were not salts of three isomeric phosphoric anhydrides, as Berzelius assumed with respect to the two known to him, but were derived from three definite acids, containing water replaceable in whole or in part by bases. " I suspect that the modifications of phosphoric acid, when in what we call a free state, are in combination with their usual proportion of base, and that base is water." Thus the three modifications of phosphoric acid may be composed as follows :

Phosphoric Acid	$\dot{H}^3\overset{\cdot\cdot\cdot\cdot\cdot}{P}$	$3H\underline{O}+P\underline{O}_5$	$3H_2O.P_2O_5$
Pyrophosphoric Acid	$\dot{H}^2\overset{\cdot\cdot\cdot\cdot\cdot}{P}$	$2H\underline{O}+P\underline{O}_5$	$2H_2O.P_2O_5$
Metaphosphoric Acid	$\dot{H}\overset{\cdot\cdot\cdot\cdot\cdot}{P}$	$H\underline{O}+P\underline{O}_5$	$H_2O.P_2O_5$

The first set of symbols (due to Berzelius, see p. 197) are those used by Graham ; the second are rewritten in ordinary notation ($H=1$; $\underline{O}=8$; $P=31$) ; the third are the modern symbols. It was clear that the individual properties of the acids could not be expressed if they were regarded as anhydrides ; they must contain chemically combined water essential to their composition. Graham formulated the phosphates in a similar way, e.g. ordinary phosphate of soda is $\dot{Na}^2\dot{H}\overset{\cdot\cdot\cdot\cdot\cdot}{P}$, the metaphosphate $\dot{Na}\overset{\cdot\cdot\cdot\cdot\cdot}{P}$.

Polybasic Acids

Liebig, after referring to Graham's work, pointed out that by loss of water the phosphoric acids change their capacity for saturating bases. He showed that there were organic acids which, like phosphoric acid, could combine with more than one base to form salts which were not double salts, i.e. there were acids one molecule of which could combine with several molecules of base. Such were cyanuric, meconic, tannic and citric acids. On heating, the acids sometimes lost water and diminished in basicity. These acids " contain three atoms of water, which can be substituted by bases ". More simply, the results could be explained by the hypothesis of Davy* and Dulong (1815), that "acids are particular compounds of hydrogen, in which the hydrogen can be replaced by metals. Neutral salts are those compounds of the same class in which the hydrogen is replaced by the equivalent of a metal ".

Davy, according to Liebig, " makes the capacity of saturation of an acid dependent on the hydrogen which it contains ". This agreed with a theory of salts put forward in 1840 by J. F. Daniell, professor in King's College, London.† In the electrolysis of solu-

* Davy, *Phil. Trans.*, 1815, p. 203 ; *Works*, vol. v, pp. 492, 510.

† *Phil. Trans.*, 1839, 1840, 1844 ; *Elements of Chemical Philosophy*, 2nd ed., 1843, p. 533.

tions of salts, besides the acid and base, there were also liberated hydrogen and oxygen in the same proportions as from water. On Berzelius's theory the current seems here to do double work, in decomposing the salt, e.g. Na_2O,SO_3, into base and acid, Na_2O and SO_3, and the water into hydrogen and oxygen. But if the salt is regarded as Na_2SO_4, then it may be decomposed by the current into $2Na$ and SO_4. The first reacts with water to form soda and liberates hydrogen :

$$2Na + H_2O = Na_2O + H_2$$

and the latter acts with water to form sulphuric acid and oxygen :

$$SO_4 + H_2O = H_2SO_4 + O.$$

" These secondary actions are found to be in equivalent proportions to the primary, but have no influence upon the current force." Sulphuric acid, according to Daniell, could be represented as a compound of the radical SO_4 with hydrogen, H_2SO_4, in accordance with Davy's theory.

Laurent

August Laurent was born at La Folie, near Langres, in 1808 and became a pupil and for a short time assistant of Dumas. Laurent spent most of his life in poverty. He was professor in Bordeaux, where he had a very inadequate laboratory, for eight years. Then he came to Paris to work with Gerhardt, but lack of means drove him to a position as assayer to the Mint, where he had a damp cellar as a laboratory and could afford only common chemicals. He died of consumption in 1853.* Both Laurent and Gerhardt were ostracized by their French colleagues. Hofmann, who visited Paris in 1851 with Graham, says they were entertained by the French chemists, " allerdings moins les deux " (i.e. Laurent and Gerhardt), who " verkehrten nur wenig mit ihren Fachgenossen ; es war, als ob eine Art von Interdikt auf ihnen gelegen hätte." It is very painful indeed to see how badly these two chemists, the most brilliant of their day, were treated by some of their colleagues. Gerhardt received overdue recognition a year before his death.

* Yorke, *J. Chem. Soc.*, 1855, vii, 149 ; Grimaux, *Revue Scientifique*, 1896, vi, 161, 203, who gives the date of birth as 14 September, 1808—the date usually given is 14 November, 1807 ; de Milt, *Chymia*, 1953, iv, 85.

Laurent was a skilful experimenter and a keen critic, and in theory had a passion for classification. His famous *Méthode de chimie* appeared in 1854, a year after his death. In the introduction, Biot says: "Cet ouvrage, rempli d'idées nouvelles . . . vous offre les convictions intimes d'un homme qui a enrichi la science de

FIG. 90.—A. LAURENT, 1808-1853.

découvertes nombreuses et inattendues . . . il attachait tant d'intérêt à laisser après lui cet héritage, qu'il a travaillé à le finir, jusque dans les bras de la mort." The earnestness and fire of the author shine from the pages of the book; his logical mind did not shrink, in new systems of nomenclature, from the Oriental *atolan-telmin-ojafin-weso* for ammonia alum, or the Slavonic *lifavinaf* for mercaptan.

In 1832 Laurent carried out an investigation on naphthalene and its derivatives and discovered anthracene, and in 1836 he obtained anthraquinone by the oxidation of anthracene. In 1836 he obtained phthalic acid by the oxidation of naphthalene, and found that on heating it readily loses water to form an anhydride. An investigation of phenol and its derivatives, made in 1841, showed that this substance is identical with " carbolic acid ", discovered in coal tar by Runge in 1834, but is different from creosote, discovered in wood tar by Reichenbach in 1832.

When Berzelius accused Dumas of putting forward the " absurd theory " that chlorine plays the same rôle as hydrogen in substitution products, Dumas replied : " I am not responsible for the exaggerated extension of my theory by Laurent ", but when his discovery of trichloracetic acid had lent probability to the supposition, Dumas seemed to claim it as his own. " Si la théorie tombe, j'en serai l'auteur, si elle réussit, un autre l'aura faite ", says Laurent, rather bitterly. The reason why chlorine can play the part of hydrogen, he explains, is that there is a strong tendency for a radical acted upon by chlorine to conserve its type. Such a stable grouping of atoms he called first a radical, afterwards a nucleus (" noyau "), and he spoke of " fundamental " and " derived nuclei " *, the latter derived from the former by substitution. The idea was not taken seriously by Berzelius and Liebig, but Laurent showed in his Thesis of 1837 that it could form the basis of a method of classification of organic compounds, and it was so used by Gmelin in his *Handbook*. The fundamental nucleus is a prism, in the angles of which are carbon atoms, whilst the edges are occupied by hydrogen atoms. These edges can be taken away, but to prevent the collapse of the molecule they must be replaced by other atoms, e.g. of chlorine, forming derived nuclei. Hydrogen, halogen or oxygen could also be added, forming " hyperhydrides ", etc.

Dumas' Theory of Types

As a result of his experiments on the chlorination of acetic acid, and probably influenced by Laurent's ideas, Dumas in 1839 con-

* *Annales de chimie*, 1836, lxi, 125 ; *Chemical Method*, transl. by Odling, 1855.

cluded that: "in organic chemistry there exist certain Types which persist even when in place of the hydrogen they contain an equal volume of chlorine, bromine, or iodine is introduced." In 1840 * he distinguished between chemical types, "substances which contain the same number of equivalents united in the same way and showing the same fundamental chemical properties", and mechanical types, substances having similar chemical formulae, produced by substitution, but essentially different in their most salient chemical properties. The mechanical type was borrowed from Regnault; the germ of the chemical type is contained in Laurent's work. As examples may be quoted:

Chemical Types		*Mechanical Types*	
Acetic acid	$C_4H_8O_4$	Marsh gas	$C_2H_2H_6$
Chloracetic acid	$C_4H_2Cl_6O_4$	Methyl ether	C_2OH_6
		Formic acid	$C_2H_2O_3$
		Chloroform	$C_2H_2Cl_6$
		Carbon chloride	$C_2Cl_2Cl_6$

A satirical letter, really written by Wöhler, signed "S.C.H. Windler" (i.e. "Swindler"), but in French and dated from "Paris", appeared in Liebig's *Annalen* for 1840. In this, the undue extensions of the theory of substitution by Dumas are ridiculed. The ingenious author had replaced one by one all the atoms of manganous acetate by chlorine, so that from $MnO,C_4H_6O_3$ he had arrived at $Cl_2Cl_2,Cl_8Cl_6Cl_6$, yet the substitution product was faithful to type and showed all the salient properties of manganese acetate, although it consisted wholly of chlorine. A footnote adds, impudently enough, that bleached fabrics consisting entirely of spun chlorine were on sale in London, and "much sought after"!

Gerhardt's Theory of Residues

Closely associated with Laurent was Charles Gerhardt †, born in Strasburg in 1816, who studied under Liebig and other German chemists, and finally worked in Paris. He became, in 1844, pro-

* *Annalen*, 1840, xxxiii, 179, 259; *Mémoires de chimie*, Paris, 1843.

† E. Grimaux and Ch. Gerhardt, Junr., *Charles Gerhardt, sa vie, son œuvre, sa correspondance*, etc., Paris, 1900; Wurtz, *Moniteur Scientifique*, 1862, iv, 477; Thorpe, *Nature*, 1918, ci, 165.

fessor in Montpellier, where he found little in the way of a laboratory, and in 1855 in Strasburg, where he died in 1856. Gerhardt carried out fundamental researches in organic chemistry, the most noteworthy of which was on the anhydrides of organic acids in

FIG. 91.—C. GERHARDT, 1816-1856.

1852, and wrote two original textbooks: *Précis de chimie organique*, 2 vols. Paris, 1844-5, and *Traité de chimie organique*, 4 vols. Paris, 1853-56, and 1860-62. The first was badly received in France and Germany, but Gregory in England praised it. Gerhardt was rather outspoken in his criticism of work with which he did not agree, in spite of a warning from Liebig, and this aroused resentment and interfered with his popularity and success.

Although in his theoretical views Gerhardt had the misfortune to differ from most of the great authorities of his day, he was ultimately admitted to be in the right, and his influence on the more independent thinkers such as Williamson and Kekulé was very important.

In 1839 Gerhardt drew attention to a new type of reaction which he at first called copulation ("accouplement") and afterwards double decomposition.* This was complementary to the processes of addition (Dalton) and substitution (Dumas), and is found when two molecules react with the elimination of a part of each in combination as a simple compound (water, hydrochloric acid, etc.) whilst the "residues" or "radicals" combine together:

$$\underset{\text{benzene}}{C_6H_5H} + \underset{\text{nitric acid}}{HO.NO_2} = \underset{\text{water}}{HOH} + \underset{\text{nitrobenzene}}{C_6H_5NO_2}$$

These "residues" need not be capable of existence in the free state: they merely express possible modes of chemical reaction. The radical had for Gerhardt a purely formal significance; his equations represent merely: "les relations qui rattachent les corps entre eux sous le rapport des transformations." In particular: "chemical formulae are not destined to represent the arrangement of the atoms."† He repeats this over and over again in his writings. One and the same substance could have several purely hypothetical formulae according to the particular reactions in which it played a part. It would be going too far to read into Gerhardt's meaning a foreshadowing of the theory of tautomeric substances, since the symbols were not intended to represent structure.

Gerhardt's Two Volume Formulae and Atomic Weights

Berzelius had taken as the equivalents of organic acids those weights combining with one molecule of silver oxide. Since he assumed the formula AgO ($Ag = 2 \times 108$) for the latter, his molecular weights of the acids, and their formulae, were double the true ones. Again, Berzelius assumed the existence of anhydrides in the formulae of organic acids, and as these had to contain a molecule of water added to the anhydride, the formula of acetic acid was double the true formula, and so on. It should also be noted that "equivalent" long meant what we now call molecular weight

* *Annales de chimie*, 1839, lxxii, 198; *Traité*, iv, 566.

† *Traité de chimie*, iv, 566; cf. pp. 563, 568, 576, 580, etc.

or *some multiple* of the molecular weight. Liebig (and Gerhardt until 1842) had used what were called " four volume formulae " (i.e. double the modern " two volume " formulae) representing the quantities of substances occupying the same space as " four volumes " (H_4) of hydrogen. They made confusion worse by calling these amounts " equivalents ". Berzelius had used the correct two volume formulae for alcohol and ether (p. 233). Dumas attached great importance to vapour density determinations, but his atomic weights ($\underline{C}=6$, $H=1$, $O=16$) are not based on two volume formulae.

In 1843 Gerhardt pointed out * that if the formulae of organic compounds were written in the manner then customary, then water, hydrochloric acid, carbon dioxide and ammonia were always eliminated in reactions as double molecules : H_4O_2, H_2Cl_2, C_2O_4 and N_2H_6, or multiples of these. He drew the conclusion that, if the formulae of Berzelius for the simple inorganic compounds (H_2O, HCl, CO_2, NH_3) were correct, then his formulae of organic compounds should be halved. Thus, acetic acid is $C_2H_4O_2$ ($C=12$, $H=1$, $O=16$), and not $C_4H_8O_4$, or $C_4H_6O_3+H_2O$ (p. 244), or $\underline{C}_8H_4O_2+H_4O_2$ ($\underline{C}=6$). The latter formula had been adopted by Liebig to introduce the " acetyl " radical, and many formulae of organic acids really represented anhydrides, and, as we now know, the anhydride of a monobasic acid is derived from *two* molecules of the acid. The formula of chloracetic acid is $C_2HCl_3O_2$, and it therefore could not consist of water+anhydride, nor could $C_2AgH_3O_2$, silver acetate, contain silver oxide (Ag_2O according to Gerhardt) : " there is no water in our acids and no oxide in the salts."

The doubled formulae could represent weights made up of equivalents, $\underline{C}=6$, $\underline{O}=8$, etc., and if these equivalents were doubled, and the atomic weights, $C=12$, $O=16$, $S=32$, etc., used, then the formulae fell into line with inorganic formulae. This comes to the same thing as halving Berzelius's formulae for organic compounds but retaining his atomic weights. Gerhardt is, in fact, using Berzelius's atomic weights, but he does not say so. His correct atomic weights of the alkali metals and silver were only half those of

* *Annales de chimie*, 1843, vii, 129 ; viii, 238.

Berzelius, since he assumed that the oxides of metals were usually formed after the type of water, Me_2O. His atomic weights of other metals such as zinc and copper were, therefore, only half the true values, so that his system of atomic weights was less accurate than the discarded one of Berzelius. With Gerhardt's "new" atomic weights, the formulae of volatile compounds agreed with the weights of "two volumes", i.e. the volume of 2 grams of hydrogen, H_2. Gerhardt did not make this result his sole guiding principle, and he nowhere emphasized the value of Avogadro's hypothesis. The laws of atomic heat and isomorphism were also used again, as they had been long before by Berzelius, but they were not systematically applied to the metals.

In 1846 Laurent adopted Gerhardt's atomic weights as the smallest quantities of elements present in a molecular weight of any of their compounds. He defines "equivalents" as weights which have the same value in reactions, or fulfil the same functions in forming compounds or in decompositions.* He pointed out that there were some exceptions to Gerhardt's definition of the molecular weight as that occupying two volumes, and in such cases the weight of four volumes must be taken.† These supposed exceptions (H_2SO_4, NH_4Cl, PCl_5, etc.) are substances which dissociate on heating, as was afterwards recognized by Cannizzaro.

In his *Traité*, however, Gerhardt used the old formulae, except in the last volume, published after his death. This was necessary, as he explained, in order that the book should find purchasers, since chemists in general would not use his new atomic weights.

At that time an almost intolerable confusion reigned in chemical circles: various atomic weights, and multitudes of formulae for the same things, were in use. It was apparently considered a sign of independence of thought for every chemist to have his own set of formulae. In 1861 Kekulé gives nineteen different formulae for acetic acid, and "in looking back on these discussions", says Schorlemmer only twenty years later, "we seem to enter a bygone age".‡

* *Chemical Method*, tr. Odling, 1855, 7, 80.

† *ibid.*, 81.

‡ Roscoe and Schorlemmer, *Treatise on Chemistry*, vol. iii, part 1, p. 22.

Cannizzaro

The use of barred symbols by Berzelius to represent double atoms produced formulae very similar to those based on Gmelin's equivalents, e.g. $\underline{H}O$ and $H\underline{O}$, and when the bars were omitted, as was often the case, it was not known which system of atomic weights was in use. The later use of barred symbols by Kekulé and Wurtz to represent Gerhardt's atomic weights added to the confusion. This did not matter when every chemist had his own set of symbols, but after the work of Williamson and Gerhardt, it seemed desirable to arrive at some kind of agreement.

The main workers in different countries accordingly met in Karlsruhe in December 1860. As might have been expected, no agreement was reached. Dumas expressed the opinion that there were two chemistries, inorganic and organic, and Cannizzaro pointed out that different atomic weights were used in the two branches. In the end it was decided that each chemist should continue to use what system he pleased. At the close of the conference a small pamphlet written by Cannizzaro was distributed, and Lothar Meyer says that after reading this, " the scales fell from my eyes, doubts vanished, and the feeling of calm certainty came in their place ". The pamphlet was in Italian, with the title : *Sunto di un corso di filosofia chimica* ; it appeared in 1858 in the *Nuovo Cimento* and was republished in 1859 at Pisa and 1880 at Rome. The German translation by Lothar Meyer was published in 1891.* The views which are expressed in it were made the basis of Lothar Meyer's classical *Modernen Theorien der Chemie*, first published in 1864, from which most chemists derived their information on the theory of Avogadro and the views of Cannizzaro.

Stanislao Cannizzaro was born in Palermo in 1826.† He studied there, then in Naples, and then became Piria's assistant in Pisa in 1850. His first investigation, on cyanogen chloride, made jointly with Cloëz in Chevreul's laboratory in Paris, was published in 1851. In 1851 he obtained an appointment in Alessandria, and in 1853 published an important research on the action of potash on benzaldehyde, obtaining benzyl alcohol (" Cannizzaro's reaction "). In

* Ostwald's *Klassiker* No. 30 ; English in *Alembic Club Reprint* No. 18.

† Tilden, Memorial Lecture, *J. Chem. Soc.*, 1912, p. 1677.

1855 he became professor in Genoa, where there was at first no laboratory. In 1861 he went to Palermo and in 1871 to Rome, where he became Senator; he died in 1910.

Cannizzaro was the first to see the full significance of the work of his countryman Avogadro.* The molecular weight of a volatile

FIG. 92.—S. CANNIZZARO, 1826-1910.

substance can be found by a determination of vapour density, the ratio of the latter to the density of a standard substance giving the ratio of the molecular weights. As standard, hydrogen, the lightest gas, was chosen, but since the hydrogen molecule contains *two* atoms (as was clear from the reasoning used by Avogadro in 1811, see p. 209), the density relative to hydrogen must be multiplied by two to give the molecular weight of the vapour as the sum of the atomic weights, the latter being referred to H = 1.

* Graebe, *J. prakt. Chem.*, 1913, lxxxvii, 145, on the development of the molecular theory.

The atomic weight of an element is then the least weight of it contained in a molecular weight of any volatile compound. In the case of a series of volatile compounds, the weights of an element contained in a molecular weight are always whole multiples (including unity) of a number which must be taken as the atomic weight. The atomic weights so found are in agreement with the law of atomic heat, and the latter is therefore used in cases where vapour densities cannot be found.

The real service of Cannizzaro was to show conclusively, as Tilden puts it, that: "There is, in fact, but one science of chemistry and one set of atomic weights." Kekulé said in 1859 * that when "the chemical molecular weights are compared with the specific gravities in the form of vapour it is found that the two for nearly all compounds, and for all carbon compounds, are identical." Gerhardt, although puzzled by the supposed exceptions to his "two volume" formulae, was not deterred from writing the formulae as H_2SO_4, PCl_5, NH_4Cl, etc., since these followed from the chemical evidence. The explanation of the difficulty, viz. the dissociation of these compounds by heat, was first given by Cannizzaro in 1857 (see p. 329).

Gerhardt's Unitary Theory

In 1841-2 Gerhardt put forward a scheme of classification of organic compounds based on their empirical formulae, arranging them on what he calls a "combustion ladder" ("échelle de combustion"), with animal matter at the top and carbon dioxide, water and ammonia at the foot, but he later made this idea more serviceable.

In his *Précis de chimie organique* (1844-5) Gerhardt used empirical formulae, pointing out that there were in use seven different constitutional formulae for alcohol, "each of which is only the expression of one or two reactions." As he says: "autant de réactions, autant de formules rationelles." He classified organic compounds in families, each containing substances with the same number of atoms of carbon. The result was that acids and their esters appeared in different groups, and even Laurent objected to this arbitrary system: "a system of empirical formulae (formules

* *Lehrbuch*, i, p. 233.

brutes) is too absolute, and if it is adopted it obscures a crowd of interesting relations."

Gerhardt drew attention to what he named *homologous series* of compounds, successive members differing by multiples of CH_2.* "These substances undergo reaction (se metamorphosent) according to the same equations and it is only necessary to know the reactions of one in order to predict those of the others." This constitutes one of the charms of organic chemistry both in theory and practice.

As examples of such series he gives † :

CH_4O	CH_2	CH_2O_2	CH_4SO_4	CH_3Cl
C_2H_6O	C_2H_4	$C_2H_4O_2$	$C_2H_6SO_4$	C_2H_5Cl
. .	. .	. .	. .	. .
$C_{16}H_{34}O$	$C_{16}H_{32}$	$C_{16}H_{32}O_2$	$C_{16}H_{34}SO_4$	$C_{16}H_{33}Cl$

The existence of the homologous series of alcohols, containing the radicals nR + H, where $R = \underline{C}_2H_2$ ($\underline{C} = 6$), had been pointed out by Schiel in 1842, and Dumas had shown that a similar relation exists among fatty acids, but it was Gerhardt who generalized the regularity and gave to it the name *homology*.

In Gerhardt's small *Introduction à l'étude de la chimie par le système unitaire* (Paris, 1848), he says: "all bodies are considered as unique molecules, the atoms of which are disposed in an order which the chemical reactions indicate only in a relative manner. Each simple or compound body is considered as an edifice, as a system formed by an assemblage of atoms. This system is called the molecule of a body." A similar point of view seems to have been held by Berzelius many years before. In a letter to Wöhler ‡ he remarks that "in benzoic acid as an acid neither benzene nor benzoyl nor a compound inflammable radical, differentiated from the rest of the compound, is present; rather it depends entirely on the conditions under which the body is decomposed as to the parts into which it breaks down. One must know into what parts it can be decomposed, but it is not correct to say that it is composed of these parts". Berzelius, however, did not consistently adopt this view.

* *Précis*, 1845, ii, 492.

† *Introduction à l'étude de la chimie par le système unitaire*, Paris, 1848, 291.

‡ Quoted by Hjelt, *Gesch. der organ. Chem.*, 1916, p. 78.

Hofmann

August Wilhelm von Hofmann was born in Giessen in 1818. After some years of study of philosophy and law he turned to chemistry and became assistant to Liebig. After a short

FIG. 93.—A. W. VON HOFMANN, 1818-1892.

stay in Bonn he was called in 1845 by Prince Albert to the newly founded College of Chemistry in London, which became a Government institution in 1853. He succeeded Mitscherlich in 1865 in Berlin, where he died in 1892. Hofmann's work was so extensive that several chemists were commissioned to write his memorial lecture for the Chemical Society.* He was a brilliant

* *J. Chem. Soc.*, 1896, p. 575.

lecturer and his stay in London was very beneficial for English chemistry.

The Ammonia Type

Dumas' idea that the properties of a compound were determined principally by its type, and that the electrochemical functions of the atoms were of little importance, received some modification as a result of the work of Hofmann on the chlorine and bromine derivatives of aniline (1845). These, although basic, were less so than the parent substance aniline, owing to the negative or acid-forming characters of the halogen substituents. Even Gerhardt was forced to recognize that, although both nitric acid and potash were derived from neutral water as a type, the first contains an electronegative radical, NO_2, conferring acidic properties, whilst the second contains an electropositive atom, K, conferring basic properties. The ethyl radical, C_2H_5, is so nearly allied to hydrogen that the molecule of alcohol is almost neutral. Thus, not only the atomic weights of Berzelius, but even parts of his electrochemical system, were incorporated into the views of one of his chief opponents.

In 1849 Wurtz had accidentally obtained methylamine and ethylamine by the action of alkali on the cyanic and cyanuric esters, and said they *could* be represented as "ammonia in which one equivalent of hydrogen is replaced by methyl or ethyl", NH_2Me and NH_2Et. The possible existence of such compounds containing the amide radical, $NH_2 = Ad$, had been predicted about ten years before by Liebig. In 1850 Hofmann * showed that by the action of ethyl iodide on ammonia, all the hydrogen atoms of the latter could be successively replaced by ethyl, forming the three amines, which he represented as follows:

$$N\begin{cases}H\\H\\H\end{cases} \qquad N\begin{cases}C_2H_5\\H\\H\end{cases} \qquad N\begin{cases}C_2H_5\\C_2H_5\\H\end{cases} \qquad N\begin{cases}C_2H_5\\C_2H_5\\C_2H_5\end{cases}$$

These were called primary, secondary and tertiary amines by Gerhardt in the fourth volume of his *Traité*. Berzelius formulated organic bases as ammonia copulated with hydrocarbons, e.g

* *Phil. Trans.*, 1850 and 1851.

aniline as $\bar{N}\bar{H}^{3}+C^{12}H^{8}=$ *an* Ak, but in formulating them as above Hofmann created the *Ammonia Type*.

The correctness of the formulae was established by the preparation of ethyl aniline and diethyl aniline from aniline and ethyl bromide, and ethyl-amyl aniline from ethyl aniline and amyl bromide:

$$\left.\begin{array}{r}C_6H_5\\H\\H\end{array}\right\}N \qquad \left.\begin{array}{r}C_6H_5\\C_2H_5\\H\end{array}\right\}N \qquad \left.\begin{array}{r}C_6H_5\\C_2H_5\\C_2H_5\end{array}\right\}N \qquad \left.\begin{array}{r}C_6H_5\\C_2H_5\\C_5H_{11}\end{array}\right\}N$$

aniline — ethyl aniline — diethyl aniline — ethyl-amyl aniline

The substances were all basic, forming hydrochlorides which combined with platinic chloride to form double compounds in the same way as ammonium chloride.

In 1855 Hofmann and Cahours prepared similar compounds derived from phosphine, PH_3. The tertiary ethyl amine was found to combine with a molecule of ethyl iodide to form a salt, $N(C_2H_5)_4I$, analogous to ammonium iodide, from which the powerful base $N(C_2H_5)_4OH$, a substituted ammonium hydroxide, was obtained by the action of moist silver oxide. Corresponding phosphonium salts, and phosphonium iodide (PH_4I), were prepared.

The Water Type

In 1846 Laurent had represented alcohol and ether as analogous to water, potassium hydroxide and potassium oxide:

OHH OEtH OEtEt OHK OKK,

and this idea of a "water type" was extended by Williamson.

Alexander William Williamson was born in Wandsworth in 1824, studied in Heidelberg and from 1844 to 1846 with Liebig in Giessen. After a period of study in Paris he succeeded Fownes in 1849 as professor of analytical chemistry in University College, London, and in 1855 he succeeded Graham as professor of general chemistry there. In the period between 1850 and 1860 he carried out important researches *, but from 1860 to his resignation in 1887 he published nothing of importance. He died in 1904.

* *Alembic Club Reprint* No. 16; papers of 1850-3.

Williamson's most important work was on the formulae ot alcohol and ether and of related compounds. The formulae of alcohol and ether had been written ($\underline{C}=6$, $C=12$):

	Dumas	Berzelius	Liebig
Alcohol	$\underline{C}^8H^8, H^4O^2$	C^2H^6O	$C_4H_{10}O, H_2O$
Ether	$\underline{C}^8H^8, H^2O$	$C^4H^{10}O$	$C_4H_{10}O$

FIG. 94.—A. W. WILLIAMSON, 1824-1904.

About 1850 Berzelius's formulae (which were correct) had been replaced by those of Liebig, who regarded alcohol as a hydrate of ether, but Williamson showed that Liebig's formula for alcohol was incorrect. He thus supplied an experimental proof for Gerhardt's view that the molecules of alcohol and ether occupied equal volumes in the state of vapour, each having a "two volume" formula, whilst according to Dumas and Liebig ether had a two volume, and alcohol a four volume, formula. Williamson started with the idea of substituting hydrogen in known alcohols by hydrocarbon radicals, with the object of preparing higher alcohols. By the action of ethyl iodide on potassium ethylate, obtained by the action of potassium on alcohol, he expected to get ethylated alcohol, but: "to my astonishment the compound thus formed had none of the properties of an alcohol, it was nothing else than common ether."

Ether must therefore contain two ethyl radicals. "Thus alcohol is $\begin{matrix}C_2H_5\\H\end{matrix}O$, and the potassium compound $\begin{matrix}C_2H_5\\K\end{matrix}O$; and by acting upon this with iodide of aethyl we have:

$$\begin{matrix}C_2H_5\\K\end{matrix}O + C_2H_5I = IK + \begin{matrix}C_2H_5\\C_2H_5\end{matrix}O.\text{"} *$$

"Alcohol is therefore water in which half the hydrogen is replaced by carburetted hydrogen, and aether is water in which both atoms of hydrogen are replaced by carburetted hydrogen: thus

$$\begin{matrix}H\\H\end{matrix}O \quad \begin{matrix}C_2H_5\\H\end{matrix}O \quad \begin{matrix}C_2H_5\\C_2H_5\end{matrix}O.\text{"}$$

That alcohol and ether belong to the water type was proved by the production of a mixed ether, containing two different hydrocarbon radicals, from potassium methylate and ethyl iodide:

$$\begin{matrix}CH_3\\K\end{matrix}O + C_2H_5I = IK + \begin{matrix}CH_3\\C_2H_5\end{matrix}O.$$

The ordinary process of the formation of ether (formerly regarded as catalytic) was now explained by the alternate formation of sulphovinic acid and its decomposition by alcohol:

$$\begin{matrix}H\\H\end{matrix}SO_4 + \begin{matrix}C_2H_5\\H\end{matrix}O = \begin{matrix}C_2H_5\\H\end{matrix}SO_4 + \begin{matrix}H\\H\end{matrix}O.$$

$$\begin{matrix}C_2H_5\\H\end{matrix}SO_4 + \begin{matrix}C_2H_5\\H\end{matrix}O = \begin{matrix}H\\H\end{matrix}SO_4 + \begin{matrix}C_2H_5\\C_2H_5\end{matrix}O.$$

* Williamson does not use brackets in his formulae. These were used by Gerhardt.

This view was again shown to be correct by the preparation of mixed ethers, e.g.

$$\begin{matrix} CH_3 \\ H \end{matrix} SO_4 \quad \text{and} \quad \begin{matrix} C_2H_5 \\ H \end{matrix} O \quad \text{gave} \quad \begin{matrix} CH_3 \\ C_2H_5 \end{matrix} O.$$

" The alternate formation and decomposition of sulphovinic acid is to me, as to the partisans of the chemical theory, the key to explaining the process of aetherification." Liebig's theory that alcohol was a hydrate of ether, and Mitscherlich's theory that the action of sulphuric acid in forming ether was only catalytic, were thus shown by Williamson to be incorrect.

Williamson recognized that acetic acid also belongs to the water type, the radical C_2H_3O, *othyl* (now called acetyl), replacing hydrogen :

$$\begin{matrix} C_2H_3O \\ H \end{matrix} O.$$

" I believe throughout inorganic chemistry and for the best known organic compounds, one simple type will be found sufficient ; it is that of water." " If the two atoms of hydrogen in water were replaced by . . . othyl we should have anhydrous acetic acid

$$\begin{matrix} C_2H_3O \\ C_2H_3O \end{matrix} O."$$

Gerhardt, who at first thought that anhydrides of monobasic acids could not exist (since the molecule of the acid could not contain a molecule of water, p. 254), in spite of the preparation of the anhydride of nitric acid by Deville in 1849, prepared in 1852 the compound already predicted by Williamson, viz. acetic anhydride, by the action of acetyl chloride (discovered by Gerhardt in 1852) on anhydrous sodium acetate :

$$\left.\begin{matrix} C_2H_3O \\ Cl \end{matrix}\right\} + \left.\begin{matrix} C_2H_3O \\ Na \end{matrix}\right\} O = \left.\begin{matrix} C_2H_3O \\ C_2H_3O \end{matrix}\right\} O + \left.\begin{matrix} Na \\ Cl \end{matrix}\right\}$$

Other, mixed, anhydrides could similarly be prepared. This research * of Gerhardt's was recognized by all chemists as of the utmost importance.

Williamson derived the dibasic acids from two molecules of water; e.g. sulphuric acid from $2H_2O$ in which 2H give place to the

* " Recherches sur les acides organiques anhydres " ; *Compt. rend.*, 1852, xxxiv, 755, 902 ; *Ann. Chim.*, 1853, xxxvii, 285 (type theory) ; *Annalen*, 1853, lxxxvii, 57, 149.

radical sulphuryl, SO_2; and carbonic acid from $2H_2O$ in which the radical CO replaces 2H:

$$\begin{matrix}SO_2\\H_2\end{matrix}O_2 \quad \begin{matrix}SO_2\\HK\end{matrix}O_2 \quad \begin{matrix}SO_2\\K_2\end{matrix}O_2 \quad \begin{matrix}CO\\H_2\end{matrix}O_2 \quad \begin{matrix}CO\\KH\end{matrix}O_2 \quad \begin{matrix}CO\\K_2\end{matrix}O_2$$

"1 atom of carbonic oxide is here equivalent to 2 atoms of hydrogen, and by replacing them, holds together the 2 atoms of hydrate in which they were contained." He showed that besides sulphuryl chloride, SO_2Cl_2, an intermediate compound, chlorosulphonic acid ("chloro-hydrated sulphuric acid"), is formed by the action of PCl_5 on sulphuric acid:

$$\begin{matrix}H\\SO_2\\H\end{matrix}\begin{matrix}O\\ \\O\end{matrix} \qquad \begin{matrix}H\\SO_2\\Cl\end{matrix}O \qquad \begin{matrix}Cl\\SO_2\\Cl\end{matrix}$$

in which he recognizes the replacement of OH by Cl.

In 1852 Williamson obtained the correct formulae of acetone and aldehyde: aldehyde is "the hydruret of othyle, as acetone is its methyle-compound":

$$\begin{matrix}C_2H_3O\\H\end{matrix} \text{ and } \begin{matrix}C_2H_3\\CH_3\end{matrix}O \text{ or } \begin{matrix}CH_3\\H\end{matrix}CO \text{ and } \begin{matrix}CH_3\\CH_3\end{matrix}CO.$$

Odling * in 1854 extended the idea of multiple types and introduced dashes over a symbol, e.g. SO_2'', Bi''', to indicate how many atoms of hydrogen the substance could replace, and in 1857 Kekulé and Odling introduced mixed types.

Examples of Odling's formulae are:

$$\left.\begin{matrix}K'\\K'\end{matrix}\right\}O'' \quad \left.\begin{matrix}NO_2'\\K'\end{matrix}\right\}O'' \quad \left.\begin{matrix}3NO_2'\\Bi'''\end{matrix}\right\}3O'' \quad \left.\begin{matrix}PO'''\\3H'\end{matrix}\right\}3O''$$

and sodium thiosulphate is a mixed type derived from water and sulphuretted hydrogen:

$$\left.\begin{matrix}H_2\\H_2\end{matrix}\right\}O''+S'' \text{ give } \left.\begin{matrix}SO_2''\\Na_2'\end{matrix}\right\}O''+S''.$$

Gerhardt's Theory of Types

In his memoir of 1853, on the anhydrides of organic acids, Gerhardt formulated *four inorganic types* from which he thought all organic compounds were derivable. These types were "used

* *J. Chem. Soc.*, 1855, vii, 1.

to represent double decomposition", and occupied equal volumes (2 vols., H_2) in the state of vapour. They did not represent the positions of the atoms in the molecules, which Gerhardt appears to have considered as inaccessible to experiment. "In deriving a body from the water type I intend to express that to this body, considered as an oxide, there correspond a chloride, a bromide, a sulphide, a nitride, etc., susceptible of double decompositions, or resulting from double decompositions, analogous to those presented by hydrochloric acid, hydrobromic acid, sulphuretted hydrogen, ammonia, etc., or which give rise to the same compounds. The type is thus the unit of comparison for all the bodies which, like it, are susceptible of similar changes or result from similar changes." *

The four types were water, hydrochloric acid, ammonia and hydrogen. Their molecules, and those of their derivatives, were looked upon as complete units, and the whole arrangement was spoken of as a "système unitaire":

Type	*Ethyl derivative*	*Benzoyl derivative*
$O\left\{\begin{matrix}H\\H\end{matrix}\right.$	$O\left\{\begin{matrix}C_2H_5\\H\end{matrix}\right.$	$O\left\{\begin{matrix}C_7H_5O\\H\end{matrix}\right.$
water	alcohol	benzoic acid
$\left\{\begin{matrix}H\\Cl\end{matrix}\right.$	$\left\{\begin{matrix}C_2H_5\\Cl\end{matrix}\right.$	$\left\{\begin{matrix}C_7H_5O\\Cl\end{matrix}\right.$
hydrochloric acid	ethyl chloride	benzoyl chloride
$N\left\{\begin{matrix}H\\H\\H\end{matrix}\right.$	$N\left\{\begin{matrix}C_2H_5\\H\\H\end{matrix}\right.$	$N\left\{\begin{matrix}C_7H_5O\\H\\H\end{matrix}\right.$
ammonia	ethylamine	benzamide
$\left\{\begin{matrix}H\\H\end{matrix}\right.$	$\left\{\begin{matrix}C_2H_5\\H\end{matrix}\right.$	$\left\{\begin{matrix}C_7H_5O\\H\end{matrix}\right.$
hydrogen	ethyl hydride	benzaldehyde

The sulphides, tellurides, oxides, acids, bases, salts, alcohols, ethers, etc., belong to the water type; chlorides, bromides, iodides, cyanides and fluorides to the hydrochloric acid type (Gerhardt does not see that these really belong to the hydrogen type); nitrides, phosphides, arsenides, etc., belong to the ammonia type; metallic hydrides, hydrocarbons and metals to the hydrogen

* *Traité*, iv, 586 f.; this volume was completed in 1856.

type. More complicated compounds were formed by the substitution of radicals for hydrogen in the types. Unknown compounds could be predicted in large numbers by this scheme of classification.*

Gerhardt † clearly distinguished between the hydrogen and chlorine radicals, H and Cl, and free hydrogen and chlorine, H_2 and Cl_2. Many so-called addition reactions, he says, are really double decompositions :

$$\mathrm{HH + ClCl = HCl + HCl} ;$$

but the generalization of this mode of reaction was carried too far, e.g. in the equations :

$$\underline{\mathrm{Zn}}\underline{\mathrm{Zn}} + \mathrm{ClH} = \underline{\mathrm{Zn}}\mathrm{H} + \underline{\mathrm{Zn}}\mathrm{Cl} ;$$
$$\underline{\mathrm{Zn}}\mathrm{H} + \mathrm{ClH} = \underline{\mathrm{Zn}}\mathrm{Cl} + \mathrm{HH}.$$

($\underline{\mathrm{Zn}}$ was half the modern atomic weight.)

An important step was the combination of the type and radical theories.‡ The fatty acids may be regarded as derived from the water type by substitution of hydrogen by radicals :

$$\begin{array}{ccc} \mathrm{O}\left\{\begin{matrix}\mathrm{H}\\ \mathrm{H}\end{matrix}\right. & \qquad \mathrm{O}\left\{\begin{matrix}\mathrm{CHO}\\ \mathrm{H}\end{matrix}\right. & \qquad \mathrm{O}\left\{\begin{matrix}\mathrm{C_2H_3O}\\ \mathrm{H}\end{matrix}\right. \\ \text{water} & \qquad \text{formic acid} & \qquad \text{acetic acid} \end{array}$$

Gerhardt calls these radicals monatomic, diatomic, etc., according as they replace 1, 2, or more atoms of hydrogen. But the monatomic acid radicals, $C_nH_{2n-1}O$, may themselves be regarded as composed of the carbonyl radical, CO, combined with hydrogen in formic acid and with a hydrocarbon radical, C_nH_{2n+1}, in the higher fatty acids. Thus acetyl, C_2H_3O, may be regarded as carbonyl combined with methyl : $CO + CH_3$. This is confirmed by the formation of methane on heating an acetate with alkali. Potassium acetate, as Kolbe had shown, on electrolysis gives carbon dioxide, hydrogen, and " methyl " (really ethane, C_2H_6). Acetic acid is not only

$$\mathrm{O}\left\{\begin{matrix}\mathrm{C_2H_3O}\\ \mathrm{H}\end{matrix}\right., \text{ but also } \mathrm{O}\left\{\begin{matrix}\mathrm{CO(CH_3)}\\ \mathrm{H}\end{matrix}\right.,$$

in which the acetyl group, C_2H_3O, is what Gerhardt calls a conjugated radical, i.e. one composed of two or more radicals, " each of which reminds one of a particular system of double decompositions ". The same idea was applied to nitro-compounds.§

* See the tables in *Traité*, iv, 612-13.

† *Traité*, iv, 568, 593. ‡ *Traité*, iv, 604-610. § *Traité*, iv, 644.

Summary and Supplement

Theory of Substitution and the Type Theory

Jean Baptiste Dumas (1800-1884), professor in Paris, carried out investigations on vapour densities (1826), and with Boullay put forward the " etherin " theory (1828)—alcohol, ether and esters are derived from ethylene. Dumas discovered oxamide (1830) and devised the combustion method for the determination of organic nitrogen (1830). The recognition that wood spirit contains an alcohol (methyl alcohol), and of the methyl radical (Dumas and Péligot, 1834), and that ethal, discovered in spermaceti by Chevreul in 1818, is an alcohol (now called cetyl alcohol), established the general series of the alcohols (Dumas and Péligot, 1836). The cinnamyl derivatives were investigated (Dumas and Péligot, 1834); the substitution of hydrogen by chlorine was recognized (from 1834); the hydrolysis of nitriles (cyanides of organic radicals) to acids and amines; acetamide (Dumas, Malaguti and Leblanc, 1847), and Dumas made accurate determinations of atomic weights (1840-59). After about 1840, Dumas and Liebig divided the authority which formerly belonged to Berzelius.

Auguste Laurent (1808-1853), for a time professor at Bordeaux, but later assayer in the Paris Mint, worked at first with Dumas. He discovered anthracene (Dumas and Laurent, 1832), chrysene and pyrene (1837), benzil (1836), anthraquinone (1836, published in 1840), phthalic acid (1836), adipic and pimelic acids (1837), and isatin (1841). Laurent identified picric acid (Woulfe, 1771; Haussmann, 1788) with trinitrophenol (1841), and devised methods of classification of organic compounds; he put forward the " Nucleus theory " (1837), suggested that alcohol and ether correspond with KHO and K_2O (1846), and in his posthumous *Chemical Method* (1854) put forward a powerful criticism of the dualistic theory, the foundations of which were then already seriously undermined.

Charles Gerhardt (1816-1856), professor in Montpellier (1844) and Strasburg (1855), was a disciple of Laurent's. In 1843 he proposed a reform in the atomic weight system. The molecular weight was defined as the weight occupying in the vapour state the same volume as 2 gm. of hydrogen (H_2); formulae so derived were called " two-volume formulae ". Since he assumed the water type, Me_2O, for oxides of most metals, his atomic weights of metals were generally less accurate than those of Berzelius, but he corrected Berzelius's atomic weights of the alkali metals and silver by dividing them by 2. Gerhardt's other contributions to chemistry were his theory of residues (1839), his recognition of homologous series of compounds

(1844), and his theory of four types (H_2O, NH_3, HCl, H_2) (1853); these were in conformity with Laurent's "unitary theory", but also included some features of the radical theory. Gerhardt discovered quinoline (1842), the anilides (1845), acetyl chloride and acid chlorides (prepared with $POCl_3$) (1852), and especially the anhydrides of monobasic organic acids, e.g. acetic anhydride (1852). His text-book of organic chemistry (1853-56) was important; the last volume contained the formulation of organic compounds with the new atomic weights.

Auguste Cahours (1813-1891), professor in Paris, discovered amyl alcohol (1839), anisol and its derivatives (1841), methyl salicylate (1843), the preparation of organic acid chlorides by means of PCl_5 (1846), tin tetraethyl (1853), allyl alcohol (1857), alkyl sulphonium bases (1865); and investigated abnormal vapour densities.

Thomas Anderson (1819-1874), professor in Glasgow, discovered pyridine, picoline, lutidine and collidine in bone oil (1846-55), the constitution of piperidine (1855), the preparation of pure pyrrol (1858), and the constitution of anthracene (1861).

August Wilhelm von Hofmann (1818-1892) was first an assistant to Liebig, then professor in the College of Chemistry, London (1845), and in Berlin (1865); he founded the German Chemical Society (1868). His researches on the constitution of aniline (1843) and on the preparation of amines from ammonia and alkyl halides (1850), led to the Ammonia Type. He discovered phenyl isocyanate (1849-58), hydrazobenzene (1863), diphenylamine (1864), isonitriles (1866-7) and formaldehyde (1867); investigated myrosin and mustard oil and discovered phenyl mustard oil (1858), worked out a vapour density method (1868), and improved the methods of organic analysis and manipulation. In 1873 Hofmann and Martius discovered methyl aniline and dimethyl aniline. The action of bromine and alkali on amides was found in 1881 to be a general method ("Hofmann reaction") for the preparation of amines. Hofmann did valuable work on the aniline dyes (alkyl rosanilines, 1863 f., etc.) and alkaloids. He was an eminent teacher and introduced the new "two-volume" atomic weights into his lectures, as well as many new lecture experiments.

Thomas Graham (1805-1869), professor in Glasgow and London, investigated the absorption of gases (1826) and vapours (1828) by liquids, solubilities of salts (1827), supersaturation (1828), the diffusion of gases (1828, with the suggestion that the rate of diffusion is inversely as some function of the density, apparently the square root; 1833, "Graham's law" that "the diffusion or spontaneous intermixture of two gases in contact is effected by an interchange of position of indefinitely minute volumes of the gases, which volumes are not necessarily of equal magnitude, being in the case of each gas, inversely proportional to the square root of the density of that gas"). He investigated the oxidation of phosphorus (1829, effect of various substances in stopping the glow), adsorption of dissolved salts on charcoal (1830), alcoholates of salts (1831), the arsenates and phosphates (1833, leading to the theory of polybasic acids), hydrated salts and oxides (1834-7), theory of the voltaic circle (1839), heats of reaction (1842-3), effusion and transpiration of gases (1845-9), liquid diffusion, dialysis and osmose (1849-61, crystalloids and colloids), viscosity of liquids (1861), diffusion of gases through septa (1863-6), colloidal silica (1864), occlusion of hydrogen by metals (1866-9). Graham was one of the founders of physical chemistry.

Alexander William Williamson (1824-1904), professor in University College, London, investigated the formation of ether and the constitution of alcohol and ether (1850-52), putting forward the Water Type, and recognizing the acetyl radical and the structure of acetone. He prepared aldehydes and ketones (including "mixed ketones") by distilling calcium salts of organic acids (1852), and discovered chlorosulphonic acid (1856). Williamson put forward a dynamical view of chemical equilibrium (1850).

Stanislao Cannizzaro (1826-1910), professor in Genoa (1855), Palermo (1861) and Rome (1871), explained (1858) how the atomic weight of an element (as distinguished from the equivalent) may be found without ambiguity by an application of Avogadro's hypothesis supplemented by Dulong and Petit's law of atomic heats. He also carried out important researches in organic chemistry, e.g. the production of benzyl alcohol and benzoic acid from benzaldehyde by treatment with potash (1853; "Cannizzaro's reaction").

CHAPTER XII

THE THEORY OF VALENCY

Frankland and Kolbe

At the time when Frankland and Kolbe began their work the ideas of Berzelius had fallen into discredit, and their later investi-

FIG. 95.—E. FRANKLAND, 1825-1899.

gations coincided in point of time with the development of the theory of types. In respect of the latter, chemists felt, according to Lothar Meyer, that " in the stereotyped patterns (*Schablonen*) of the theory something deeper lay concealed". It was precisely the work of Frankland and Kolbe which revealed this deeper meaning of the types. Their researches were guided mainly by the older teachings of Berzelius, and when they were completed the type theory as such had ceased to have any interest for chemists.

The radicals for Berzelius and his school were realities: they existed in compounds and in some cases, as with cyanogen, and Bunsen's cacodyl, they could be isolated and prepared in the free state. For Gerhardt and his school, on the other hand, the radicals were mere phantoms, "the ghosts of departed reactions", incapable of isolation and existing only in the imagination.

Edward Frankland was born in Churchtown, Lancashire, in 1825. He was a pupil of Playfair, Bunsen and Liebig, and occupied the chairs of chemistry in Owens College, Manchester (1851-57), St. Bartholomew's Hospital (1863), the Royal Institution, and the Royal School of Mines (1865). He died in 1899.*

Hermann Kolbe, born near Göttingen in 1818, studied under Wöhler in 1838. From 1842, when his first research was published, he was assistant to Bunsen in Marburg, then (1845) to Playfair in London, and he succeeded Bunsen in Marburg in 1851. In 1865 he became professor in Leipzig, where he died in 1884.

Kolbe was very successful as a teacher, his original work was of great importance, and his text-book of Organic Chemistry (3 vols., 1854-78) is clear and well arranged. As editor of the *Journal für praktische Chemie* he was a keen critic of the work and ideas of his contemporaries, sometimes overstepping the bounds of what would now be considered good taste. In particular, his criticisms of the type theory, of van't Hoff's stereochemistry, and of Kekulé's structural formulae, were very virulent.

In 1843 Kolbe † showed that carbon disulphide on chlorination gave carbon tetrachloride, which had been discovered by Regnault in 1839. By the action of chlorine and water in presence of sunlight on liquid carbon chloride (tetrachlorethylene, C_2Cl_4, discovered by Fara-

* Memorial Lecture, *Proc. Chem. Soc.*, 1901, p. 193; *Sketches from the Life of Sir Edward Frankland*, edited by his two daughters, 1902. Frankland's papers were re-published (not very accurately) in 1877 as *Researches in Pure, Applied, and Physical Chemistry.*

† *Annalen*, 1843, xlv, 41; 1844, xlix, 339; 1845, liv, 145.

day in 1821) Kolbe in 1844 obtained trichloracetic acid, hexachlorethane, C_2Cl_6, being an intermediate product. This was an important experiment, since according to Berzelius chloracetic acid contained carbon chloride conjugated with oxalic acid (p. 244). In 1845 Kolbe showed that C_2Cl_4 is formed (with chlorine) when CCl_4 vapour is passed through a red-hot tube, and that Melsens' reduction of chloracetic acid to acetic acid (p. 244) could be effected by electrolytic hydrogen.

In 1848 Frankland and Kolbe* obtained acetic acid by heating methyl cyanide with potash; Dumas, Malaguti and Leblanc in the same year showed that propionic acid was obtained from ethyl cyanide, and thus an important general method (hydrolysis of the cyanides of radicals) for the synthesis of organic acids was evolved. This again was in accordance with Berzelius's ideas, since cyanogen itself on hydrolysis gave oxalic acid, and acetic acid was regarded as methyl conjugated with the latter. In announcing this discovery, Frankland and Kolbe state that they had been able actually to isolate the alcohol radicals from the corresponding organic acids by electrolysis.

In 1849, in a paper on the electrolysis of organic compounds, Kolbe † announced that potassium acetate on electrolysis gave hydrogen at the positive pole, and at the negative pole carbon dioxide and the "free methyl radical". The latter was really ethane, C_2H_6. Kolbe expected that the electrolysis of the acetate "would bring about a decomposition . . . into its two constituents in such a way that, in consequence of the decomposition of water, carbonic acid as the oxidation product of oxalic acid would appear at the positive pole, and at the negative pole a compound of methyl with hydrogen, viz. marsh gas", should collect. Instead of this, hydrogen and "methyl" itself collected at opposite poles. This was in excellent agreement with Berzelius's theory ($\underline{C}=6$, $\underline{O}=8$):

$$H\underline{O}\,.\,(\underline{C}_2H_3)\underline{C}_2\underline{O}_3+\underline{O}=\underline{C}_2H_3+2\underline{CO}_2+H\underline{O}.$$

Kolbe remarked that the electrolysis of organic compounds promised to throw light on their chemical constitution. In his textbook ‡ Kolbe represented the above decomposition as follows:

$$K\underline{O}\,.\,(\underline{C}_2H_3)\underline{C}_2\underline{O}_3+H\underline{O}=\underline{C}_2H_3+\underline{CO}_2+K\underline{O}\,.\,\underline{CO}_2+H.$$

* *Annalen*, 1848, lxv, 288.

† *Annalen*, 1849, lxix, 257; *Alembic Club Reprint* No. 15.

‡ *Lehrbuch der organischen Chemie*, 1854, i, 234.

The preparation of acetic acid from carbon chloride, mentioned above, made it possible to obtain this acid from its elements, and Kolbe in 1845 remarks on " the interesting fact that acetic acid, hitherto known only as an oxidation product of organic matter, can be almost immediately composed by synthesis from its elements." According to Graebe * this is the first use of the word " synthesis " in a memoir on organic chemistry.

The Alcohol Radicals

It will be recalled that Bunsen had obtained the cacodyl radical by heating cacodyl chloride with zinc. In 1848 Frankland and Kolbe by the action of potassium on ethyl cyanide in the cold obtained a gas having the empirical composition of free methyl. It did not, however, combine with chlorine to form methyl chloride, as would be expected. Frankland then heated methyl and ethyl iodides with zinc in sealed tubes at 150°. Gases were obtained having the empirical formulae CH_3 and C_2H_5 ($C=12$), and were regarded as the free methyl and ethyl radicals.

Gerhardt in 1849-50, however, pointed out that the densities showed that these formulae should be doubled, giving C_2H_6 and C_4H_{10}, and the gases would then be homologues of marsh gas. Hofmann and Laurent also pointed out, in 1850, that unless the formulae were doubled the increments of boiling point were twice those known in other homologous series. Frankland, however, considered that : " The isolation of four of the compound radicals (methyl, ethyl, valyl and amyl) belonging to the alcohol series now excludes every doubt of their existence, and furnishes a complete and satisfactory proof of the correctness of the theory proposed by Kane, Berzelius, and Liebig." This means the ethyl theory (p. 233). H. Rose, in his memorial lecture on Berzelius (1852), says : " It is to be regretted that Berzelius was not spared to see how many of his hypothetical radicals were actually prepared, and that in so short a time after his death." Gerhardt's view was, however, shown to be correct by Wurtz's preparation, in 1855, of " mixed radicals ", e.g. methyl + ethyl, by the action of sodium on

* *Geschichte der organischen Chemie,* p. 149.

the alkyl iodides: the molecules all contained two radicals, which might be the same or different.

By heating zinc with excess of methyl or ethyl iodide in a sealed tube under pressure, and distilling the white solid product in a stream of hydrogen, Frankland obtained compounds of the metal with the

Deutsches Museum, Munich.

FIG. 96.—H. KOLBE, 1818-1884.

methyl and ethyl radicals, viz. zinc methyl and zinc ethyl, as spontaneously inflammable liquids. On treatment with water these gave marsh gas, and what he called "methyl" and "ethyl", respectively, as gases.* Further investigation of "methyl" convinced Frankland

* *J. Chem. Soc.*, 1849, ii, 263, 297; *Experimental Researches*, 1877, p. 144.

in 1850 that it was really the "hydride of ethyl", $\underline{C}_4H_6$ ($\underline{C}=6$), whilst the gas from the electrolysis of acetate was regarded as free methyl, $\underline{C}_2H_3$. It was not until 1864 that Schorlemmer proved them to be identical, and to be ethane, C_2H_6 ($C=12$).

Kolbe on the Structure of Carbon Compounds

In 1855 Kolbe made the very important suggestion that organic acids could be regarded as derived from carbonic acid, $\underline{C}_2\underline{O}_4$ ($\underline{C}=6$, $\underline{O}=8$). In 1853 he and Frankland had suggested that the so-called copulated substances were formed by the substitution of oxygen by hydrocarbon radicals. In 1855 acetic acid was regarded as a molecule of carbonic acid with one *equivalent* ($\underline{O}=8$) of oxygen replaced by methyl:

$$\underline{C}_2\underline{O}_3(\underline{C}_2H_3) + H\underline{O}.$$

It is important to remember that in Kolbe's formulae hydrogen and hydrocarbon radicals replace oxygen, since equivalent weights are used, and that the double atom of carbon, $\underline{C}_2$, always behaves as a single atom. Organic compounds are thus formed from inorganic compounds by substitution. It will be remembered that Berzelius had suggested (p. 242) that organic compounds should be formulated in the same way as inorganic. In this way Kolbe obtained real types, as distinct from Gerhardt's *, and Kolbe's theory is sometimes called the "newer type theory". Thus, there are three type theories, Dumas', Gerhardt's and Kolbe's.

The next step was to break up the radicals into their immediate constituents; this step had already been formally taken by Gerhardt (p. 268):

$$\underbrace{\underline{C}_2\underline{O}_3(\underline{C}_2H_3)}_{\text{acetyl}} = \underbrace{\underline{C}_2\underline{O}_2}_{\text{carbonyl}} + \underline{O} + \underbrace{\underline{C}_2H_3}_{\text{methyl}}.$$

This representation was in agreement with Wanklyn's experiment (1858) of passing carbon dioxide through zinc ethyl mixed with sodium, when sodium propionate was formed:

$$\underline{C}_4H_5Na + \underline{C}_2\underline{O}_4 = \underline{C}_4H_5\frown\underline{C}_2\underline{O}_3 + \underline{O}Na.$$

(In 1860 Kolbe obtained sodium salicylate from carbon dioxide, phenol and sodium, and in 1873 from carbon dioxide and sodium

* *Annalen*, 1850, lxxv, 211; lxxvi, 1.

phenate.) The link between $\underline{C}_4H_5$ and $\underline{C}_2\underline{O}_3$ was a copula, and expressed Berzelius's view that in organic acids the radicals were conjugated with oxalic acid (p. 244).

Kolbe thus arrived in 1860 * at the following formulae, all derived from the inorganic carbonic acid ($\underline{C}_2\underline{O}_4$) by substitution of oxygen by *equivalents* of hydrogen or hydrocarbon radicals:

Carbonic acid - - -	$(\underline{C}_2\underline{O}_2)\underline{O}_2$
Formic acid - - -	$(\underline{C}_2\underline{O}_2)\underline{O}H + H\underline{O}$
Acetic acid - - -	$(\underline{C}_2\underline{O}_2)\underline{O}\underline{C}_2H_3 + H\underline{O}$
Acetone - - - -	$\underline{C}_2H_3(\underline{C}_2\underline{O}_2)\underline{C}_2H_3$
Aldehyde - - -	$H(\underline{C}_2\underline{O}_2)\underline{C}_2H_3$
Methyl alcohol - -	$H\underline{O} + H_3\underline{C}_2, \underline{O}$
Ethyl alcohol - - -	$H\underline{O} + \left.\begin{matrix}H_2\\ \underline{C}_2H_3\end{matrix}\right\}\underline{C}_2, \underline{O}.$

After Kekulé's recognition of the quadrivalency of carbon (p. 288), Kolbe pointed out that the *double* atom of carbon, $\underline{C}_2$, in these formulae (he called it "carbonyl") is always combined with four *equivalents* of hydrogen, oxygen, or alcohol radicals. He claimed that he had anticipated Kekulé, but it is difficult to admit that his claim was reasonable. Kolbe did not adopt Gerhardt's atomic weights, C = 12, O = 16, etc., until 1870.

In the first four formulae there is substitution *outside* the brackets of what is now called the carbonyl radical ($\underline{C}_2\underline{O}_2$); in the last two there is substitution in the radical itself, *inside* the brackets. (These brackets are important.) Water, $H\underline{O}$, is sometimes added to the formula, sometimes not, and no reason is given for the difference. Kolbe freely admits substitution in radicals: "The facts compel us at the present moment almost forcibly to the view that the organic radicals are variable groups of atoms, in which chlorine, bromine, nitrogen peroxide, etc., can enter in place of hydrogen equivalents, in such a way that the molecular grouping (*molekulare Gruppirung*) of their atoms remains unchanged, and that secondary radicals arise, which are in part endowed with similar properties to the primary." This was, of course, a significant step for a supporter of the electrochemical theory.

* *Annalen*, 1860, cxiii, 293; Ostwald's *Klassiker* No. 92.

In 1855 Kolbe compared the three formulae :

Acetic acid - - -	$H\underline{O}+\underline{C}_2H_3(\underline{C}_2\underline{O}_2)\underline{O}$
Aldehyde - - -	$\left.\begin{matrix}\underline{C}_2H_3\\ H\end{matrix}\right\}(\underline{C}_2\underline{O}_2)$
Alcohol - - - -	$H\underline{O}+\left.\begin{matrix}\underline{C}_2H_3\\ H_2\end{matrix}\right\}\underline{C}_2,\ \underline{O}$

and pointed out that only those hydrogen atoms which stand by themselves, i.e. outside the methyl group $\underline{C}_2H_3$, are capable of oxidation : they are more readily attacked than the hydrogens of the methyl group, which are firmly held to carbon.

In 1856 Frankland and Kolbe showed that an acid could be converted into an aldehyde by way of reduction of the cyanide by nascent hydrogen.

The following formulae are considered by Kolbe :

Carbonic acid - - -	$2H\underline{O}+(\underline{C}_2\underline{O}_2)\underline{O}_2$
Acetic acid - - -	$H\underline{O}+\underline{C}_2H_3(\underline{C}_2\underline{O}_2)\underline{O}$
Aldehyde - - -	$\left.\begin{matrix}\underline{C}_2H_3\\ H\end{matrix}\right\}(\underline{C}_2\underline{O}_2)$
Alcohol - - - -	$H\underline{O}+\left.\begin{matrix}\underline{C}_2H_3\\ H_2\end{matrix}\right\}\underline{C}_2,\ \underline{O}$
Substance A - - -	$H\underline{O}+\left.\begin{matrix}(\underline{C}_2H_3)_2\\ H\end{matrix}\right\}\underline{C}_2,\ \underline{O}$
Substance B - - -	$H\underline{O}+(\underline{C}_2H_3)_3\underline{C}_2,\ \underline{O}.$

The basicity of an acid is determined by the number of oxygen atoms *outside* the $\underline{C}_2\underline{O}_2$ radical ; to express this, water equivalents, $H\underline{O}$, are added in equal numbers.* Substance A is ordinary alcohol with one hydrogen substituted by methyl ; substance B is alcohol with two hydrogens substituted by methyls. Now compare the oxidation of substance A with that of ordinary alcohol :

$$\underbrace{H\underline{O}+\left.\begin{matrix}\underline{C}_2H_3\\ H_2\end{matrix}\right\}\underline{C}_2,\ \underline{O}}_{\text{ethyl alcohol}}\ \text{gives}\ \underbrace{\left.\begin{matrix}\underline{C}_2H_3\\ H\end{matrix}\right\}\underline{C}_2\underline{O}_2}_{\text{acetaldehyde}}$$

$$\underbrace{H\underline{O}+\left.\begin{matrix}\underline{C}_2H_3\\ \underline{C}_2H_3\\ H\end{matrix}\right\}\underline{C}_2,\ \underline{O}}_{\text{substance A}}\ \text{gives}\ \underbrace{\left.\begin{matrix}\underline{C}_2H_3\\ \underline{C}_2H_3\end{matrix}\right\}\underline{C}_2\underline{O}_2}_{\text{acetone}}$$

* *Annalen*, 1860, cxiii, 293.

The substance A should yield acetone on oxidation. Kolbe named it dimethyl carbinol; it is a secondary alcohol, whilst ordinary alcohol is a primary alcohol, yielding aldehyde on oxidation. The substance A (isopropyl alcohol) was prepared by Friedel shortly after its prediction by Kolbe; its modern formula is

$$\left.\begin{matrix}CH_3\\CH_3\end{matrix}\right> CH\,.\,OH.$$

The substance B contains no hydrogen atom attached to the carbonyl, $\underline{C}_2$; it cannot, therefore, yield a simple oxidation product. It is a tertiary alcohol, and was discovered by Butlerow in 1864. Its modern formula is $(CH_3)_3C\,.\,OH$, tertiary butyl alcohol.

Dibasic organic acids are formed from two molecules of carbonic acid by replacement of two oxygens outside the radical and addition of $2H\underline{O}$:

$$\begin{matrix}2H\underline{O}+(\underline{C}_2\underline{O}_2)\underline{O}_2\\2H\underline{O}+(\underline{C}_2\underline{O}_2)\underline{O}_2\end{matrix}\ \text{give}\ 2H\underline{O}+(\underline{C}_4H_4)\left.\begin{matrix}\underline{C}_2\underline{O}_2\\\underline{C}_2\underline{O}_2\end{matrix}\right\}\underline{O}_2.$$

In this example (succinic acid) two oxygens are replaced by the single group $\underline{C}_4H_4$ (i.e. C_2H_4 if $C = 12$) and $2\underline{O}H$ disappear with the oxygens outside the $\underline{C}_2\underline{O}_2$ radicals. Citric acid was derived from three molecules of carbonic acid:

$$3H\underline{O}+(\underline{C}_6H_5\underline{O}_2)\left.\begin{matrix}\underline{C}_2\underline{O}_2\\\underline{C}_2\underline{O}_2\\\underline{C}_2\underline{O}_2\end{matrix}\right\}\underline{O}_3.$$

It will be seen that Kolbe in his papers showed a deep insight into the constitution of organic compounds; his memoirs, although difficult, are worthy of careful study. Many of his pupils carried out important researches. The remarkable series of diazo-compounds obtained by the action of nitrous acid in the cold on solutions of salts of aromatic amines such as aniline, were discovered in 1860 by Peter Griess in Kolbe's laboratory. They form the starting point in the preparation of numerous organic compounds, including azo-dyes, the first member of which (aminoazo-benzene, "aniline yellow") was obtained by Mène in 1861.

Wurtz

The masterly investigations of Berthelot on glycerin (1854) showed that it "gives rise to three distinct series of neutral combinations " with acids ; it is a " tribasic alcohol ", and it was formulated by Wurtz (1855) as derived from six water molecules ($\underline{C}=6$; $\underline{O}=8$) :

$$\left.\begin{matrix}\underline{C}_6H_5\\H_3\end{matrix}\right\}\underline{O}_6.$$

The corresponding monatomic alcohol is propyl alcohol, $(\underline{C}_6H_7)H\,.\,\underline{O}_2$, and the intermediate diatomic alcohol would be

FIG. 97.—A. WURTZ, 1817-1884.

$(\underline{C}_6H_6)H_2\,.\,\underline{O}_4$, a representative of which type of compound was discovered by Wurtz in 1856, viz. $(\underline{C}_4H_4)H_2\,.\,\underline{O}_4$, by the action of ethylene iodide on silver acetate and saponification of the resulting glycol diacetate. He named it glycol, since it resembles in its properties at the same time both an alcohol properly so-called and glycerin.*

* *Compt. rend.*, 1856, xliii, 199 ; *Ann. Chim. Phys.*, 1859, lv, 400 ; Ostwald's *Klassiker* No. 170.

Adolphe Wurtz, born in Strasburg in 1817, was a pupil of Liebig and Dumas. He was professor in the École de Médecine and in the Sorbonne in Paris, and died in 1884. He made important investigations on amines (p. 261) and on polyatomic alcohols. His early researches on phosphorous and hypophosphorous acids are also interesting. Wurtz was the first to teach Gerhardt's views in France.*

Berthelot

Marcellin Berthelot † was born in Paris in 1827, and became professor in the Collège de France. He later held the posts of

FIG. 98.—M. BERTHELOT. 1827-1907.

Minister of Education and Foreign Minister, and carried out, with Vieille and alone, important researches on explosives. He

* Hofman, *Berichte*, 1887, vol. iii, p. 815.

† Memorial Lecture by Dixon, *J. Chem. Soc.*, 1911, p. 2353.

died in 1907. His first work dealt with organic synthesis.* After the work on glycerin in 1854 came the synthesis in 1854 (published in January, 1855) of alcohol from ethylene and sulphuric acid, followed by hydrolysis of the ethylsulphuric acid (this had been carried out by Hennell in 1825-8), and in 1855 of formic acid from carbon monoxide and heated alkali : $CO + KOH = H.COOK$. In 1856 he prepared methane (and ethylene) by passing a mixture of carbon disulphide vapour and sulphuretted hydrogen over heated copper. In 1857 methyl alcohol was obtained from methane by way of the chloride. Camphol and camphor were obtained from turpentine in 1859, borneol from camphor in 1859, acetylene from carbon and hydrogen in 1862, benzene from acetylene in 1866, and hydrocyanic acid from acetylene and nitrogen in 1868. In 1870 Berthelot used hydriodic acid in sealed tubes for reduction.

Berthelot also carried out important work in physical chemistry. His fundamental investigations with St. Gilles on reaction velocity (1862-3) will be referred to later ; that on the explosion wave, developed simultaneously and later by Dixon, opened out an important branch of investigation, and the masterly researches on thermochemistry † provided a wealth of accurate data in that science.

Berthelot's work on the history of alchemy first gave us a nearly complete and critical edition of the writings of the oldest chemists of the Alexandrian period ; his other publications in historical chemistry are classical.

The Theory of Valency

Kolbe was of the opinion that the radical or element with which a substance is conjugated has only a subordinate influence on the nature of the compound. Thus, acetic acid, $H\underline{O} + \underline{C}_2H_3{}^\frown(\underline{C}_2\underline{O}_2)\underline{O}$ is similar to carbonic acid, $2H\underline{O} + (\underline{C}_2\underline{O}_2)\underline{O}_2$ ($\underline{C} = 6$; $\underline{O} = 8$). Frankland‡ in 1852 showed that this idea is not correct. If the organo-metallic compounds are considered, it is seen that when arsenic is conjugated

* *Chimie organique fondée sur la synthèse*, 2 vols., 1860 : " La chimie crée son objet. Cette faculté créatrice, semblable à l'art lui-même, la distingue essentiellement des sciences naturelles et historiques."

† *Mécanique chimique fondée sur la thermochimie*, 2 vols., 1879 ; *Thermochimie*, 2 vols., 1897.

‡ *Phil. Trans.*, 1852, vol. cxlii, 417.

with methyl the former changes its saturation capacity. Whereas arsenic can combine with five equivalents of oxygen ($\underline{O}=8$) to form arsenic acid, $As\underline{O}_5$, cacodyl can combine with only three to form cacodylic acid, $(C_2H_3)_2As\underline{O}_3$, which resists the action of strong oxidising agents.

The conjugation of metals with hydrocarbon radicals thus alters their saturation capacities. Frankland goes on to say that: "When the formulae of inorganic compounds are considered, even a superficial observer is struck with the general symmetry of their construction; the compounds of nitrogen, phosphorus, antimony and arsenic especially exhibit the tendency of these elements to form compounds containing 3 or 5 equivs. of other elements, and it is in these proportions that their affinities are best satisfied; thus in the ternal groups we have NO_3, NH_3, NI_3, NS_3, PO_3, PH_3, PCl_3, SbO_3, SbH_3, $SbCl_3$, AsO_3, AsH_3, $AsCl_3$, &c.; and in the five-atom group, NO_5, NH_4O, NH_4I, PO_5, PH_4I, &c. Without offering any hypothesis regarding the cause of this symmetrical grouping of atoms, it is sufficiently evident, from the examples just given, that such a tendency or law prevails, and that, no matter what the characters of the uniting atoms may be, the combining power of the attracting element, if I may be allowed the term, is always satisfied by the same number of these atoms." It must be noted that Frankland used the equivalent weights in his formulae.

This "combining power" was afterwards called quantivalence, or valency. (American chemists still use the older name "valence".) Frankland proceeded to point out, with respect to Gerhardt's views: "The formation and examination of the organo-metallic bodies promise to assist in effecting a fusion of the two theories which have so long divided the opinions of chemists, and which have too hastily been considered irreconcilable; for whilst it is evident that certain types of series of compounds exist, it is equally clear that the nature of the body derived from the original type is essentially dependent upon the electrochemical character of its single atoms, and not merely on the relative position of these atoms."

This was an important step, and for two reasons. In the first place, the older theory of radicals (as distinct from Gerhardt's residues) was fused with the type theory, and in the second place, attention was once more focused on the atoms as the real determining factor in chemical changes, instead of radicals or types. The vague idea of copulae also disappeared.

Kekulé

Frankland's idea was developed and extended by Kekulé. Friedrich August Kekulé was born in 1829 in Darmstadt, and in

FIG. 99.—F. A. KEKULÉ, 1829-1896.

1847 studied architecture in the University of Giessen. He was destined, however, to be master of a more refined architecture, and the influence of Liebig turned his attention to chemistry. He also studied under Dumas and corresponded with Gerhardt. Whilst in London as Stenhouse's assistant, he thought out his structure theory (molecular architecture)—he says on the top of an omnibus—

and in this period he was a friend of Williamson. As he said later : "Originally a pupil of Liebig, I became a pupil of Dumas, Gerhardt and Williamson : now I no longer belong to any school." He was professor in Ghent in 1858, and whilst there developed his benzene theory ; in 1867 he was called to Bonn, where he died in 1896.*

Kekulé grasped the true significance of valency and made it the principal guide in his theory of the structure of organic molecules. In 1853, when in London, he discovered that phosphorus pentasulphide gives thioacetic acid with acetic acid, and contrasted the reaction with that of phosphorus pentachloride :

$$5\left.\begin{matrix}C_2H_3O\\ H\end{matrix}\right\}O + P_2S_5 = 5\left.\begin{matrix}C_2H_3O\\ H\end{matrix}\right\}S + P_2O_5$$

$$5\left.\begin{matrix}C_2H_3O\\ H\end{matrix}\right\}O + 2PCl_5 = \frac{5C_2H_3OCl}{5HCl} + P_2O_5.$$

Whereas in the second case the product breaks up, in the first it does not, "because the quantity of sulphur which is equivalent to two atoms of chlorine is not divisible." This is explicable with Gerhardt's atomic weights, "since sulphur, like oxygen itself, is dibasic (*zweibasisch*) so that one atom is equivalent to two atoms of chlorine."

In 1857 Kekulé added to Gerhardt's types that of marsh gas, but applied it only to a few compounds ; he had now returned to Germany and had adopted the equivalent weights there in use ($\underline{C}=6$, $\underline{O}=8$, $N=14$, $\underline{Hg}=100$) :

Marsh gas	$\underline{C}_2HHHH$
Methyl chloride	$\underline{C}_2HHHCl$
Chloroform	$\underline{C}_2HClClCl$
Chloropicrin	$\underline{C}_2(N\underline{O}_4)ClClCl$
Mercury fulminate	$\underline{C}_2(N\underline{O}_4)\ \underline{Hg}\ \underline{Hg}\ (\underline{C}_2N)$.

Odling, in a lecture at the Royal Institution in 1855, had already used the marsh gas type, but this did not attract attention.

In 1858 † Kekulé extended the marsh gas type to all carbon compounds and re-adopted Gerhardt's atomic weights, which he

* Japp, Memorial Lecture, *J. Chem. Soc.*, 1898, p. 97.

† *Annalen*, 1858, cvi, 129 ; a fundamental paper.

represented by barred symbols, ~~C~~, ~~O~~, ~~H~~, etc. "The quantity of carbon which chemists have recognized as the atom always unites with four atoms of a monatomic or two atoms of a di-atomic element; in general the sum of the chemical units of the elements combined with one atom of carbon is 4."

Kekulé had now clearly recognized that carbon is quadrivalent (*vierbasisch oder vieratomig*) * and he also introduced the fundamentally important conception of the linking of carbon atoms with one another. "In the cases of substances which contain several atoms of carbon, it must be admitted that at least some of the atoms are held in the compound by the affinity of carbon, and that the carbon atoms attach themselves to one another, whereby a part of the affinity of one is naturally engaged with an equal part of the affinity of the other. The simplest and therefore most probable case of such an association of carbon atoms is that in which one affinity unit of one is bound by one of the other. Of the 2×4 affinity units of the two carbon atoms, two are used up in holding the atoms together; six remain over, which can be bound by atoms of other elements. In other words, a group of two carbon atoms, C_2, will be sexatomic, and will form a compound with six atoms of a monatomic element, or generally with so many atoms that the sum of the chemical units of these is six."

Kekulé says: "I must repeatedly emphasize that I do not consider a great number of these views as originating with myself, but am rather of the opinion that besides the earlier named chemists (Williamson, Odling, Gerhardt), from whom detailed considerations of these matters are available, others, especially Wurtz, have shared in the outlines of these views." The omission of Frankland's name is remarkable; Kekulé later said: "Unless I am mistaken I am the one who introduced the idea of the atomicity of the elements into chemistry." He was, of course, mistaken. Perhaps the circumstance that Kekulé felt that so much of his theory was due to others led him, further on in the memoir, to say that he "attaches but a subordinate value to such considerations."

Kekulé particularly emphasized the necessity of going back to the atoms in considering combining capacities: "the nature of the

* This is first stated in a footnote in *Annalen,* 1857, civ, 133.

radicals and of their compounds must be derived from those of the elements." He points out that Gerhardt's double decomposition as a typical reaction is not sufficient to cover all the cases, and proposed the following types of chemical reactions:

1. Direct addition: $CO + Cl_2 = COCl_2$.

2. Combination of several molecules by rearrangement (*Umlagerung*) of a polyatomic radical:

$$SO_3 + H_2O = \left.\begin{matrix} SO_2 \\ H_2 \end{matrix}\right\} O_2.$$

3. Changes in which the numbers of molecules (and in the case of gases the volumes) remain the same, i.e. double decompositions, usually regarded as an exchange of radicals but considered by Kekulé as addition followed by rearrangement:

before		*during*		*after*	
a	b	a	b	a	b
a_1	b_1	a_1	b_1	a_1	b_1

He points out that mass action and catalytic force have an important influence on chemical reactions. Usually the force which brings about the association of the molecules (*Aneinanderlagerung*) also causes their decomposition, but there are also cases where the latter does not occur under suitable conditions, and the intermediate product may be isolated.

The quadrivalency of carbon was assumed in 1858 independently of Kekulé by Archibald Scott Couper, born in Scotland in 1831.* He also made use of bonds (for which he used dotted lines) joining the atoms, and of structural formulae.† Thus ($C = 12$, $\underline{O} = 8$):

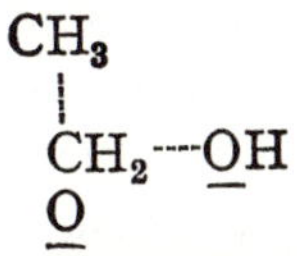

alcohol

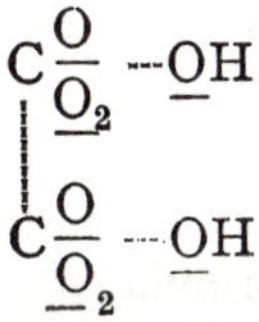

oxalic acid

* Anschütz, *Proc. Roy. Soc. Edin.*, 1909, xxix, 193.

† *Compt. Rend.*, 1858, xlvi, 1157; *Phil Mag.*, 1858, xvi, 104; Dobbin, *J. Chem. Educ.*, 1934, xi, 331; *Alembic Club Reprint* No. 21.

Couper's formula of alcohol may be compared with Kolbe's and the modern formula :

$$\underline{O}, \underline{C}_2\left\{\begin{matrix}\underline{C}_2H_3\\H_2\end{matrix}\right. + H\underline{O} \qquad\qquad \begin{matrix}CH_3\\|\\CH_2\text{—}OH\end{matrix}$$

Kolbe's formula | Modern formula

Couper recognized two valencies of carbon, one in carbon monoxide and one in carbon dioxide : " the highest known power of combination of carbon is that of the second degree, i.e. 4." Frankland also admitted that the valency of an element could vary, e.g. phosphorus could have valencies of 3 and 5, whilst Kekulé considered that the valency was constant, PCl_5, for example, being a molecular compound of PCl_3 and Cl_2. An illness on his return to Scotland prevented Couper from developing his ideas, and this was left to Kekulé.

The Formula of Benzene

In 1858 Kekulé had suggested that the carbon atoms of benzene and especially naphthalene are more densely arranged than in most organic compounds (" eine dichtere Aneinanderlagerung des Kohlenstoffes "). The formula of benzene was first satisfactorily written by Kekulé in 1865. He also gave the clue to the constitution of the so-called " aromatic compounds ", which contain closed rings of carbon atoms, as contrasted with open chains in the so-called " fatty " or " aliphatic " compounds. (The names aromatic and aliphatic are due to Kekulé and to Hofmann respectively.) The origin of the discovery has been described by Kekulé : in a doze in his study in Ghent he saw the long chains of carbon atoms like snakes twisting and curling until one " gripped its own tail and the picture whirled scornfully before my eyes." *

The six atoms of carbon in the benzene ring were regarded as joined alternately by single and double linkages, so as to preserve the quadrivalency of carbon. Kekulé at first made use of curious symbols † which were satirized by Kolbe, but in 1865 ‡ he used the hexagon formula for benzene without bonds (I) and in

* *Berichte*, 1890, xxiii, 1306 ; Walker, *Annals of Science*, 1939, iv, 34.

† See Lowry, *Historical Introduction to Chemistry*, p. 439.

‡ *Bull. Acad. Roy. Belg.*, 1865, xix, 551.

1866 * he gave a sketch (II) of a space model with single and double bonds equivalent to the modern formula (III):

I

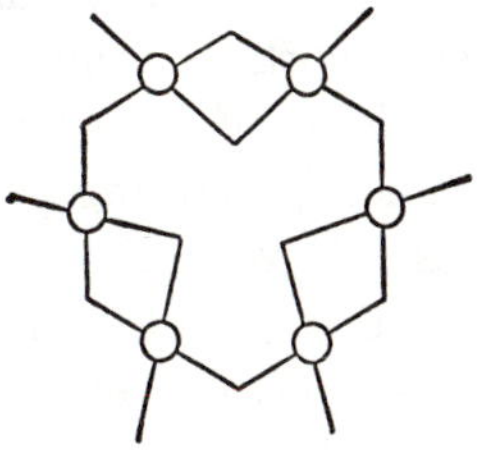

II

III

The existence of multiple linkings between carbon atoms was vaguely recognized by Kekulé in 1859 † and in 1864 the formula of ethylene was written as follows by Crum Brown:

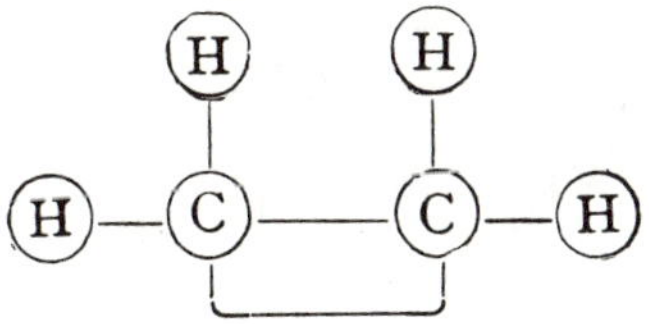

and in this a double bond appears. Lothar Meyer in the same year proposed to regard ethylene as containing two unsaturated affinities, since the molecule readily combines with two atoms of chlorine or bromine by addition, but Kekulé's method of double bonds was generally preferred, although carbon monoxide was a difficulty. The existence of triple linkages in acetylene was recognized by Erlenmeyer in 1862. The formulation of multiple linkages is contained in the graphic formulae of Wilbrand‡ (the modern formulae are shown below):

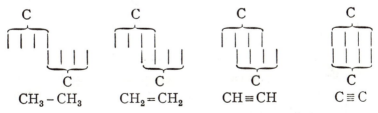

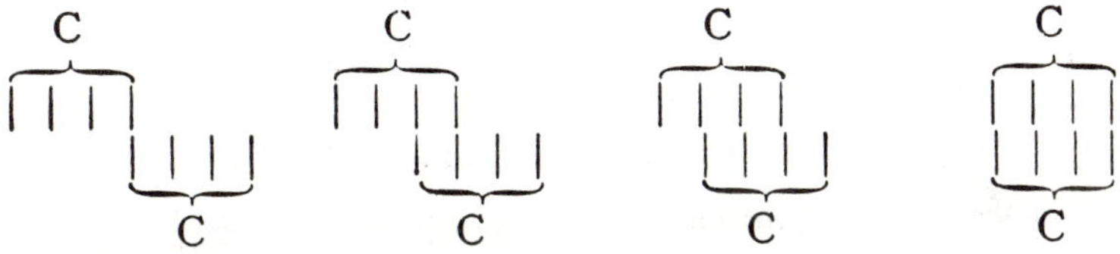

CH_3-CH_3 $CH_2=CH_2$ $CH\equiv CH$ $C\equiv C$

* Kekulé, *Lehrbuch*, 1866, ii, 496.

† *Lehrbuch*, i, 156.

‡ *Zeitsch. f. Chemie*, 1865, viii, 685.

The immense service of Kekulé's benzene formula in organic chemistry cannot be over-emphasized. It explained cases of isomerism in the derivatives of benzene and has proved sufficient during years of intensive work following its inception. The study of the structure of aromatic compounds by X-rays has also confirmed the existence of the flat hexagonal ring of six carbon atoms, but the distribution of valencies still remains a difficulty.

It was recognized before Kekulé's theory that other aromatic hydrocarbons are homologues of benzene, e.g. toluene was synthesized by Fittig and Tollens in 1864 from bromobenzene and methyl iodide by the action of sodium, and is thus methyl-benzene. Xylene, discovered in crude wood-spirit by Cahours in 1849, is also present in coal tar and was shown to be dimethyl-benzene. In this case three isomers are possible, according as the two methyl groups are related in the *ortho*-, *meta*-, or *para*-positions :

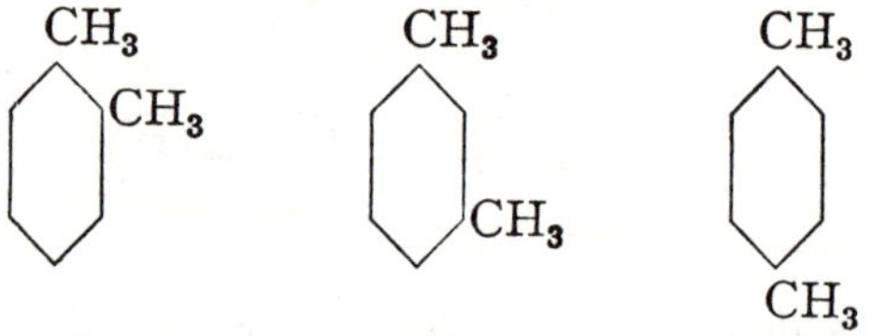

o-xylene *m*-xylene *p*-xylene

The determination of orientation in such cases was worked out in 1874 by Wilhelm Körner, Kekulé's assistant and later professor in Milan.*

Kekulé pointed out that isomerism may also arise by substitution according as the latter occurs in the ring or in the side-chain :

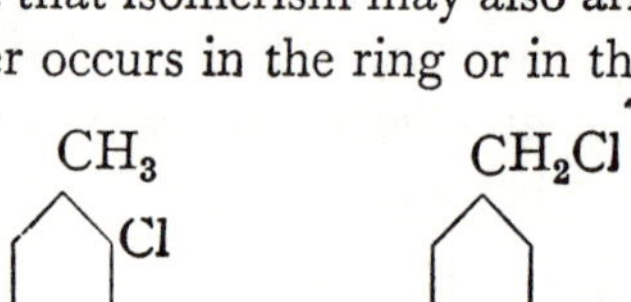

o-chlorotoluene benzyl chloride

Chlorine attached to carbon in the ring is very firmly bound and cannot be removed by boiling alkali, as can that in the side chain or in aliphatic compounds.

* *Gazzetta*, 1874, iv, 305 ; *J. Chem. Soc.*, 1876, xxix, 204 ; Ostwald's *Klassiker* No. 174.

In 1866 Erlenmeyer suggested that naphthalene, discovered by Garden * and by Kidd † in coal tar, consists of two fused benzene rings having two carbon atoms in common. By the oxidation of naphthalene the important phthalic acid was obtained by Laurent in 1836; its constitution was established by Graebe, since it very easily forms an anhydride by loss of water, and thus probably contains the two carboxyl groups in two adjacent positions:

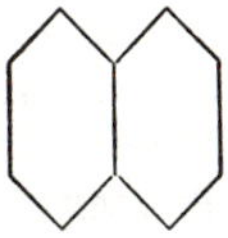

naphthalene

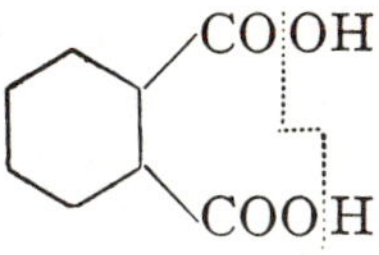

phthalic acid

Another hydrocarbon present in coal tar is anthracene, discovered in 1832 by Dumas and Laurent. Its constitution was elucidated in 1868 by Graebe and Liebermann, who showed that it consists of three fused benzene rings:

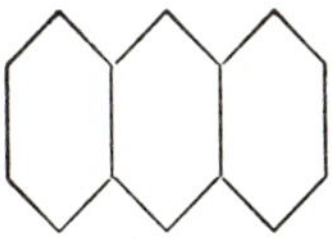

anthracene

Examples of heterocyclic compounds (containing atoms other than carbon in the ring) soon came to light, e.g. pyridine:

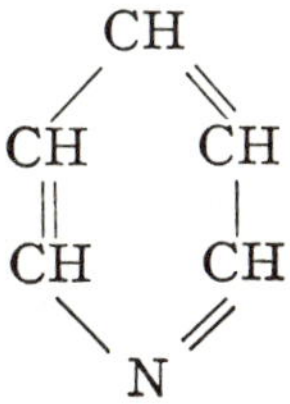

(Körner, 1869; independently by Dewar, 1871). They are parent substances of the alkaloids.

The further development of synthetic organic chemistry will be briefly mentioned in the next chapter.

* *Ann. Phil.*, 1820, xv, 74. † *Phil. Trans.*, 1821, cxi, 209.

SUMMARY AND SUPPLEMENT

The Theory of Valency

Hermann Kolbe (1818-1884), professor in Marburg (1851) and Leipzig (1865), developed the views of Berzelius in a period when most other chemists had abandoned them. He synthesized acetic acid (1844), obtained methyl sulphonic acid (1845), studied the electrolysis of salts of fatty acids (1850), synthesized salicylic acid (1860), investigated taurine (1862), malonic acid (1864) and aliphatic nitro-compounds (1872). His theoretical views (1854-65) were important: they foreshadowed modern structural formulae and valency, and enabled him to predict the existence of compounds afterwards synthesized, such as secondary and tertiary alcohols.

Edward Frankland (1825-1899), professor in Manchester (1851) and London (1863), discovered organo-metallic compounds of zinc, etc. (1849-64) by an extension of Bunsen's cacodyl researches; obtained hydrocarbons from zinc alkyls (1849), put forward the theory of valency (1852-60), and (after Crum Brown, 1864) devised modern structural formulae (1867). He recognized that the valency of an element could vary. With Duppa he investigated (1864 f.) the reactions and constitution of acetoacetic ester (independently discovered by Geuther in 1863).

The varying valencies of an element in different compounds, e.g. Cl 1 in HCl and 7 in $HClO_4$; S 2 in H_2S, 4 in SO_2, and 6 in H_2SO_4, were clearly expressed by Christian Wilhelm Blomstrand (1826-1897), professor in Lund (*Chemie der Jetztzeit*, Heidelberg, 1869), who also investigated complex ammine compounds of metals (addition compounds of ammonia with metal salts, e.g. cobaltammines) and (1869) put forward the "diazonium" formula for diazo-salts, also put forward independently by Strecker (1871) and Erlenmeyer (1874). Blomstrand's theoretical views were based on those of Berzelius, suitably adapted to the theory of valency.

August Kekulé (1829-1896), professor in Ghent (1858) and Bonn (from 1865), added the Marsh Gas Type (1857) to those of Gerhardt, and recognized (simultaneously with Couper) the quadrivalency of carbon (1857-8) and the linking of carbon atoms. He put forward the hexagon formula for benzene (1865-6; 1872), and carried out special researches in different branches of organic chemistry (e.g. on diazo-compounds). He believed that the valency of an element was constant. Kekulé's experimental investigations are less important than his theoretical publications. He obtained acetylene by the electrolysis of fumaric acid (1864), recognized crotonaldehyde (1872) and discovered triphenylmethane (1872). Kekulé's *Lehrbuch der*

Organischen Chemie oder der Chemie der Kohlenstoffverbindungen (vols. i-iii and part 1 of vol. iv, incomplete, Erlangen and Stuttgart, 1866-87) contains a valuable (if rather prejudiced) historical introduction (vol. i, pp. 1-94).

Charles Adolphe Wurtz (1817-1884), professor in Paris, investigated the lower oxy-acids of phosphorus (1844-9), discovered phosphorus oxychloride, $POCl_3$ (1846), the production of amines from alkyl isocyanates (1848), the synthesis of hydrocarbons from alkyl halides and sodium (1855), glycol (1856), ethylene oxide (1859), the reduction of aldehydes to alcohols with sodium (1862), and the aldol condensation (1872). His advocacy of Gerhardt's views and his text-books had an important influence on the development of chemical theory.

Marcellin (or Marcelin) Berthelot (1827-1907), professor in the Collège de France, Paris, was one of the most distinguished chemists of the nineteenth century. His interests were extraordinarily wide, and his work, all of which was highly original and of a fundamental character, may be divided into four periods : (I) on the constitution and synthesis of polyatomic aliphatic alcohols (including glycerol), and organic acids (1850-60) ; (II) on the synthesis of hydrocarbons, including the synthesis of acetylene in the electric arc and its polymerization to benzene, the reduction of organic compounds by heating in sealed tubes with hydriodic acid, and on the velocity of reaction (with Sainte-Gilles) (1861-70) ; (III) thermochemical researches, work on explosives (discovery of the "detonation wave"), the specific heats of gases at high temperatures, and syntheses by the silent electric discharge (1869-85) ; (IV) researches in historical and agricultural chemistry (1885-1907).

A Sketch of the Earlier Development of Organic Chemistry

I. Berzelius (1814) improved Gay-Lussac and Thenard's method of organic analysis (1811), in which the substance was heated with potassium chlorate and the water and carbon dioxide collected. Liebig, from 1823, used combustion with copper oxide, collecting the carbon dioxide in "potash bulbs". Dumas (1830) determined nitrogen by collecting it as gas from a combustion experiment. In 1815 Gay-Lussac discovered cyanogen and recognized it as a radical. It was the prevailing belief (expressed e.g. by Gmelin in 1817) that organic compounds could be produced only by the operation of a mysterious "vital force".

II. Wöhler and Liebig in 1832 published their research on the benzoyl radical. They found that oil of bitter almonds (now called benzaldehyde), benzoic acid, benzoyl chloride and benzamide (the last two compounds being discovered by them) all contained a group of atoms, C_7H_5O, which they wrote $C_{14}H_{10}O_2$ and called benzoyl. Berzelius, who at first welcomed this theory, soon opposed it, since he would not sanction radicals containing oxygen. Liebig in 1838 gave a general definition of a radical.

Dumas and Boullay (1827) put forward the "etherin" theory, regarding alcohol and ether as compounds of ethylene and water.

Liebig in 1834 adopted the ethyl radical, C_4H_{10}, and regarded ether as its oxide, $C_4H_{10}O$, and alcohol as the hydrate of ether, $C_4H_{10}O.H_2O$. Berzelius, on the other hand, considered alcohol to be the oxide of a radical, C^2H^6, viz. C^2H^6O, whilst ether was the oxide of ethyl, $C^4H^{10}O$.

III. As a result of the experiments of Dumas and Laurent, Dumas in 1839 put forward his Theory of Substitution and of Types. He found that hydrogen in an organic compound could be replaced by an equivalent of chlorine, of oxygen, etc., and he proposed to give up the dualistic formulation of organic compounds, replacing it by a Unitary Theory, in which the molecule was regarded as a whole, its parts being capable of substitution.

Laurent pointed out that the compound after such substitution still retained its essential chemical properties, or preserved its chemical type, and he suggested that chlorine "takes the place" of hydrogen in substitution. At first Dumas did not accept this, but after his research on the formation of trichloracetic acid from acetic acid, he adopted Laurent's views.

These ideas were violently attacked by Berzelius, who could not admit that electronegative chlorine could replace electropositive hydrogen, and he proposed a formula for trichloracetic acid which made it completely different in structure from acetic acid:

acetic acid = hydrated trioxide of acetyl:

$$C^4H^6.O^3 + H^2O;$$

trichloracetic acid = carbon chloride + oxalic acid:

$$C^2Cl^6 + C^2O^3 + H^2O.$$

In 1842, however, Melsens showed that trichloracetic acid is easily converted into acetic acid by reduction with nascent hydrogen, so that Berzelius was forced to regard acetic and trichloracetic acids as closely analogous compounds. He did this by writing their formulae as oxalic acid (C^2O^3) "copulated" with methyl or with carbon chloride:

acetic acid: $C^2H^6 + C^2O^3 + H^2O$
trichloracetic acid: $C^2Cl^6 + C^2O^3 + H^2O$

and explained that the nature of the substance copulated with oxalic acid had very little influence on the chemical properties. He had now tacitly admitted substitution, but he never accepted the unitary or type theories.

IV. Gerhardt (1839) modified the Older Radical Theory and combined it with the Type Theory by postulating substitution of radicals in the types. He supposed that radicals were incapable of independent existence, but were "residues" formed by double decomposition, e.g. C_6H_5 and NO_2 are residues in the reaction between benzene and nitric acid:

$$C_6H_6 + HNO_3 = C_6H_5NO_2 + H_2O.$$

Bunsen's investigations on cacodyl compounds (1837-43) showed that these contained the radical $As_2(CH_3)_4$ (cacodyl) which he obtained in the free state. (Actually the cacodyl *radical* is $As(CH_3)_2$.) This work pleased Berzelius, since it supported his theory that organic compounds were similar to inorganic compounds, except that elements in the latter were replaced by radicals in the former.

Kolbe and Frankland in 1848 thought they had achieved the isolation of the alcohol radicals, e.g. free methyl, CH_3. Actually they obtained hydrocarbons, e.g. ethane, C_2H_6, which are homologues of methane. They were largely influenced by Berzelius's ideas, especially Kolbe, who later developed his views on the structure of organic compounds, and came near the recognition of valency.

V. Wurtz in 1849 discovered the amines predicted by Liebig, i.e. organic compounds containing the radical NH_2, and Hofmann in 1850 proved that they could be formulated as NH_3, in which one or more atoms of hydrogen are replaced by organic radicals, and thus created the Ammonia Type:

$$N\begin{cases}H\\H\\H\end{cases}\qquad N\begin{cases}CH_3\\H\\H\end{cases}\qquad N\begin{cases}CH_3\\CH_3\\H\end{cases}\qquad N\begin{cases}CH_3\\CH_3\\CH_3\end{cases}$$

Williamson's researches from 1850 proved by experiment that alcohol, ether, and many other compounds, inorganic and organic, can be derived from the Water Type:

$$\begin{matrix}H\\H\end{matrix}O\qquad\begin{matrix}C^2H^5\\H\end{matrix}O\qquad\begin{matrix}C^2H^5\\C^2H^5\end{matrix}O$$

It was an important step to formulate the molecules of alcohol and ether as occupying the same volume ("two volumes") in the vapour state.

Gerhardt in 1856 proposed four types (water, ammonia, hydrogen, and hydrochloric acid) :

$$\left.\begin{matrix}H\\H\end{matrix}\right\}O \qquad \left.\begin{matrix}H\\H\\H\end{matrix}\right\}N \qquad \left.\begin{matrix}H\\H\end{matrix}\right\} \qquad \left.\begin{matrix}H\\Cl\end{matrix}\right\},$$

to which Kekulé in 1857 added the marsh gas type :

$$\left.\begin{matrix}H\\H\\H\\H\end{matrix}\right\}C$$

stating that carbon is " tetratomic ", i.e. 1 at. C is equivalent to 4 at. H.

VI. Frankland in 1852 had put forward the idea of valency, based on his study of organo-metallic compounds, and in 1858 Kekulé showed that the two assumptions (1) carbon is quadrivalent, (2) atoms of carbon can link together, can explain the structures of a great many organic compounds. In 1865 Kekulé assumed that in aromatic compounds the atoms of carbon are linked together into closed rings (e.g. hexagon-rings of benzene). The theory of types was then no longer necessary, and modern structural chemistry began.

CHAPTER XIII

THE DEVELOPMENT OF ORGANIC CHEMISTRY

Pasteur

Louis Pasteur * was born in Dôle in 1822, his father being an old soldier of the First Empire, decorated by Napoleon on the field of battle, but become a tanner. Louis studied in Arbois, then in Besançon, and then, when his talents became obvious, he proceeded to Paris to study in the École normale, which he entered in 1843. He attended the lectures of both Balard and Dumas; his great energy found an outlet in research and his attention was directed to the study of crystals by Delafosse, a former pupil of Haüy, whilst he became Balard's assistant. In 1848 he became professor of physics at the Lycée in Dijon, in 1852 he was appointed professor of chemistry in Strasbourg, and in 1857 he became Director of the École normale in Paris. He there carried out, in a private laboratory, those researches on fermentation which founded microbiology, and in 1889 the Pasteur Institute was founded by a large subscription. He worked there until his death in 1895.

In 1815-35 Biot observed that a number of naturally occurring organic compounds, such as turpentine, camphor, sugars, and tartaric acid, possess the property of rotating the plane of polarized light in one sense or the other. This property is inherent in the molecules, since it is exhibited by solutions. In 1844 Mitscherlich observed that, although tartaric and racemic acids are isomeric, the former is optically active, as well as its salts, whilst racemic acid is inactive.

In 1848 Pasteur found that the crystals of tartaric acid are characterized by small facets which had unaccountably been overlooked by Mitscherlich, similar to the hemihedral facets observed by Haüy on optically active quartz. Quartz crystals may be divided into

* Frankland, *J. Chem. Soc.*, 1897, 683; Patterson, *Annals of Science*, 1938, iii, 431.

two groups, right and left handed, according to the position of the facets, the two forms having the relation of an object and its image in a mirror. Herschel in 1821 had pointed out that this was probably connected with Biot's observation that some quartz

FIG. 100.—L. PASTEUR, 1822-1895.

crystals rotated the plane of polarized light to the right and some to the left.

There was, it will be seen, a considerable amount of information available when Pasteur began his work on the tartaric acids, and the connection between crystalline form and optical activity was as good as established. His important discovery was that by the slow crystallization of a solution of sodium ammonium racemate, crystals were obtained with facets, some on the right and some on the left.

"I carefully separated the crystals which were hemihedral to the right from those which were hemihedral to the left, and examined their solutions separately in the polarizing apparatus. I then saw with no less surprise than pleasure that the crystals hemihedral to the right deviated the plane of polarization to the right, and that those hemihedral to the left deviated it to the left; and when I took an equal weight of each of the two kinds of crystals, the mixed solution was indifferent towards the light in consequence of the neutralization of the two equal and opposite individual deviations."* Pasteur remarks that: "For more than thirty years Biot had striven in vain to induce chemists to share his conviction that the study of rotatory polarization offered one of the surest means of gaining a knowledge of the molecular constitution of substances."

The two tartaric acids, separated from the salts, gave out heat on mixing and from the solution their compound, optically inactive racemic acid, crystallized out. The interpretation of Pasteur's results in terms of chemical constitution was given independently in 1874, by van't Hoff and LeBel.

Van't Hoff

Jacobus Henricus van't Hoff † was born in Rotterdam in 1852. He was educated in the Polytechnic in Delft, in the University of Leyden, in Bonn under Kekulé (whom he found unsympathetic), and under Wurtz at the École de Médecine in Paris, where he met LeBel. In 1874 van't Hoff received his doctorate at Utrecht for a straightforward piece of routine work in organic chemistry, the presentation of which, rather than the *Chemistry in Space*, showed good judgment on his part. In 1876 he obtained a post as assistant in the Veterinary College at Utrecht, where he wrote the first part of his *Ansichten über die organische Chemie* ‡, a book which is very original and deals with general principles, although according to Walker it "is almost unreadable". In 1878 he became professor in Amsterdam, where he spent eighteen years teaching inorganic and organic chemistry, crystallography, mineralogy, geo-

* *Alembic Club Reprint* No. 14; Delépine, *Bull. Soc. Chim.*, 1925, xxxvii, 197.

† Walker, Memorial Lecture, *J. Chem. Soc.*, 1913, 1127.

‡ 2 vols., Brunswick, 1878-1881.

logy and palaeontology, and conducting practical classes for medical students. It was during this period that he laid the foundations of modern physical chemistry.

In 1884 van't Hoff's *Études de dynamique chimique* appeared, in which the general treatment of reaction velocity and the

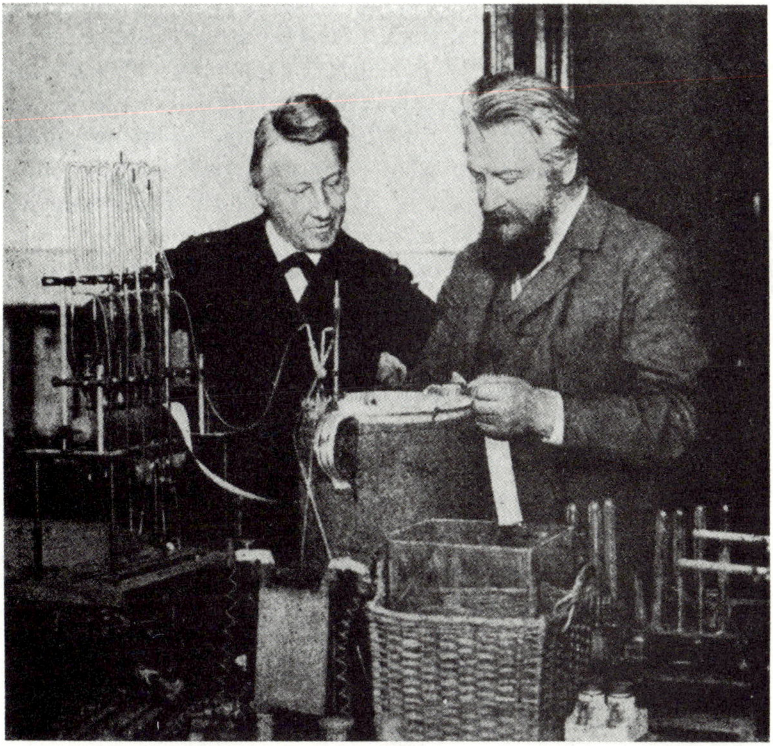

FIG. 101.—J. H. VAN'T HOFF (1852-1911) AND W. OSTWALD (1853-1932).

application of thermodynamics to chemistry are expounded. The Principle of Mobile Equilibrium is enunciated: "Every equilibrium between two systems is displaced by fall of temperature in the direction of that system in the production of which heat is evolved." The importance of the "transition point" in heterogeneous systems is shown. The last section is devoted to the study of affinity as measured by the diminution of available energy, e.g. by electromotive forces. The equation connecting heat of reaction with change of equilibrium constant with temperature is applied. Arrhenius said of this book that the author "succeeds with relatively scanty experimental material in developing an imposing and harmonious scheme for the whole subject of chemical influence and action . . . an enormous perspective has been opened out for future investigation."

Van't Hoff's fundamental work on the theory of dilute solutions was published in 1886 as "L'équilibre chimique dans les systèmes gazeux ou dissous à l'état dilué", in the *Archives néerlandaises* and in the Transactions of the Swedish Academy. The new ideas first became widely known when they were published, together with the fundamental memoir of Arrhenius, in the first volume of Ostwald's *Zeitschrift für physikalische Chemie* in 1887.

In 1896 van't Hoff moved to Berlin as a professor in the University with nominal teaching duties and a research laboratory. He carried out there, in conjunction with Meyerhoffer and others, investigations of the Stassfurt salt deposits, but this work was of a routine character. He died in 1911. "With no great mathematical or experimental attainment," says Walker, "with no striking gift as a teacher, van't Hoff yet influenced and moulded the current thought, and even much of the practice, of chemistry for decades."

Stereochemistry

Van't Hoff's pamphlet on Pasteur's results was published at Utrecht in September 1874 (*Voorstel tot Uitbreiding der . . . structuurformules in de ruimte*); an enlarged French translation (*La chimie dans l'espace*) appeared in 1875. In November 1874 the paper of LeBel appeared in the *Bulletin de la Société Chimique* of Paris. The modes of reasoning of the two authors differ somewhat: van't Hoff seems to have been more influenced by Kekulé than by Pasteur, but the results are practically the same. LeBel's argument is more abstract, not accompanied by figures and less easy to understand, but more general.

According to van't Hoff: "In the case where the four affinities of an atom of carbon are satured by four different univalent groups, two and only two different tetrahedra can be obtained, of which one is the mirror image of the other . . . , two structural formulae of isomers in space." Kekulé in 1867 had described the tetrahedral model of the carbon atom, with a central sphere as the atom and four wires pointing to the corners of a tetrahedron as the valencies. Further, says van't Hoff: " Every carbon compound which in solution impresses a deviation on the plane of polarization possesses an asymmetric carbon atom ", i.e. one in which the four groups attached to the carbon are all different. In this case two tetrahedra, related as object and mirror image, are possible.

An example of this type of isomerism is furnished by the optically active lactic acids studied by Wislicenus, who adopted van't Hoff's ideas:

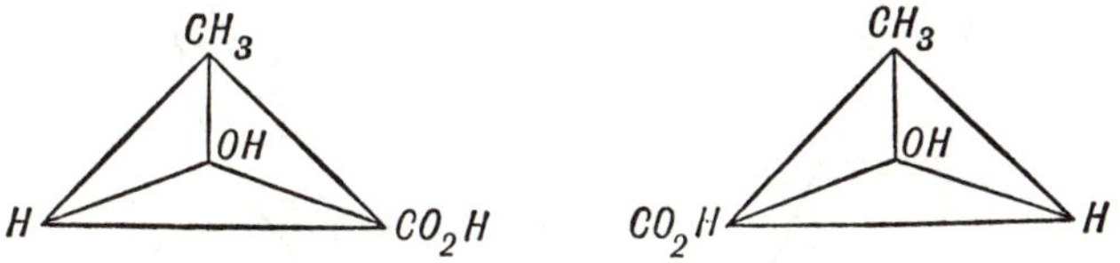

FIG. 102.—STEREOISOMERIC LACTIC ACIDS.

The optically active tartaric acids contain two asymmetric carbon atoms, and a third isomer is also possible, in which the two rotations due to the two asymmetric carbons compensate and the molecule is inactive as a whole. This form, known as mesotartaric acid, was prepared by Pasteur; it differs from racemic acid, which is a compound of *d*- and *l*-tartaric acids, in being *internally* compensated, and is irresolvable:

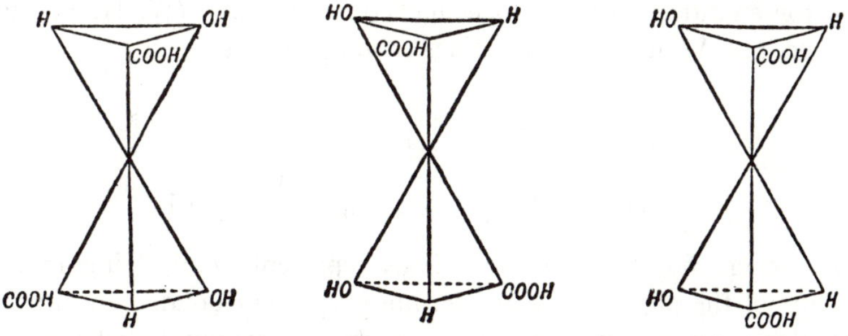

d-tartaric acid *l*-tartaric acid *meso*-tartaric acid

FIG. 103.—STEREOISOMERIC TARTARIC ACIDS.

The work of Wislicenus, Emil Fischer, Baeyer, Wallach, and others, placed the theory of the asymmetric carbon atom on a firm foundation.

Van't Hoff in 1874 was also able to explain the isomerism of fumaric and maleic acids by *geometrical isomerism.* The tetrahedra surrounding the two carbon atoms in the molecules are supposed to be united along an edge, representing a double bond, and incapable of rotation, and the H and COOH groups can then be arranged in two models, not superposable. These are not optical isomers, and differ in physical properties :

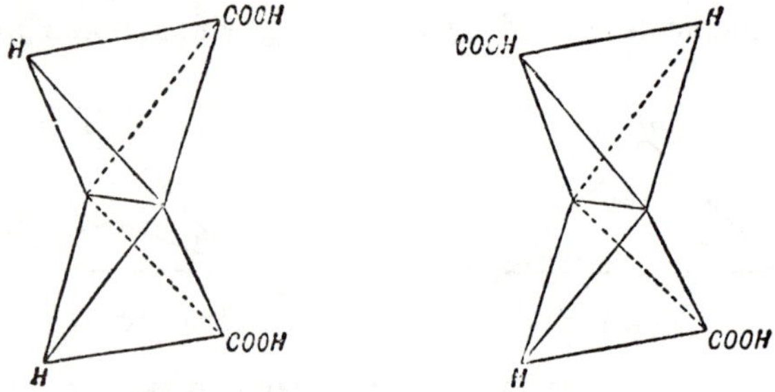

maleic acid · · · · · · · · · · · · · fumaric acid

FIG. 104.—GEOMETRICAL ISOMERISM.

It has been proved that optically active compounds can exist which contain no asymmetric carbon atom, but can form two non-superposable molecular configurations : " optical activity is not due to the presence of an asymmetric carbon atom, but originates in the enantiomorphous molecular configuration " (W. H. Perkin, Pope, and Wallach, 1909). Such a compound is :

$$\overset{9}{CH_3}\diagdown \diagup CH_2{-}CH_2 \diagdown \diagup H$$
$$\qquad C \diagdown \diagup \qquad\quad C{=\!=\!=}C$$
$$H \diagup {}_1 \diagdown CH_2{-}CH_2 \diagup {}_4 \quad {}_7 \diagdown CO_2H$$

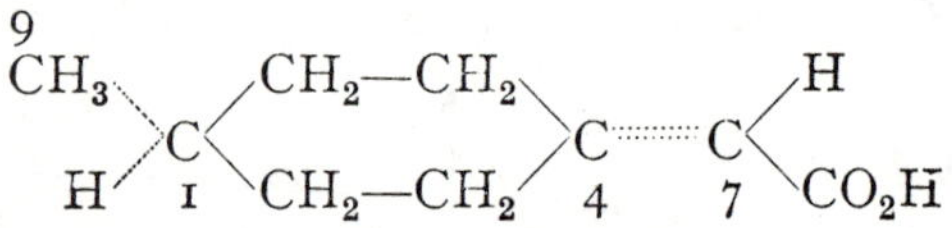

The ring and the continuous lines represent bonds lying in the plane of the paper ; the broken lines represent bonds lying in the vertical plane passing through the carbon atoms numbered 1, 4 and 7. The plane of the paper is not a plane of symmetry, nor is any other plane a plane of symmetry. There are no axes of symmetry nor a centre of symmetry. The substance is therefore asymmetric, although it does not contain an asymmetric carbon atom, and it is optically active.

Optical activity has been established in the case of compounds of selenium, tin, sulphur, chromium, etc., some of the compounds being quite free from carbon.*

SYNTHETIC ORGANIC CHEMISTRY

Perhaps one of the most striking features of chemistry in the second half of the nineteenth century was the extraordinarily rapid growth of synthetic organic chemistry. When the distinction between inorganic and organic compounds was removed, attempts to prepare in the laboratory those compounds which are built up in plant and animal organisms were made in increasing numbers, although the methods used are very different from those employed by the living cells. In addition, numerous drugs and dyestuffs have been prepared which are not found in the storehouse of Nature. It is not possible to follow this branch of chemistry in any detail here, but we may see something of its progress by considering briefly the work of three of the leading organic chemists of last century: Adolf von Baeyer, Emil Fischer and Victor Meyer.

Baeyer

Adolf Baeyer was born in 1835 in Berlin † and in 1853 proceeded to Heidelberg to study under Bunsen. He also worked with Kekulé and in 1858 presented a thesis on cacodyl compounds. In the same year he followed Kekulé to Ghent and worked on derivatives of uric acid. In 1860 he returned to Berlin and taught for twenty years in a modest position in a technical school, where he continued his work on uric acid. In 1872 he was called to Strasburg, where Emil Fischer joined him. Work on the condensation of aromatic hydrocarbons and phenols with aldehydes was carried out. In 1875 Baeyer went to Munich as successor to Liebig and remained there until his resignation in 1915. He died in 1917.

In 1870 Baeyer and Emmerling carried out the first synthesis of indigo, a blue dyestuff the empirical formula of which, $C_{16}H_5NO_2$, had been determined by Dumas in 1840. The synthesis was effected

* See the summary by Walden, *Berichte*, 1925, p. 237; Lowry, *Optical Rotatory Power*, London, 1935.

† Memorial Lecture, Perkin, *J. Chem. Soc.*, 1923, p. 1520; *Nature*, 1935, cxxxvi, 669; Baeyer, *Gesammelte Werke*, 2 vols., 1905.

by heating isatin chloride with phosphorus and hydriodic acid. Between 1878 and 1884 Baeyer carried out those researches on

FIG. 105.—A. VON BAEYER, 1835-1917.

indigo and its derivatives which constitute his best known work. He showed that the formula of indigo is :

$$\begin{array}{cccc} C_6H_4\text{—}CO & & CO\text{—}C_6H_4 \\ | \quad\quad\quad | & & | \quad\quad\quad | \\ NH - C & = & C - NH \end{array} .$$

It is now manufactured in large quantities. The methods of the German dye firms with which Baeyer was associated gave him a distaste for the subject, and he turned his attention to a study of the derivatives of acetylene.   He prepared a number of curious compounds, e.g. tetra-acetylene dicarboxylic acid :

$$CO_2H . C{\equiv}C{-}C{\equiv}C{-}C{\equiv}C{-}C{\equiv}C . CO_2H,$$

some of which are highly explosive.   The unstable nature of such compounds was assumed to be due to the distortion and strain set up in the molecule by the multiple linkages.   " Baeyer's strain theory " has recently been modified in its application to ring compounds by the discovery of non-planar " strainless rings ".

Baeyer showed that benzene can be reduced, by the addition of six atoms of hydrogen, to hexahydrobenzene, a substance no longer possessing any of the properties of aromatic compounds, and he also worked on the terpenes, a series of compounds including oil of turpentine, various essential oils and camphor.

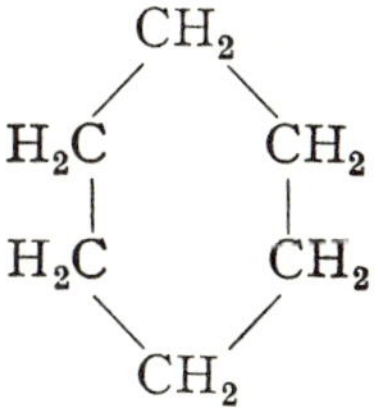

hexahydrobenzene   centric formula
(cyclohexane)   for benzene

The " centric " formula of benzene, in which six valencies forming double bonds in Kekulé's formula (p. 291) are directed to the centre of the ring, was adopted by Baeyer ; it had been given by Lothar Meyer* in 1884.

Baeyer's work with indigo derivatives led him in 1882 to conclude that the same substance may behave towards different reagents as though it possessed two different structural formulae, e.g. in the case of isatin :

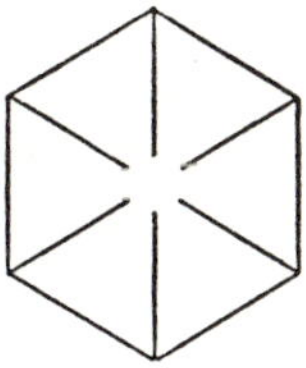

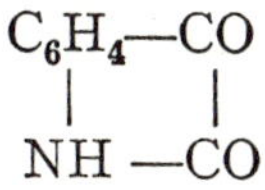

$$\begin{array}{cc} C_6H_4{-}CO & \qquad C_6H_4{-}CO \\ | \qquad | & \qquad | \qquad | \\ NH{-}CO & \qquad N{=\!=}C(OH) \\ \text{lactam-form} & \qquad \text{lactim-form} \end{array}$$

This phenomenon was called *tautomerism*.   It was at first explained

---

* *Die modernen Theorien der Chemie*, 5 edit., 1884, p. 268 ; Armstrong, *J. Chem. Soc.*, 1887, li, 264 ; Baeyer, *Annalen*, 1888, ccxlv, 123.

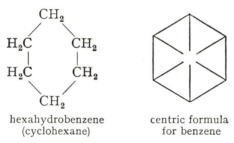

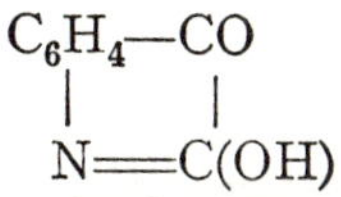

by an oscillation of the linkages in the molecule (Laar, 1885), but afterwards it was suggested that a tautomeric substance is a mixture of two forms in equilibrium, these being very readily converted into one another (dynamic isomerism). This was shown to be the case for acetoacetic ester by the isolation of the two forms by Knorr and by Kurt Meyer in 1911 and 1920:*

$CH_3 . CO . CH_2 . COOEt$	$CH_3 . C(OH) = CH . COOEt$
keto-form	enol-form

Baeyer later extended the work of J. N. Collie (professor of organic chemistry in University College, London) and Tickle, who had shown that oxygen may function apparently as a quadrivalent and base-forming element in the oxonium salts, e.g. of dimethylpyrone, which may be compared with ammonium salts. Baeyer's last research, published in his eightieth year, dealt with the reduction of the pyrones, substances allied with colouring matters of plants and the haemoglobin of blood.

Emil Fischer

Emil Fischer † was born in 1852 at Euskirchen, near Bonn, and became a pupil of Kekulé in 1871. He graduated under Baeyer in Strasburg in 1874, was professor at Erlangen in 1882 and Würzburg in 1885. In 1892 he succeeded Hofmann in Berlin, where he remained until his death in 1919. Fischer represents the culmination of the type of high specialization in synthetic organic chemistry which had been gathering momentum in the second half of the nineteenth century.

The beginning of Fischer's work was his discovery in 1875 of phenyl hydrazine, $C_6H_5 . NH . NH_2$, which in his hands became a general reagent for the carbonyl group, $=CO$, condensing with it to the grouping $:C:N.NHC_6H_5$. In 1884 he found that the sugars reacted in this way, and the presence of carbonyl groups in their molecules was inferred. In 1887 Fischer and Tafel found that from acrolein, $CH_2 : CH . CHO$, an aldehyde, two synthetic

* On tautomerism see Perkin, *J. Chem. Soc.*, 1914, p. 1176; Lowry, *B.A. Report* 1904, p. 193.

† Memorial Lecture, *J. Chem. Soc.*, 1920, p. 1157.

sugars could be produced, one of which is identical with natural fructose. By an application of ingenious reactions, and of methods which had been used by Pasteur, the optically active sugars dextrose, fructose and mannose were obtained. In 1891 the configuration of the sugar molecules was attacked, and gradually the

FIG. 106.—EMIL FISCHER, 1852-1919.

structural formulae of these were elucidated. Fischer's formulae of the sugars have been somewhat modified by later work.

From the sugars Fischer passed to another group of natural substances, the purine derivatives. Among these is caffeine, the stimulating principle of tea and coffee, and uric acid. Some derivatives of uric acid had been investigated by Liebig and Wöhler.

Fischer showed that this group of substances may be derived from a base, *purine* :

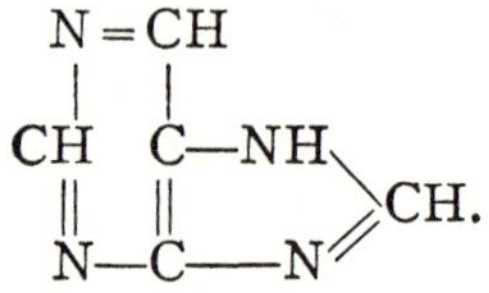

About 130 purine derivatives were investigated by Fischer and his students prior to 1900.

From the purine derivatives Fischer turned to problems of increasing complexity, and in 1899 his researches on amino-acids began. The simplest amino-acid is amino-acetic acid, glycocoll or glycine: $CH_2.NH_2.COOH$. This on condensation yields glycylglycine: $H_2N.CH_2 . CO . NH . CH_2 . CO_2H$. Once started on this road Fischer proceeded to substances of increasing complexity, until in 1907 he had arrived at the substance:

$$H_2N.CH(C_4H_9).CO(NH.CH_2.CO)_3.NH.CH(C_4H_9).CO.(NH.CH_2.CO)_3.NH.CH(C_4H_9).CO(NH.CH_2.CO)_8.NH.CH_2.CO_2H.$$

This complicated organic amino-acid resembles in some of its properties natural albumins; it is called a polypeptide and has 816 possible isomerides. There is now very little doubt that the natural proteins are related to amino-acids and polypeptides by containing the grouping —CO . NH—.

Victor Meyer

Victor Meyer * was born in 1848 in Berlin, studied under Bunsen in Heidelberg from 1865 and became his assistant; in 1868 he went to Berlin to work with Baeyer on organic chemistry. After a brief period at Würtemburg he became professor at the Zurich Polytechnic in 1872. He went to Göttingen in 1885 and Heidelberg in 1889. He was a brilliant teacher as well as an indefatigable worker and had much wider interests than Baeyer or Emil Fischer. He died in 1897.

In 1874 Victor Meyer investigated the nitro-compounds of the aliphatic series, a subject which occupied him and his students for

* Thorpe, Memorial Lecture, *J. Chem. Soc.*, 1900, p. 169.

twenty years. It was found that nitro-ethane, a true nitro-substitution product of a hydrocarbon, and isomeric with an ester of nitrous acid :

$$CH_3 . CH_2 . NO_2 \qquad CH_3 . CH_2 . O . NO$$
nitroethane   ethyl nitrite

behaves as an acid towards sodium hydroxide, a tautomeric form called a nitronic acid being formed :

$$CH_3 . CH_2 . NO_2 \rightleftharpoons CH_3 . CH : NO(OH)$$

which gives a sodium salt $CH_3 . CH : NO(ONa)$.

In 1877 the well-known vapour density method was described, and this was modified in later work so that determinations at high temperatures could be carried out. It was shown that at high temperatures bromine and iodine molecules are dissociated into atoms.

In 1882 the action of hydroxylamine on aldehydes and ketones was examined, and an important series of compounds, the oximes, discovered. In these the carbonyl group of the aldehyde or ketone condenses with the hydroxylamine :

$$: C \, O + H_2 \, N . OH = : C : N . OH + H_2O.$$

The isomerism of the oximes has given rise to much interesting work.

In 1882, in the course of a lecture, Victor Meyer found that a specimen of benzene did not give the indophenine reaction described by Baeyer, i.e. the formation of a blue colour with concentrated sulphuric acid and isatin. It was found that the benzene put out for the lecture had been prepared from benzoic acid and was very pure ; ordinary coal-tar benzene, which gave the reaction, did so because it contained a new substance, *thiophene*, a sulphur compound containing a five-membered ring, which was discovered in 1883 by Victor Meyer. It is analogous to furan (a derivative of which, furfurol, was discovered by Döbereiner in 1831 and Fownes in 1845) and pyrrol (recognized but not analyzed by Runge, 1834 ; formula by Baeyer, 1866) :

$\begin{array}{l} CH{=}CH \\ | \qquad\quad \rangle O \\ CH{=}CH \end{array}$ $\begin{array}{l} CH{=}CH \\ | \qquad\quad \rangle S \\ CH{=}CH \end{array}$ $\begin{array}{l} CH{=}CH \\ | \qquad\quad \rangle NH \\ CH{=}CH \end{array}$

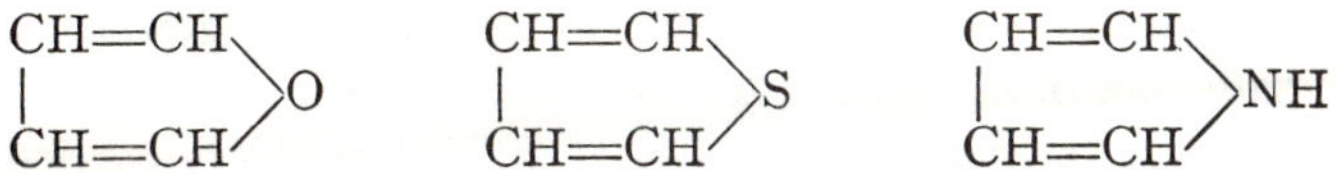

furan   thiophene   pyrrol

In 1891 Victor Meyer carried out a series of investigations on the ignition temperatures of explosive gas mixtures in which the very marked effect of the walls of the containing vessel came to light. In 1893 the dissociation of hydriodic acid was studied, and shown to be a reversible reaction.

In 1892-94 Victor Meyer and his pupils investigated the organic iodine compounds : iodosobenzoic acid, iodoxybenzoic acid, iodoso-

FIG. 107.—VICTOR MEYER, 1848-1897.

benzene, iodoxybenzene, and compounds derived from the strong base diphenyl iodonium hydroxide. These substances, in which iodine functions as a base-forming element like nitrogen in ammonium salts, are of considerable theoretical interest.

From 1894 to 1897 Victor Meyer and his students carried out investigations on what is known as steric hindrance. Whereas benzoic acid, COOH, and most of its substitution products, readily yield esters with alcohol and hydrogen chloride in the cold, trisubstituted benzoic acids, e.g.

$$CH_3 \overset{CH_3}{\underset{CH_3}{\hexagon}} COOH$$

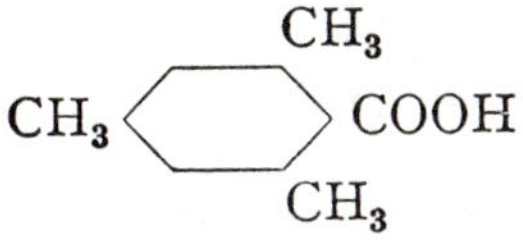

do not, unless the carboxyl group, COOH, is prolonged well beyond the ring by the interposition of a chain of carbon atoms, e.g. in

$$Br \overset{Br}{\underset{Br}{\hexagon}} —CH_2 . CH_2 . COOH.$$

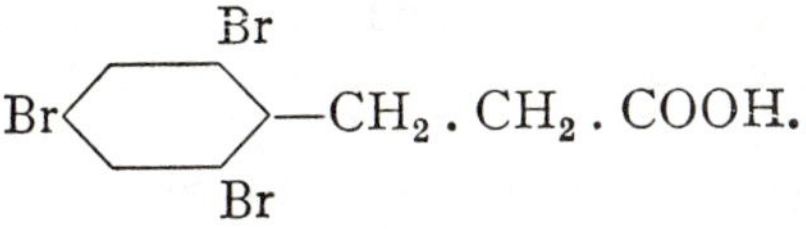

It was thought that the ortho-substituents are so near the carboxyl group in the former compound as to exert a screening effect, which is not observed when the group is extended by a chain of atoms. It is now known that this explanation is only valid in some cases and the " ortho-effect " is related to the electrochemical (" polar ") properties of the substituent groups, a fact which would have pleased Berzelius.

The later developments of synthetic organic chemistry led to the recognition of many new classes of compounds, and to special methods for introducing substituents and for analysis.

SUMMARY AND SUPPLEMENT

The Development of Organic Chemistry

Louis Pasteur (1822-95), professor in Strasburg and Paris, and from 1889 director of the Pasteur Institute, investigated optical activity, observing hemihedral facets on crystals of tartaric acid and tartrates, but not on racemates, and resolving an optically inactive compound, sodium ammonium racemate, by crystallization (1848-50). He also resolved racemic or optically inactive substances by fractional crystallization with an optically active acid or base (1853) or by the growth of a mould which preferentially uses up one form (1860). He discovered laevotartaric acid and mesotartaric acid. Pasteur provided experimental proofs for the vitalistic theory of fermentation (1857), and later carried out fundamental researches in bacteriology, disproving " spontaneous generation ".

The relation of optical activity to the presence of an asymmetric carbon atom was given independently in 1874 by van't Hoff and by

Le Bel. Researches on optical activity were carried out by Hans Landolt (1831-1901), professor in Berlin, who also investigated organic arsenic and antimony compounds (1853), and tested an empirical formula for the refractive power of substances proposed in 1858 by John Hall Gladstone (1827-1902) and J. Dale. Molecular refractions, from the point of view of a theoretical formula proposed independently in 1880 by L. Lorenz and by H. Lorentz, were studied from 1880 by J. W. Brühl (1850-1911), with special reference to the influence of structure.

Emil Erlenmeyer, senr. (1825-1909), professor in Frankfort, proposed a valency theory (1863) and a structure notation (1867), discovered isobutyric acid (1865), the constitution of naphthalene (1866), and of hydracrylic and lactic acids (1867 f.), elucidated the structure of the lactones (1880), and synthesized tyrosine (1883).

Alexander Michailowitsch Butlerow (1828-86), professor in St. Petersburg (Leningrad), prepared synthetic hexoses (" methylenitan ") (1861), and tertiary alcohols from zinc alkyls and acid chlorides (1864) ; he investigated isomeric dibutylenes, so recognizing the existence of " dynamic isomerism " (1877).

Johann Peter Griess (1829-88), chemist to Allsopp and Sons, Burton-on-Trent, discovered the aromatic diazo-compounds (1858 ; published in 1860). The aliphatic diazo-compounds were discovered in 1894 by Hans von Pechmann (1850-1904), professor in Tübingen, by the action of alkalis on nitrosamines.

Carl Schorlemmer (1834-1892), assistant (1859) and then professor (1874) of organic chemistry in Manchester (the first chair of its kind in England), worked on the paraffin hydrocarbons (1861 f.) and showed that the supposed free methyl, CH_3, and ethyl hydride, $C_2H_5.H$, regarded as isomeric, were identical and had the formula C_2H_6. He discovered a general method for the conversion of a secondary into a primary alcohol (1866), studied the boiling points of the paraffins (1872), and the constitutions of aurin (1871) and suberone (1874). He collaborated with Roscoe in the authorship of a large *Treatise on Chemistry* (1877 f.), the organic section of which was unfinished in the English edition, but was completed by Brühl in the German translation.

Rudolf Fittig (1835-1910), professor in Tübingen (1869) and Strasburg (1876), discovered the pinacone reaction (1859), and diphenyl (1862), synthesized aromatic hydrocarbons by the action of sodium on a mixture of alkyl and aryl halides (" Fittig's reaction ") (with Tollens, 1864), synthesized mesitylene (1867), discovered isophthalic acid (1867), investigated complex salts (e.g. $K_4Mn(CN)_6$ and $K_3Mn(CN)_6$) (1868), worked on piperine (1869 f.), discovered phenanthrene (with Ostermeyer, 1872; Glaser and Graebe almost simultaneously obtained it), put forward the diketone formula for quinone (1873-6), worked on unsaturated carboxylic acids (1877-1904), discovered coumarone (1883), synthesized α-naphthol (1885), investigated ketonic esters (1887 f.), discovered diacetyl (1887), and demonstrated the genetic relationship of the γ-hydroxy-acids and lactones (1881-1904).

William Henry Perkin, senr. (1838-1907), produced the first "aniline" dye, mauve (1856), synthesized glycocoll (Perkin and Duppa, 1858), tartaric acid (Perkin and Duppa, 1861), coumarin (1868), and cinnamic acid (1877: the "Perkin reaction"), investigated the constitution of saligenin (1868) and the relation between magnetic rotatory power and chemical constitution (1884-1907).

Johann Wislicenus (1835-1902), professor in Würzburg and Leipzig, investigated the condensation of aldehyde ammonia (1857-8), and the lactic acids (1863-75). He adopted and extended van't Hoff's stereochemical theories (from 1875). From 1874 Wislicenus and his students (Conrad, Limpach) studied the use of acetoacetic ester (discovered by Geuther in 1863, and independently by Frankland and Duppa in 1865) in organic syntheses, e.g. synthesizing glutaric acid (1878), methyl β-butyl ketone (1883), etc. Conrad and Limpach (1874-7) worked out the details of the method of synthesis depending on the interaction of the sodium derivative of acetoacetic ester and an alkyl iodide. Wislicenus discovered vinyl ether (1878) and vinyl acetic acid (1899). His discovery of the simple synthesis of hydrazoic acid (1892) and his papers on the geometrical isomerism of unsaturated compounds (1887-9) are important.

Adolf Baeyer (1835-1917), professor in Berlin and Munich, investigated purine derivatives (1861-70), discovered indole and the constitution and synthesis of indigo (1866-85); he put forward the strain theory of carbon rings (1885), investigated the reduction products of benzene and its derivatives (1888-93), adopted the "centric formula" for benzene (1888), and worked on the terpenes (1893-99). The chemistry of the terpenes was also studied by Albin Haller (1849-1925) in Paris, Otto Wallach (1847-1931) in Göttingen, William Henry Perkin, junr., and Gustav Komppa (b. 1867) in Helsingfors.

Wilhelm Körner (1839-1925), professor in Milan, put forward the formula for pyridine (1869), devised a method for determining the orientation of substitution products of benzene (1874), and synthesized asparagine (with Menozzi, 1887).

Albert Ladenburg (1842-1911), professor in Breslau, put forward the prism formula for benzene (1869), investigated the aliphatic (1867-72) and aromatic (1873-4) compounds of silicon (partly in conjunction with Charles Friedel, 1832-99, professor in Paris), synthesized optically active coniine (1886), and worked on the constitution of other alkaloids (from 1879).

Jacob Volhard (1834-1910), professor in Halle, synthesized sarcosine (1862) and creatine (1868), invented the thiocyanate silver titration (1874), and synthesized thiophene (1885).

Carl Graebe (1841-1927), professor in Geneva, showed that alizarin is reduced to anthracene by distilling with zinc dust (with Liebermann, 1868), synthesized alizarin from anthraquinone (with Liebermann, 1869), determined the constitution of anthracene (with Liebermann, 1869), discovered acridin (with Caro, 1870), isolated and synthesized carbazole (with Glaser, 1872), synthesized quinoline derivatives (1878) and acenaphthenes (1893). He also investigated dyestuffs.

Carl Liebermann (1842-1914), professor in Berlin, investigated the action of nitrous acid on phenols and secondary amines ("Liebermann's reaction") (1874), the constitution of anthraquinone and alizarin (with Graebe, 1868-82), of anthracene and phenanthrene (with Graebe, 1869-70), of quercitrin and rhamnetin (1879), and discovered the so-called Liebermann dyes, which belong in part to the indophenols.

Victor Meyer (1848-97), professor in Zurich and Heidelberg, discovered the aliphatic nitro-compounds (1872), nitroso-compounds, isonitroso-compounds and nitrols (1875-82), thiophene (1882), the oximes and their isomerism (1883-90), iodoso-, iodoxy- and iodonium-compounds (1892-7) and steric hindrance (1894-7). In physical chemistry he invented the displacement method of determining vapour densities (1878-80), studied reaction velocities and equilibria, and carried out researches at high temperatures.

Emil Fischer (1852-1919), professor in Berlin, investigated the constitution of rosaniline (1878), discovered phenyl hydrazine (1875) and applied it as a reagent in extensive researches on the constitution, synthesis and configuration of the sugars (1884-1900); investigated and synthesized several members of the purine group (1882-1901); synthesized polypeptides (1900-06); investigated the Walden inversion (1908-12), and worked on the constitution and synthesis of tanning substances (1912-19). The chemistry of polypeptides and proteins was studied by Emil Abderhalden (b. 1877).

Theodor Curtius (1857-1928), professor in Heidelberg, discovered diazoacetic ester (1883), hydrazine (1887), hydrazoic acid and its derivatives (1890 f.), pyrazoline derivatives (1891), benzaldazine and aldazines (1900), and so-called tetrazine derivatives (1906). He worked on glycollic esters (1883) and on polypeptides (1904).

Ernst Beckmann (1853-1923), professor in Erlangen and Leipzig, improved the freezing point (1888) and boiling point (1889) methods for the determination of molecular weights in solution, and invented useful apparatus; in organic chemistry he discovered the "Beckmann rearrangement" of oximes (1886 f.), worked on the production of thymol from menthol (1896), and examined the possible technical uses of furfurol (1919); in inorganic chemistry he investigated the higher chlorides of sulphur (1909 f.).

Ludwig Claisen (1851-1930), professor in Aix-la-Chapelle and Kiel, prepared aromatic ketonic esters (1881), benzalacetone (1881) and benzoylacetic ester (1887). He explained the synthesis of aceto-acetic ester as due to the intermediate formation of an ortho-compound (1887-8, 1897); prepared β-diketones by the condensation of ketones and acid esters in presence of sodium, sodium ethoxide or sodamide (1887 f.), the so-called Claisen condensation of carbonyl compounds with the group —CH_2.CO—; synthesized pyrazole (after Knorr, 1885, and Emil Fischer, 1885) and dipyrazole derivatives from 1, 3-diketones and oxalyldiacetone with phenyl hydrazine (1894); and investigated cases of isomerism involving the transformation of R.CO.CH_2.CO— into R.C(OH):CH.CO—, and tautomerism of such *keto-* and *enol* forms, respectively, as exhibited for example by acetyldibenzoylmethane (1892 f.), so recognizing the influence of negative groups.

Eugen Bamberger (1857-1932), professor in Zurich, investigated guanidine derivatives (guanamines, etc.) (1880), discovered retene (1885) and pyrene (1887), introduced the use of sodium and amyl alcohol as a reducing agent (1887), investigated the reduction products of naphthylamines (1888 f.), cyanuric acid (1890), cyanamines (1892), the reduction of nitrobenzene to nitrosobenzene and phenyl hydroxylamine (1894 f.), benzazimide (1899), dimethylaniline oxide (1899), and the isomerism of nitroso-compounds (1903). He engaged in a long controversy (1894-1900) with Hantzsch on the structure of the diazo-compounds.

Arthur Hantzsch (1857-1935), professor in Leipzig, worked on the synthesis of pyridine from acetoacetic ester and aldehyde ammonia (1882), coumarone (1886), thiazole (1889), the stereochemistry of nitrogen and the structure of the oximes (with Werner, 1890) and diazo-compounds (1894 f.), isolated hyponitrous acid (1896), investigated the tautomeric behaviour of phenyl nitromethane and nitrophenols as pseudo-acids (1899; 1905 f.), the structure of cyanuric acid and cyamelide (1905 f.), applied physico-chemical methods (e.g. electrical conductivity) to organic chemistry, and advanced theories of indicators (1907) and of the structure of acids (1917-25).

Ludwig Knorr (1859-1921), professor in Jena, investigated pyrazolone and isopyrazolone and their derivatives (1883-1911), discovering the important drug antipyrine (phenyl dimethyl pyrazolone) (1887). His work on compounds exhibiting tautomerism, such as diacetyl succinic ester (1896 f.), acetoacetic ester (1904), and acetyl acetone (with H. Fischer, 1911), led him to regard a tautomeric substance as a mixture of two forms in mobile equilibrium (" allelotropic mixture "), and he was able to furnish experimental proofs for this view. He devised a colorimetric method for the determination of the proportion of the enol-form in a tautomeric mixture of this with the keto-form (with H. Schubert, 1911). He devised a method for preparing acetonylacetone (1900), discovered aminoethyl ether (with G. Meyer, 1905), and worked on piperazine derivatives and alkaloids (1904).

William Henry Perkin, junr. (1860-1929), professor at Heriot Watt College, Edinburgh (1887-92), Manchester (1892-1913), and Oxford (1913-29), synthesized polymethylene rings (1883-94), ethyl benzoylacetate and derivatives (1884-6), anthraquinone by heating orthobenzoylacetic acid with sulphuric acid (1891), indene and hydrindene and derivatives (1894 f.), di- and tri-methyl glutaric acids and derivatives (1895 f.), camphor and its derivatives (1897 f.; synthesis of camphoronic acid with J. F. Thorpe, 1897), alkaloids (1890-1929: berberine, narcotine, cotarnine, cryptopine, protopine, strychnine); terpenes (1904-21; synthesis of terpineol, 1904), brazilin and haematoxylin (1901-28), harmine and harmaline (1912-27), and isoquinoline derivatives (1925-6). His work on polymethylene rings led to Baeyer's strain theory (1885, p. 309).

Johannes Thiele (1865-1918), professor in Munich and Strasburg, discovered semicarbazide (1894), investigated tetrazole derivatives (1895) and nitramide (1895), discovered derivatives of fulvene (a coloured hydrocarbon) (1900), investigated guanidine derivatives (1892 f.) and unsaturated lactones (1899-1902) and put forward (1899) a theory of " partial valencies " to explain addition reactions to double bonds and the constitution of benzene.

Victor Grignard (1871-1935), professor in Nancy and Lyons, developed the so-called Grignard reaction (1900)—in part discovered by his teacher P. A. Barbier—which replaced the use of metal alkyls in synthetic reactions by magnesium alkyl hadides (MgRX) in solution in ether.

Richard Willstätter (1872-1942), professor in Munich and Berlin, investigated alkaloids and their derivatives (tropic acid, tropine, atropine, ecgonine and cocaine) (1896-1903), the synthesis of betaïne (1902), lecithin (1904), ortho-quinones, quinone-imines, and pyrones

(1904 f.), chlorophyll (1906 f.), the colouring matters of flowers (anthocyanin, etc.), of blood, etc. (1913 f.), the assimilation of carbon dioxide by plants (1917 f.) and enzymes (1921 f.).

Paul Sabatier (1854-1941) obtained pure hydrogen disulphide (1886) and introduced the method of catalytic hydrogenation (from 1897).

Henry Edward Armstrong (1848-1937), pupil of Hofmann and Frankland in London and Kolbe in Leipzig (1867), professor at the London Institution, Finsbury Circus (1871), and the City and Guilds Institute (1879-1911) ; researches on naphthalene derivatives (1881-1900), terpenes and camphor (1878-1902), and quinonoid theory of colour of dyestuffs (1888, 1892); wrote on inorganic chemistry, on organic chemistry, and on the teaching of science. He was a keen critic of the theory of electrolytic dissociation.

William Jackson Pope (1870-1939), professor in the Manchester School of Technology (1901) and at Cambridge (1908), worked mostly in the field of stereochemistry, particularly optically active compounds (see p. 306), including elements other than carbon; introduced camphorsulphonic acids in the resolution of optically active bases, and hydroxyhydrindamine (with Read) in the resolution of acids; with C. S. Gibson prepared dichlorodiethyl sulphide (" mustard gas ") by the action of ethylene on sulphur chloride (1920).

Gilbert Thomas Morgan (1870-1940), assistant at the Royal College of Science, professor in Dublin (1912) and Birmingham (1919), and head of the Chemical Research Laboratory, Teddington (1927) ; investigations related to dyestuffs ; organic compounds of arsenic, antimony, selenium and tellurium ; coordination compounds ; catalytic hydrogenation ; phenol-formaldehyde condensation products.

Arthur Lapworth (1872-1941), head of the chemistry department at Goldsmiths' Institute, London (1900), lecturer in inorganic and physical chemistry (1909), professor of organic (1913) and physical and inorganic chemistry (1922), in Manchester ; a pioneer of organic chemistry based on the electronic theory of valency ; cyanohydrin formation (1907 f.) ; mechanism of esterification (1908 f.) ; enol-keto change in the bromination of acetone (1913) ; basic properties of water and alcohol and a general theory of acids and bases (1908), later developed by Lowry and Brønsted.

Jocelyn Field Thorpe (1872-1940) at first studied engineering, but later chemistry in London and Heidelberg ; lecturer in Manchester (1897), and (1913) professor at Imperial College, London ; camphor and terpenes (Perkin and Thorpe, synthesis of camphoronic acid, 1897, and camphoric acid, 1903, independently of Komppa, 1903) ; imino-compounds and hydrindene derivatives (1904 f.) ; glutaconic acids (1923 f.) ; tautomerism and bridged rings (1911 f.).

CHAPTER XIV

THE HISTORY OF PHYSICAL CHEMISTRY

Affinity and Mass Action

One of the branches of physical chemistry first studied was the theory of affinity. The name (*affinitas*) is used in the sense of

FIG. 108.—GEOFFROY'S AFFINITY TABLE.

chemical relation by Albertus Magnus (*c.* 1250). Boyle, Mayow, Glauber, Newton and Stahl put forward ideas on elective affinity.

In 1718 E. F. Geoffroy arranged substances, including acids and bases, in a table of affinities. At the head of a column is a substance with which all the substances below can combine. The latter are so placed that any substance replaces all others lower in the column from their compounds with that at the head of the table. This method was extended by Bergman (1775), although it was necessary, as pointed out by Baumé in 1773, to have two sets of tables, one for reactions in solution at the ordinary temperature and another for reactions of fusion. Bergman also pointed out that in some cases it is necessary to use an excess of a reagent in order to bring a reaction to completion.

Boyle in 1674, in speaking of the action of the atmosphere on bodies, says *: "I have long thought, that, in divers cases, the quantity of a menstruum may much more considerably compensate its want of strength, than chemists are commonly aware of." Wenzel in 1777 had an idea that the concentration with which substances are present in solution has an influence on the affinities, and he showed that the rate of solution of metals in acids is proportional to the concentration of the acid. The "effect of mass" was, however, first clearly pointed out by Berthollet.† Increasing concentration causes a reaction to proceed further; reactions are not usually complete, but a state of equilibrium is set up in which the products of reaction tend to pass back into the initial substances: "affinities do not act as absolute forces by means of which a substance is displaced from its compound, but rather in all combinations and decompositions . . . the substance on which two other substances act with opposing forces is divided between them, and the ratio of partition depends not only on the inner strength of the affinity but also on the quantities of the acting bodies present, so that to produce an equal degree of saturation the mass can make good what is wanting in the strength of the affinity."

Berthollet's views led him to the incorrect theory of combination in variable proportions, which was disproved by Proust, and what was true in his theory was not put to much service. Gay-Lussac supported the law of action of mass, Heinrich Rose in 1842 made use of it to explain certain reactions, and Berzelius pointed out that the criticism that the action of mass was incompatible with the law of constant proportions was based on a misunderstanding.

* *Suspicions about the Hidden Qualities of the Air*; *Works*, ed. Birch, 1744, iii, 464.

† *Recherches sur les lois de l'affinité*, 1801; *Essai de statique chimique*, 1803; Ostwald's *Klassiker* No. 74.

Most textbooks at least mentioned the theory, and some favourably. Thus Henry *, after giving five objections to the theory, remarks that: " Notwithstanding these objections to the theory of Berthollet, when carried so far as has been done by its author, in the explanation of chemical phenomena, it must still be admitted that the extraneous forces, pointed out by that acute philosopher, have great influence in modifying the effects of chemical affinity." Berthollet had assumed that the production of volatile or insoluble products could modify considerably the course of reactions.

In 1853 Bunsen † exploded mixtures of carbon monoxide and hydrogen with quantities of oxygen insufficient to burn both gases, and obtained the curious result that the proportion of each gas uniting with oxygen did not vary in a uniform manner with the alteration in the ratio of the combustible gases, but changed *per saltum* in whole multiples at certain compositions. This was, of course, incompatible with Berthollet's theory. The results were shown to be erroneous by Horstmann ‡ (1877) and especially by Dixon § (1884): if the condensation of steam on the walls of the eudiometer is prevented, the partition is continuous and in agreement with the law of mass action.

In 1850 Ludwig Wilhelmy || investigated for the first time the rate of progress of a chemical reaction, the inversion of cane sugar by water in presence of acid. This is particularly suited to experimental investigation because the progress of the reaction can be followed by the polarimeter, without disturbing the conditions of the reacting system. Wilhelmy found that, in the presence of a large and practically constant mass of water, the amount of sugar changed in a small interval of time, dt, is proportional to the amount, M, actually present:

$$-dM/dt = kM, \text{ where } k \text{ is a constant, } \therefore -\log M = kt + \text{const.}$$

The dynamical character of a chemical equilibrium, viz. that it is the result of two opposite changes taking place at equal rates, was pointed out by Williamson in 1850.

* *Elements of Experimental Chemistry*, 9th ed., vol. i, p. 64 (1823).

† *Annalen*, 1853, lxxxv, 137.

‡ *Annalen*, 1878, cxc, 228.

§ *Phil. Trans.*, 1884, clxxv, 617.

|| *Ann. Physik*, 1850, lxxxi, 413, 499; Ostwald's *Klassiker* No. 29.

In 1855 J. H. Gladstone * investigated the reaction:

$$Fe(NO_3)_3 + 3KCNS = Fe(CNS)_3 + 3KNO_3,$$

the extent of which can be determined by the depth of colour of the solution due to the $Fe(CNS)_3$ produced. A state of equilibrium is set up: "a change in the mass of one of the binary compounds brings about a change in the amount of each of the other binary compounds, and that in a regularly progressive ratio."

A very important investigation is that of Berthelot and Péan de Sainte-Gilles † (1862-3), who studied the reaction:

$$\text{alcohol} + \text{acid} \rightleftharpoons \text{ester} + \text{water}.$$

This is never complete, but slowly approaches a limit corresponding with equilibrium. The same equilibrium is attained whether alcohol and acid, or ester and water, are mixed in the first instance. "The amount of ester formed in each moment is proportional to the product of the reacting substances and inversely proportional to the volume." Although they succeeded in deducing mathematical expressions for the velocity of reaction, Berthelot and Sainte-Gilles did not arrive at the law of mass action as later formulated by Guldberg and Waage, because they neglected to take into account the velocity of the reverse reaction between the ester and water, although they knew that this occurred.

In 1866-7 Harcourt and Esson, ‡ in Oxford, attempted to obtain information on the course of chemical change by studying the reaction: $2HI + H_2O_2 = 2H_2O + I_2$, in presence of dilute sulphuric acid.

Harcourt and Esson also investigated the reduction of potassium permanganate by oxalic acid in presence of dilute sulphuric acid and manganous sulphate. This is a consecutive reaction, in which a substance A is converted into an intermediate substance M which then forms a final substance B. They set up and solved the differential equations for the reactions studied.

The law of mass action was first announced with full generality by Guldberg and Waage, professors of applied mathematics and of chemistry, respectively, in the University of Christiania (now

* *Phil. Trans.*, 1855, cxlv, 179.

† *Annales de chimie*, 1862, lxv, 385; lxvi, 5; 1863, lxviii, 225.

‡ *Phil. Trans.*, 1866, clvi, 193; 1867, clvii, 117.

Oslo). In 1864 their first paper appeared in Norwegian; the full memoir was published in French in 1867.* They point out that, as in mechanics, " we must study the chemical reactions in which the forces which produce new compounds are held in equilibrium by other forces . . . where the reaction is not complete, but partial ".

FIG. 109.—C. M. GULDBERG, 1836-1902, AND P. WAAGE, 1833-1900.

They defined " active mass " as the number of molecules in unit volume. When " two substances A and B are transformed by double substitution into two new substances A' and B', and under the same conditions A' and B' can transform themselves into A and B . . . the force which causes the formation of A' and B' increases proportionally to the affinity coefficients of the reaction

* *Études sur les affinités chimiques*, Christiania, 1867; Ostwald's *Klassiker* No. 104.

$A + B = A' + B'$, but depends also on the masses of A and B. We have learned from our experiments that the force is proportional to the product of the active masses of the two substances A and B. If we designate the active masses of A and B by p and q, and the affinity coefficient by k, the force $= k.p.q$."

This, however, is not the only force acting and here the work of Berthelot and St. Gilles is completed. "Let the active masses of A' and B' be p' and q', and the affinity coefficient of the reaction $A' + B' = A + B$ be k'. The force causing re-formation of A and B will be $k'.p'.q'$. This force is in equilibrium with the first force, and consequently $kpq = k'p'q'$. By determining experimentally the active masses p, q, p' and q' we can find the ratio between the coefficients k and k'. On the other hand, if we have found this ratio k/k' we can calculate the result of the reaction for any original condition of the four substances."

Van't Hoff (1877) replaced the indefinite "force" by velocity of reaction. The velocity with which A and B react to form A' and B' will be $v = kpq$, and the velocity with which A' and B' react to form A and B will be $v' = k'p'q'$. The actual velocity of the complete reaction will be $V = v - v' = kpq - k'p'q'$. In equilibrium, $V = 0$, and $p'q'/pq = k/k'$.

Thermochemistry

The study of affinity from another aspect was represented in the attempts to find a measure of the chemical forces by the amount of heat given out in a chemical reaction. The importance of thermal phenomena in chemical reactions was clearly realized by Lavoisier and Laplace, who laid the foundations of thermochemistry.* They showed that the heat evolved in a reaction is equal to the heat absorbed in the reverse reaction. They investigated the specific and latent heats of a number of substances, and amounts of heat evolved in combustion. In 1840 Hess enunciated the law that the evolution of heat in a reaction is the same whether the process is accomplished in one step or in a number of stages. Thus, the heat of formation of CO_2 is the sum of the heat of the formation of CO and the heat of oxidation of CO to CO_2. Only the amounts of

* *Mémoires de l'Académie*, 1780 [1783]; *Œuvres de Lavoisier*, tome ii, p. 287.

heat given out in the first and third of these changes can be measured experimentally, but Hess's law enables us to calculate that in the second stage.*

With the advent of the mechanical theory of heat, Hess's law was seen to be a consequence of the law of conservation of energy. The study of thermochemistry was taken up by Andrews † and by Favre and Silbermann ‡, but particularly by Berthelot in Paris (see p. 284) and by Julius Thomsen in Copenhagen. § Both considered the heat evolved in the formation of a compound as a measure of the affinity, or the work done by the chemical forces, but a consideration of the second law of thermodynamics, enunciated by Carnot in 1824, led Helmholtz in 1882 to a more correct estimate of the work done by the chemical forces. This is not the heat evolved in the reaction but the largest quantity of work which can be gained when the reaction is carried out in a reversible manner, e.g. electrical work in a reversible cell. This maximum work is regarded as the diminution of the free, or available, energy of the system, whilst the heat evolved is usually a measure of the diminution of the total energy of the system. ||

The application of the second law of thermodynamics to chemistry, which was begun by Horstmann ¶ in 1873, is important in the study of dissociation phenomena. Although isolated examples were known of chemical changes brought about by heat which were reversed on cooling, it was mainly the work of Deville which focused attention on this important type of chemical reaction. Deville ** was able by means of ingenious apparatus to study the dissociation of steam, carbon dioxide, sulphur dioxide, hydrogen chloride and carbon monoxide at high temperatures. Earlier experiments †† of

* Hess, *Ann. Physik*, 1840, l, 385 ; Ostwald's *Klassiker* No. 9.

† *Scientific Papers*, London, 1889.

‡ *Recherches sur les quantités de chaleur dégagées dans les actions chimiques et moléculaires*, Paris, 1853.

§ *Thermochemische Untersuchungen*, 4 vols., Leipzig, 1882-6; on Thomsen, see *J. Chem. Soc.*, 1910, p. 161.

|| Ostwald's *Klassiker* No. 124.

¶ Ostwald's *Klassiker* No. 137.

** *Leçons sur la dissociation*, 1866; the work was begun in 1857.

†† Pebal, *Annalen*, 1862, cxxiii, 199; Than, *ibid.*, 1864, cxxxi, 138.

Pebal (1862) and of Than (1864) on the dissociation of sal ammoniac had removed a supposed exception to Avogadro's principle, since the vapour density of this substance was only half that corresponding with the molecular weight. It was shown that the vapour is really a mixture of ammonia and hydrogen chloride, occupying double the volume of the undissociated substance: $NH_4Cl = NH_3 + HCl$. Cannizzaro in 1857 and Kopp and Kekulé in 1858 pointed out that all the so-called exceptions were cases where dissociation occurred.

Deville and Debray (1867-8) showed that in many cases where a solid is dissociated, e.g. $CaCO_3 = CaO + CO_2$, the dissociation pressure depends only on the temperature and not on the quantities of the substances present. This was regarded as an exception to the law of mass action, but the case where the system is not homogeneous requires special treatment, and the exception is only apparent. Horstmann pointed out that this case, and some others studied by him, are exactly similar to physical changes of state, and that the thermodynamic equation derived by Clapeyron and Clausius for changes of state applies here also, viz. $dp/dT = Q/T(v' - v)$, where p is the dissociation pressure, Q the heat of dissociation, T the absolute temperature, and v' and v the volumes of the system after and before dissociation. If the gas or vapour formed is supposed to behave as an ideal gas, and the volume of the solids neglected in comparison with those of the gas, the equation becomes: $d \log p/dT = Q/RT^2$, where R is the gas constant for one gram molecule.

Van't Hoff in 1884-6 generalized this equation so as to make it applicable to all cases of chemical equilibrium between gases or substances in dilute solution. In this case the equilibrium constant K replaces p in the above equation.

The integration of this equation gives a means of finding Q from measurements of the effect of temperature on K, but the problem of calculating K from the heat of reaction Q is not solved, since an unknown constant of integration enters: $\log K = -Q/RT + \text{const.}$ Nernst in 1906 showed that by making another assumption it was possible to calculate the value of this constant, and the problem of finding K and thence the affinity from the heat of reaction was solved.

The Phase Rule

In systems which are not homogeneous, the Phase Rule of Willard Gibbs, enunciated * in 1876, is the most useful guide. Gibbs's memoir on equilibrium in heterogeneous systems was published in a rather obscure American journal ; it was mentioned by Clerk

FIG. 110.—J. W. GIBBS, 1839-1903.

Maxwell in his *Theory of Heat*, but was practically unnoticed until 1884 when Bakhuis Roozeboom, working in the laboratory of van Bemmelen in Leyden on hydrates of sulphur dioxide, had his attention directed to it by van der Waals, the professor of physics. The rule deals with systems containing two or more bodies called *phases*, separated by boundaries, e.g. liquid and gas, or liquid and solid. The system can be built up from a certain minimum number of

* *Trans. Connecticut Acad.*, 1875-6, iii, 152 (January, 1876).

chemical substances called the *components*. Thus water and sulphur dioxide were the components in Roozeboom's case. In order that the system shall be in equilibrium, certain conditions of temperature, pressure, and the concentrations in the various phases, must be satisfied. The number of these variables which must be fixed before there is equilibrium is called the number of *degrees of freedom* of the system. If P is the number of phases, C the number of components and F the number of degrees of freedom, then Gibbs showed that $P+F=C+2$.

Solutions

The fundamental researches on which the modern theory of solutions is based are those of Raoult and Pfeffer.

François-Marie Raoult was born in 1830 in Fournes, in the Département du Nord in France, and was of modest origin. He obtained permission to go to Paris to study, but was unable to support himself. In 1853 he obtained a post as a teacher in the Lycée at Rheims, and after taking other situations he was able to proceed to a degree. In 1862 he was appointed in Sens, and in the most unfavourable surroundings he worked for, and obtained, a doctorate of Paris in 1863. His investigation concerned the heat of reaction and electromotive forces of galvanic cells, and he noticed that the two did not correspond, as they should do on the theory (p. 328) that heat was a measure of affinity. In 1867 he entered the University of Grenoble as assistant in the chemistry department, being promoted to professor in 1870. He remained at Grenoble until his death in 1901.

In 1878 Raoult's first publication on the freezing points of solutions appeared. He used chiefly organic substances and as a result of many measurements he was able in 1882 to publish a table * from which it appeared that the product obtained by multiplying the depression of freezing point for a solution containing 1 gram of substance in 100 grams of water by the molecular weight of the substance is a constant. " This tends to prove ", he says, " that in the majority of cases the molecules of organic compounds are simply

* *Comptes rendus*, 1882, xciv, 1517; cf. Memorial Lecture by van't Hoff, *J. Chem. Soc.*, 1902, p. 969.

separated by the act of dissolution and brought to the same state, in which they exert the same influence on the physical properties of water." It was also clear that the molecular weights of dissolved substances could be found from the freezing points of solutions. If C is the depression due to P grams in 100 grams of solvent, then

FIG. 111.—F. M. RAOULT, 1830-1901.

$CM/P=K$, where M is the molecular weight and K a constant representing the value of C when $P=M$, i.e. the molecular depression of freezing point, the same for a given solvent for all dissolved substances.

In 1884 Raoult found that "contrary to what I have believed up to now, the general law of congelation does not apply to salts dissolved in water . . . on the contrary it applies to the radicals constituting the salts, almost as though these radicals were simply mixed in the solution ". Raoult also worked on the vapour pressures of solutions, and showed in 1886-7 that: " one molecule of a fixed, non-saline substance, in dissolving in 100 molecules of any volatile liquid, diminishes the vapour tension of the liquid by a nearly constant fraction of its value, and nearly 0·0105." This again is a method capable of finding molecular weights in solution, although Beckmann in 1889 showed that it is more convenient to measure the elevation of boiling point, which obeys laws exactly parallel to those for depression of freezing point. The two methods are now in daily use.

A different type of investigation was the osmotic pressure work of Wilhelm Pfeffer, later professor of botany in Leipzig. It had been known from early in the eighteenth century that when a solution is separated from pure water by a membrane such as bladder, the water permeates the membrane and dilutes the solution, setting up a pressure if the volume of the latter is prevented from increasing. This pressure was called osmotic pressure. In 1877 Pfeffer was able, by depositing copper ferrocyanide in the walls of a porous pot, to prepare membranes which would permit water to pass but were not permeable to dissolved sugar. By means of these semipermeable membranes he measured the osmotic pressures of solutions. His results were not well known except to botanists, and van't Hoff's attention was drawn to them by his colleague de Vries, professor of botany. Van't Hoff at once saw their great significance. He showed in 1886 from Pfeffer's results that the osmotic pressure exerted by a dissolved substance is equal to the pressure which it would exert as a gas confined in a space equal to the volume of the solution; "... it gradually appeared that there is a fundamental analogy, nay almost an identity, with gases, more especially in their physical aspect, if only in solutions we consider the so-called osmotic pressure instead of the ordinary gaseous pressure ... we are not here dealing with a fanciful analogy, but with one which is fundamental ".

By an application of the principles of thermodynamics, van't Hoff was able to deduce a quantitative connection between the osmotic pressure and the depression of freezing point or lowering of vapour pressure (see p. 304).

There was still the difficulty of the solutions of salts in water which had been noticed by Raoult. These exhibited higher osmotic pressures than they would be expected to do on van't Hoff's theory. By analogy with the dissociation of gases we might expect some kind of dissociation of the salts in solution, and Raoult had already gone some distance in assuming this. Planck in 1887 also assumed a dissociation into the radicals, but the full explanation was first given in the theory of electrolytic dissociation of Arrhenius. The theory of electrolytic dissociation was based on the fact that solutions which showed abnormally high osmotic pressures were conductors of electricity.

Electrolytic Dissociation

Svante Arrhenius * was born in 1859 at Wijk, near Uppsala in Sweden, and studied in Uppsala, Stockholm, and Riga (with Ostwald). He then worked in Germany, and with van't Hoff in Amsterdam. In 1891 he was appointed lecturer, and in 1895 professor, of physics in the Technical High School in Stockholm. He received the Nobel Prize in 1903 and after declining an invitation to Berlin was Director of the Nobel Institute at Stockholm from 1905 till his death in 1927.

Hittorf † in 1853-59 had shown by numerous experiments, to which Arrhenius drew attention, that the current in electrolysis is carried unequally by the two ions, which move with different speeds; the fraction carried by the anion he called the transport number n. He said: " the ions of an electrolyte cannot be combined in a firm manner to complete molecules."

Friedrich W. G. Kohlrausch in 1874 showed that the equivalent conductivity, i.e. the specific conductivity κ divided by the concentration of the solution in equivalents per c.c., $\kappa/c=\Lambda$, of a very dilute solution of a salt, is the sum of two terms, one depending only on the cation and the other only on the anion: $\Lambda=l_c+l_a$. l_c and l_a are called the mobilities of the ions. If n is Hittorf's transport number, $l_a=n\Lambda$.

Taken in conjunction with Raoult's results, this seemed to point

* *J. Chem. Soc.*, 1928, 1380; van't Hoff's and Arrhenius's memoirs on solutions are translated in *Alembic Club Reprint* No. 19.

† Ostwald's *Klassiker* Nos. 21 and 23.

to a dissociation of the salt into its two ions in the solution—a so-called ionization. The action of the electric current is then merely to direct the positively and negatively charged ions towards the electrodes of opposite charge. Since the number of particles in a given

FIG. 112.—SVANTE ARRHENIUS, 1859-1927.

volume is increased by ionization, the abnormally high osmotic pressures and the related abnormal depressions of freezing point were simply explained.

Arrhenius was able to show * that the degree of ionization calculated from the electrical conductivity, Λ/Λ_∞, where Λ_∞ is the equivalent conductivity at infinite dilution (when there is complete ionization), was very approximately the same as that calculated

* Arrhenius's experiments on conductivity, and a " chemical theory of electrolytes ", were published (in French) in the *Bihang* of the transactions of the Swedish Academy, 1884, vol. 8 (received June 1883), Ostwald's *Klassiker* No. 160, but the full statement of the theory of electrolytic dissociation was first published in the *Zeitschrift für physikalische Chemie*, 1887, vol. i, p. 631.

from the deviations from Raoult's law in the case of freezing point measurements. These are two independent methods, and the substantial correctness of the theory was thus established. Modern investigation has modified Arrhenius's theory in some respects, since strong acids, bases and salts are now supposed to be practically completely ionized in fairly dilute solutions, the deviations from the results for complete ionization being ascribed to interaction between the ions of opposite charge, which causes a reduction in the mobilities of the ions and hence a diminution of Λ (Debye and Hückel, 1923).

The theory of electrolytic dissociation was not at first received with much enthusiasm by the majority of chemists, who could not understand how sodium and chlorine, for example, could be contained in a solution of common salt. In 1885, H. E. Armstrong had proposed a theory of chemical change which regarded all changes as electrolytic in character, and as examples of "reversed electrolysis". They occurred between three bodies, one at least of which must be an electrolyte. Thus in the combination of hydrogen and oxygen, the gases acted as electrodes, and the water which must also be present acted as an electrolyte:

$$2H_2 + 2O \vdots H_2 + O_2 = 2H_2O + 2H_2O.$$

Armstrong assumed that, as pure water is almost a non-conductor of electricity, there must also be some impurity present which renders it conducting. He was one of the principal opponents of Arrhenius's theory of ionization.

One of the chief advocates of the new theories of solution was Wilhelm Ostwald (1853-1932), professor in Leipzig.* Ostwald had previously carried out important researches on affinity and mass action, and he adopted the new views of Arrhenius with enthusiasm. He was an excellent teacher and in his lectures and text-books he popularized the theory of electrolytic dissociation and showed how it could be applied in the explanation of chemical reactions, including those in analytical chemistry. He had many pupils from other countries, especially from America, where the new teachings were soon adopted and extended.

Important extensions of Arrhenius's theory were made by

* Donnan, *J. Chem. Soc.*, 1933, p. 346.

Walther Nernst (1864-1941),* first an assistant to Ostwald and then professor in Göttingen and Berlin. He showed in 1889 that the production of electromotive force in galvanic cells could be explained in terms of a "solution pressure" of the metal electrodes,

FIG. 113.—W. NERNST, 1864-1941.

tending to throw off charged ions into the solution, this tendency being balanced by the osmotic pressure of the dissolved ions. Nernst also introduced in 1889 the important theory of solubility product, explaining precipitation reactions. His text-book, *Theoretical Chemistry from the Standpoint of Avogadro's Rule and Thermodynamics* (1893, and later editions) gave a well-balanced survey of the subject.

Nernst (1906, see p. 329) assumed that the Thomsen-Berthelot principle (p. 328) that the heat of reaction *U* is equal to the maxi-

* Memorial lecture by Partington, *J. Chem. Soc.*, 1953, 2853.

mum work or affinity A is true for reactions between pure solids near the absolute zero. Lord Kelvin (1855) and Helmholtz (1882) had shown that A and U are related by the equation $A - U = T(dA/dT)$, and for such reactions the Nernst heat theorem or third law of thermodynamics asserts that dA/dT tends to zero at the absolute zero. This is equivalent to the statement that the entropy change is zero and Planck (1911) assumed that the entropy of every pure solid is zero at the absolute zero. This implies that its specific heat vanishes at $T=0$, which was confirmed experimentally by Nernst. Einstein (1907) and Debye (1913) gave equations based on the quantum theory which led to this result. Nernst (1914) assumed that the specific heat of a gas should also vanish very near the absolute zero.

Nernst (1906) showed that the heat theorem could be applied to calculate the integration constant of van't Hoff's equation for the equilibrium constant of a gas reaction (p. 329). This is now carried out by using the entropies of the gases, reckoned from those of the solids at absolute zero, and these can also be calculated from spectoscopic data.

Einstein (1912) explained photochemical reactions by assuming that a quantum ϵ of radiant energy absorbed dissociates one molecule of reacting gas. In the photochemical union of hydrogen and chlorine (studied by Bunsen and Roscoe, 1855 f.) the yield of hydrogen chloride is much larger than would be expected. Nernst (1918) explained this by assuming that a molecule of chlorine is dissociated according to Einstein's law, but the chlorine atoms set up a " chain reaction " which proceeds without the intervention of light until atoms are removed by collision with the wall of the vessel or with foreign gas molecules in the mixture :

$$Cl_2 + \epsilon = 2Cl \quad \begin{cases} Cl + H_2 = HCl + H \\ H + Cl_2 = HCl + Cl \end{cases}$$

Summary and Supplement

The Development of Physical Chemistry

Hermann Kopp (1817-1892), professor in Giessen and Heidelberg, is perhaps best known as a historian of Chemistry, but he was also one of the founders of physical chemistry. His work on atomic and molecular volumes (from 1840), crystallography (from 1841), boiling points (from 1844), specific heats (1848 f.), the influence of constitution on physical properties (1855 f.), and dissociation (1858), is characterized by the aim to establish relations between the physical properties and chemical composition and properties of substances.

August Horstmann (1842-1929), assistant professor in Heidelberg, founded chemical thermodynamics by measurements of dissociation pressures and the application of the second law of thermodynamics (1868 f.), showed that the law of mass action applied to gaseous equilibria (1877: the water-gas equilibrium), and determined the molecular volumes of liquids (1887).

Cato Maximilian Guldberg (1836-1902), professor of applied mathematics and technology in Christiania, published important memoirs on thermodynamics and chemical equilibrium (1867 f.) and collaborated with Peter Waage (1833-1900), professor of chemistry in Christiania, in the quantitative formulation of the law of mass action (1864-7).

Henri Étienne Sainte-Claire Deville (1818-1881), born in the West Indies, was professor in Paris. He discovered nitrogen pentoxide (1849), developed technical processes for the production of sodium, aluminium (1855) and magnesium (1863), fused platinum with the oxy-hydrogen blowpipe on the technical scale (1875), and worked on crystalline boron and silicon (1856), dissociation (1857 f.), and artificial minerals, developing high-temperature technique. He collaborated with Henri Jules Debray (1827-1888), professor in Paris, who worked chiefly on inorganic chemistry and metallurgy.

Josiah Willard Gibbs (1839-1903), professor of mathematical physics in Yale University, developed the application of thermodynamics to chemistry, including thermal dissociation, surface tension and electrochemistry, and stated the phase rule (1873-8). Among other equations developed by him was one connecting adsorption and interfacial tension.

Johannes Diderik van der Waals (1837-1923), professor of physics in Amsterdam, put forward the famous equation of state (1873) for imperfect gases, from which the critical constants could be calculated. Critical phenomena were discovered in 1862-3 by Thomas Andrews (1813-1885), professor in Belfast, who also worked on thermochemistry and on ozone (1857-9). Heike Kamerlingh Onnes (1853-1927), founder of the Cryogenic Laboratory in Leyden, worked on critical phenomena and low temperatures, liquefying helium in 1907.

Hendrik Willem Bakhuis Roozeboom (1854-1907) made the first practical applications of Gibbs's phase rule (1884 f.), studied triple points (1885), classified chemical equilibria and used graphical methods (1887, 1894) and studied and classified solid solutions (1891).

François-Marie Raoult (1830-1901), professor in Grenoble, investigated (1863) the electromotive forces of galvanic cells in relation to heat of reaction, and especially the freezing points (1878 f.) and vapour pressures (1886-7) of solutions. He showed that the depression of freezing point or of vapour pressure of the solvent by the dissolved substance could be used to calculate the molecular weight of the latter, and thus laid the foundations of the modern theory of solutions enunciated by van't Hoff.

Jacobus Henricus van't Hoff (1852-1911), professor in Amsterdam (1878) and Berlin (1896), developed (1874) the idea of the stereochemistry of carbon independently of Le Bel. He investigated mass action and reaction velocity, developed the modern theory of dilute solutions (1886), including the theory of osmotic pressure, freezing and boiling points and vapour pressure, worked out methods for the determination of transition points (from 1884), and applied the phase rule in the study of the crystallization of salts from solutions.

Wilhelm Ostwald (1853-1932), professor in Riga and Leipzig, improved physico-chemical methods and apparatus (1873), investigated the partition of a base between two acids by the volume method and by the refractive power (1878), the rates of hydrolysis of salts and esters (1883), the conductivities of acids (1878-87), the affinity constants of acids and bases (from 1885), the viscosities of solutions (1891), the ionization of pure water (1893), and catalysis. He adopted and advocated Arrhenius's theory of electrolytic dissociation, put forward his dilution law (1888), and by his text-books and teaching popularized the new views on physical chemistry, especially on solutions, of van't Hoff and Arrhenius.

Johann Wilhelm Hittorf (1824-1914), professor of physics and chemistry in Münster, investigated the transport of salts in electrolysis of solutions (1853-9), and cathode rays (1869). He discovered " metallic " phosphorus (1865).

Svante Arrhenius (1859-1927), born in Wijk, near Uppsala, and latterly Director of the Nobel Institute at Stockholm, is best known for his theory of electrolytic dissociation (1887). He also investigated the viscosity of solutions, the effect of temperature on reaction velocity (1889), and immunity (1902 f.), and speculated on the structure of the universe.

Walther Nernst (1864-1941), professor in Göttingen and Berlin, put forward a theory of galvanic cells (" electrolytic solution pressure ") (1889), investigated the solubility product relation (1888-9) and diffusion in solutions, including liquid contact potentials (1888-9), put forward the so-called Heat Theorem (" third law of thermodynamics ") (1906), investigated the specific heats of solids at low temperatures from the point of view of the quantum theory (1911 f.), and put forward the " atom chain-reaction " theory in photochemistry (1918).

Fritz Haber (1868-1934), professor in Karlsruhe, is best known for his investigations on the synthesis of ammonia (1905, 1915). The first experiments at higher pressures, however, were made by Nernst (1906), followed by Haber and Le Rossignol (1908). The method was developed by Karl Bosch, of the Badische Anilin und Soda Fabrik, into an industrial process. Haber also worked on chemical equilibrium in flames (1895 f.), the electrolytic reduction of nitrobenzene (1898 f.), autoxidation (1900), the synthesis of nitric oxide in the electric arc (1908 f.) and many electrochemical problems.

Georg Bredig (b. 1868), assistant in Leipzig and professor in Karlsruhe (1911), measured the dissociations of weak bases (1894), drew attention to the important group of amphoteric electrolytes (1898), and investigated catalytic action, particularly of colloidal platinum (1901), and the " poisoning " of catalysts.

Richard Abegg (1869-1910) worked on the freezing points of solutions (1894-8), the dielectric constant of ice (1898), the polyiodides of alkali metals (1906) and (with Neustadt) on potentials in non-aqueous solutions (1909). In conjunction with Bodländer he put forward a theory of valency (1899). He distinguished between homopolar and heteropolar compounds (1906).

Henry Le Chatelier (1850-1936), professor in Paris, investigated the specific heats of gases at high temperatures (1883), the dissociation of calcium carbonate (1886 f.), mass action in explosion reactions (1888), the electrical conductivity of alloys (1895), freezing point curves (1896), and the chemistry of silicates. He put forward in 1888 the important " law of reaction ", governing the effect of pressure and temperature on equilibrium.

Sir James Walker (1863-1935), professor in Dundee and Edinburgh, investigated hydrolysis (1889 f.), the strengths of acids (1889 f.), the electrosynthesis of dibasic organic acids (1891) and amphoteric electrolytes (1904-6).

Sir James Dewar (1842-1923), pupil of Kekulé (1867), assistant in Edinburgh, professor in Cambridge (1875) and the Royal Institution (1877), gave the structural formula of pyridine (1871), invented the vacuum-jacketed vessel, liquefied hydrogen (1895-98) and made extensive researches on the properties of matter at low temperatures. (Helium was first liquefied by Kamerlingh Onnes in Leyden in 1908.) Dewar produced high vacua by adsorption on charcoal cooled in liquid air (1905) and also worked on soap films.

CHAPTER XV

THE PERIODIC LAW

Atomic Weights

THE developments in general chemistry during the twentieth century originated in the Periodic Law, which first disclosed a possible genetic relationship among the chemical elements. The first suggestion in this direction, Prout's Hypothesis (p. 210), had in the meantime fallen into discredit, since the accurate determinations of atomic weights made by Turner, Penny, Dumas and Stas, Marignac, and especially by Stas from 1860, indicated that they were not really whole numbers, and some of them deviated very considerably from whole numbers, e.g. Cl = 35·46. Lothar Meyer, therefore, said in 1891 that: "Prout's hypothesis is tempting in its simplicity, and for a time was favourably received by chemists, excepting those who had made accurate atomic weight investigations." Stas, in his researches, was led to conclude that it "is only an illusion, a pure hypothesis definitely contradicted by experiment", although he had begun with "an almost complete confidence in the exactness of the law of Prout".*

Stas's researches aimed at establishing the combining ratios of the elements through the analysis and synthesis of compounds which could be obtained in a very pure state. Although he nominally took oxygen as the standard, with the atomic weight of 16·000, his values are mostly based on the atomic weight of silver, and this could not be related directly to oxygen. A revision of Stas's values by the American chemist Theodore William Richards (1868-1928) †

* See the Memorial Lecture on Stas, by Mallet, *J. Chem. Soc.*, 1893, p. 1; Stas, *Œuvres complètes*, 3 vols., Brussels, 1894.

† Memorial Lecture, *J. Chem. Soc.*, 1930, p. 1937.

disclosed small but important errors, and the physico-chemical methods used by Guye and others, making use of the limiting densities of gases (i.e. relative densities extrapolated to zero pressure,

FIG. 114.—J. S. STAS, 1813-1891.

when Avogadro's hypothesis applies strictly), also showed that some of Stas's values were less accurate than had been supposed. The deviations from Prout's hypothesis still remained.

The Periodic Law

In 1817 and 1829 Johann Wolfgang Döbereiner * (1780-1849), professor in Jena and the chemical teacher of Goethe, noticed that in certain groups of three elements, e.g. Ca, Sr, Ba, the atomic

* *Ann. Physik,* 1817, lvi, 331 (Ca, Sr, Ba) ; 1829, **xv**, 301 (other triads) ; Ostwald's *Klassiker* No. 66.

weight of the middle element was approximately the mean of those of the first and third. These were known as Döbereiner's triads. Many attempts to derive regularities among atomic weights were made by chemists, especially Dumas,* but without success. In 1865, Odling arranged the elements in a table † which is a revision

FIG. 115.—T. W. RICHARDS, 1868-1928.

of one published in 1864 ‡ and shows a close resemblance to Mendelejeff's table of 1869 (p. 348).

Newlands §, a London industrial chemist, drew up a table of the

* *Ann. Chim.*, 1859, lv, 129.

† Watts, *Dictionary of Chemistry*, 1865, vol. iii, p. 975.

‡ *Quarterly Journal of Science*, 1864, i, 642.

§ *Chem. News*, 1863, vii, 70; 1864, x, 59, 94, 240; 1865, xii, 83, 94; 1866, xiii, 113, 130; *On the Discovery of the Periodic Law*, 1884.

Odling's Table of the Elements (1865)

			Mo 96 — Pd 106·5	W 184 Au 196·5 Pt 197
L 7 G 9 B 11 C 12 N 14 O 16 F 19	Na 23 Mg 24 Al 27 Si 28 P 31 S 32 Cl 35·5	— Zn 65 — — As 75 Se 79·5 Br 80	Ag 108 Cd 112 — Sn 118 Sb 122 Te 129 I 127	— Hg 200 Tl 203 Pb 207 Bi 210 — —
	K 39 Ca 40 Ti 48 Cr 52·5 Mn 55	Rb 85 Sr 87·5 Zr 89·5 —	Cs 133 Ba 137 — V 138	Th 231

elements arranged in the order of their atomic weights, and noted that " the eighth element, starting from a given one, is a kind of repetition of the first, like the eighth note in an octave of music ". He called this the *law of octaves*. When Newlands read a paper at a meeting of the London Chemical Society and exhibited his table, he was asked by Carey Foster whether he had ever tried classifying the elements in the order of the initial letters of their names. In 1887 Newlands received the Davy Medal of the Royal Society.

Newlands's Table of Atomic Numbers (1865)

H	1	F	8	Cl	15	Co & Ni	22	Br	29	Pd	36	I	42	Pt & Ir	50
Li	2	Na	9	K	16	Cu	23	Rb	30	Ag	37	Cs	44	Tl	53
G	3	Mg	10	Ca	17	Zn	25	Sr	31	Cd	38	Ba & V	45	Pb	54
Bo	4	Al	11	Cr	19	Y	24	Ce & La	33	U	40	Ta	46	Th	56
C	5	Si	12	Ti	18	In	26	Zr	32	Sn	39	W	47	Hg	52
N	6	P	13	Mn	20	As	27	Di & Mo	34	Sb	41	Nb	48	Bi	55
O	7	S	14	Fe	21	Se	28	Ro & Ru	35	Te	43	Au	49	Os	51

Newlands first clearly arranged the elements according to the atomic number, i.e., the serial number of an element in the order of atomic weights, beginning with hydrogen as 1. Newlands also emphasized that only when the new atomic weights proposed by Cannizzaro (p. 258) were used was any regularity observed.

The Periodic Law was put forward almost simultaneously and quite independently by Julius Lothar Meyer * (1830-1895) in Germany, and Dmitri Ivanovich Mendelejeff in Russia. Mendelejeff was born in 1834 in Tobolsk, in Siberia, the fourteenth child of a

FIG. 116.—J. LOTHAR MEYER, 1830-1895.

teacher. His father became blind and the family was looked after by the mother, who directed a glass factory. In 1848 she went the thousands of miles by road to Moscow with her son, with the idea of entering him in the University, but as a Siberian he could not enter. The two set out again with their last resources for St. Peters-

* Memorial Lecture by Bedson, *J. Chem. Soc.*, 1896, lxix, 1403.

burg, where in 1850 Mendelejeff secured admission to a training college for teachers. His mother died in the same year.

In 1887, in dedicating his book on solutions to his mother, Mendelejeff said: "She instructed by example, corrected with love, and in order to devote her son to science she left Siberia with him,

FIG. 117.—D. I. MENDELEJEFF, 1834-1907.

spending her last resources and strength." In 1859 he spent two years in Heidelberg, working out his own ideas, and in 1867, after holding other appointments, he became professor in the University of St. Petersburg, remaining there until 1890, when he resigned on account of a dispute with the authorities about his too liberal views. He died in 1907.*

* Tilden, *J. Chem. Soc.*, 1909, 2077; *ibid.*, *Great Chemists*, 1921, 241; *Nature*, 1934, cxxiii, 161.

Mendelejeff's First Periodic Table (March, 1869)

			Ti 50	Zr 90	? 100
			V 51	Nb 94	Ta 182
			Cr 52	Mo 96	W 186
			Mn 55	Rh 104·4	Pt 197·4
			Fe 56	Ru 104·4	Ir 198
			Ni=Co 59	Pd 106·6	Os 199
H 1			Cu 63·4	Ag 108	Hg 200
	Be 9·4	Mg 24	Zn 65·2	Cd 112	
	B 11	Al 27	? 68	U 116	Au 197?
	C 12	Si 28	? 70	Sn 118	
	N 14	P 31	As 75	Sb 122	Bi 210?
	O 16	S 32	Se 79·4	Te 128?	
	F 19	Cl 35·5	Br 80	I 127	
Li 7	Na 23	K 39	Rb 85·4	Cs 133	Tl 204
		Ca 40	Sr 87·6	Ba 137	Pb 207
		? 45	Ce 92		
		Er? 56	La 94		
		Yt? 60	Di 95		
		In 75·6?	Th 118?		

Lothar Meyer's Periodic Table (December 1869)

I.	II.	III.	IV.	V.	VI.	VII.	VIII.	IX.
	B 11	Al 27·3	—	—		?In 113·4		Tl 202·7
	C 11·97	Si 28		—		Sn 117·8		Pb 206·4
			Ti 48		Zr 89·7		—	
	N 14·01	P 30·9		As 74·9		Sb 122·1		Bi 207·5
			V 51·2		Nb 93·7		Ta 182·2	
	O 15·96	S 31·98		Se 78		Te 128?		
			Cr 52·4		Mo 95·6		W 182·2	
	F 19·1	Cl 35·38		Br 79·75		I 126·5		
			Mn 54·8		Ru 103·5		Os 198·6?	
			Fe 55·9		Rh 104·1		Ir 196·7	
			Co & Ni 58·6		Pd 106·2		Pt 196·7	
Li 7·01	Na 22·99	K 39·04		Rb 85·2		Cs 132·7		
			Cu 63·3		Ag 107·66		Au 196·7	
?Be 9·3	Mg 23·9	Ca 39·9		Sr 87·0		Ba 136·7		
			Zn 64·9		Cd 111·6		Hg 199·8	

Mendelejeff published his periodic law in Russian in April, 1869 (presented in March) * ; Lothar Meyer's paper, dated December, 1869, was published † in German in 1870, but he had drawn up a very imperfect periodic table in 1868, which he did not publish.‡ A longer paper by Mendelejeff in German in 1871 drew some attention to his views. Chemists first took a keen interest in the Periodic Law when Mendelejeff's predictions of missing elements were confirmed by the discovery of gallium in 1875, and of scandium in 1879,

* *J. Russ. Phys. Chem. Soc.*, 1869, i, 60 ; *J. prakt. Chem.*, 1869, cvi, 251 ; *Jahresber.*, 1869, 9 ; *Ber.*, 1871, iv, 348. † *Ann.*, 1870, Suppl. vii, 354.

‡ Seubert, *Z. anorg. Chem.*, 1895, xi, 334 ; Ostwald's *Klassiker* No. 68.

and in text-books it first came into use about 1895. Mendelejeff's own text-book, *Principles of Chemistry*, in writing which he was led to the Periodic Law, appeared in Russian in 1869 and was translated into English.

Mendelejeff expressed his discovery in the "periodic law", that "the properties of the elements are in periodic dependence upon their atomic weights".

Lothar Meyer in 1870 published his well-known atomic volume curve, in which the periodic dependence of a quantitative property was clearly shown as a function of the atomic weight; as the atomic weight steadily increases, the property alternately rises and falls over definite periods of the elements. Mendelejeff had emphasized this as a general result.

Period.	Series.	Group.															
		I. a	I. b	II. a	II. b	III. a	III. b	IV. a	IV. b	V. a	V. b	VI. a	VI. b	VII. a	VII. b	VIII. a	(o) b
1	1	H 1															He 2
2	2	Li 3		Be 4			B 5		C 6		N 7		O 8		F 9		Ne 10
3	3	Na 11		Mg 12			Al 13		Si 14		P 15		S 16		Cl 17		A 18
4	4	K 19		Ca 20		Sc 21		Ti 22		V 23		Cr 24		Mn 25		Fe 26 Co 27 Ni 28	
	5		Cu 29		Zn 30		Ga 31		Ge 32		As 33		Se 34		Br 35		Kr 36
5	6	Rb 37		Sr 38		Y 39		Zr 40		Nb 41		Mo 42		Tc 43		Ru 44 Rh 45 Pd 46	
	7		Ag 47		Cd 48		In 49		Sn 50		Sb 51		Te 52		I 53		Xe 54
6	8	Cs 55		Ba 56		Rare Earths 57–71		Hf 72		Ta 73		W 74		Re 75		Os 76 Ir 77 Pt 78	
	9		Au 79		Hg 80		Tl 81		Pb 82		Bi 83		Po 84		At 85		Rn 86
7	10	Fr 87		Ra 88		Ac 89		Th 90		Pa 91		U 92					

The modern Periodic Table, in which the elements are arranged according to the atomic numbers, from hydrogen 1 to uranium 92, is shown above. It may be compared with those of Mendelejeff and Lothar Meyer. The arrangement is practically the same as that in a later table given by Lothar Meyer, but the new group O or VIII *b* was added by Ramsay.

The Periodic Law made clear a number of previously unsuspected analogies among the elements, and it stimulated the study of Inorganic Chemistry, which had been rather neglected in the second half of the nineteenth century owing to the great specialization in

FIG. 118.—H. MOISSAN, 1852-1907.

Organic Chemistry. Notable researches in Inorganic Chemistry were, however, carried out by Roscoe in Manchester, by Brauner in Prague, and by Moissan * in Paris. The latter succeeded in isolating fluorine, a problem which had taxed the ingenuity of chemists since the early efforts of Davy, and he also carried out researches at high temperatures with the electric furnace, such as the artificial

* Ramsay, *J. Chem. Soc.*, 1912, ci, 477; Lebeau, *Bull. Soc. Chim.*, 1908, iii.

production of diamonds and the synthesis of metallic carbides. Calcium carbide had been obtained by Wöhler in 1862 by heating an alloy of zinc and calcium with carbon, and he showed that it gave acetylene with water. Moissan's researches on the carbides

FIG. 119.—SIR W. RAMSAY, 1852-1916.

are described in his book *Le four électrique* (Paris, 1897). Moissan published a *Traité de Chimie minérale* in several volumes.

The study of Inorganic Chemistry was also extended by the researches of Crookes * on thallium and the rare earths, and of Ramsay,† beginning with his collaboration with Lord Rayleigh in the discovery of argon in 1894. Although the atmosphere had been analysed by many chemists, it had not previously been suspected

* *Life*, by Fournier d'Albe, 1923.

† Tilden, *Sir William Ramsay*, 1918; Partington, *Nature*, 1952. clxx, 554.

that it contains over one per cent. of an inert element, and some chemists were at first inclined to think that argon was only a polymer of nitrogen, perhaps N_3. The discovery by Ramsay and Travers of other inert gases, elements of the group, made this idea seem very improbable and it was soon admitted that several new elements were present in the atmosphere.*

SUMMARY AND SUPPLEMENT

The Development of Inorganic Chemistry

Sir Henry Enfield Roscoe, born in London in 1833 of a Liverpool family, studied under Graham and Williamson and from 1853 to 1857 under Bunsen in Heidelberg, where he carried out researches on the action of light in the union of hydrogen and chlorine which laid the foundations of quantitative photochemistry. From 1857 to 1885 he was professor in Manchester, where he was associated with Schorlemmer. In 1865 Roscoe showed that vanadium is an element related to nitrogen, not to sulphur, its highest oxide being V_2O_5 and not VO_3, as Berzelius supposed. This was of importance later in the exposition of the periodic law by Mendelejeff. Roscoe, from 1863, made contributions to spectrum analysis. He became a Privy Councillor in 1909 and died in 1915.

Roscoe was succeeded in Manchester by Harold Baily Dixon (1852-1930), born in London, who carried out important researches on flame, combustion and explosions in gases, and was an inspiring teacher. Dixon discovered (1880) that a dry mixture of carbon monoxide and oxygen was not exploded by an electric spark. His researches on the influence of water on chemical change were continued by Herbert Brereton Baker (1862-1935), a master at Dulwich School and then professor in Imperial College, South Kensington, who also investigated nitrogen trioxide (1907) and the atomic weight of tellurium (1907).

* See Ramsay, *Gases of the Atmosphere*, 4th ed., 1915; *John William Strutt, Third Baron Rayleigh*, by R. J. Strutt, Fourth Baron Rayleigh, London, 1924, 186-225; W. M. Travers, *The Discovery of the Rare Gases*, 4to, London, 1928.

Jean Servais Stas (1813-1891), professor in Brussels, at first worked with Dumas on organic chemistry (phloridzin, acids, aldehydes, esters) and on the atomic weights of carbon, hydrogen and oxygen (1840). He then developed very accurate methods for the determination of atomic weights (1860-65 f.) and for many years Stas's values were by far the best. A man of great independence, Stas incurred the displeasure of the official and clerical circles in Belgium and worked for a small salary, unmarried and in modest circumstances.

Jean Charles Galissard de Marignac (1817-1894), professor in Geneva, worked on the rare earths (from 1840), discovering ytterbium (1878) and gadolinium (1880) ; on the isomorphism of salts of niobium, tin and tungsten, and of fluostannates and fluosilicates (1858 f.), and fluozirconates (1860), in this case correcting atomic weights. He discovered silicotungstic acid (1862), worked on the thermochemistry of solutions (1870 f.) and determined atomic weights. In 1865 he suggested that deviations from Prout's hypothesis might be due to the fact that the elements consist of mixtures of atoms of different masses (now called isotopes).

Sir William Crookes (1832-1919), born in London, was for a time assistant to Hofmann and then had a private practice. He discovered thallium (1861) and investigated cathode rays, putting forward the theory that they consisted of corpuscles identical with protyle, or the fundamental matter of the elements (1886). He studied the rare-earths, including their spectra, and directed attention to the necessity of providing sources of combined nitrogen for fertilizers (1892). Crookes carried out much work of technical interest and was an expert analyst. He received the Order of Merit in 1910 and in 1913-16 was President of the Royal Society.

Sir Thomas Edward Thorpe (1845-1925), professor in the Andersonian College, Glasgow (1870), Leeds (1874), and South Kensington (1885), discovered phosphorus pentafluoride (1877), phosphoryl fluoride (with Hambly, 1889), and thiophosphoryl fluoride (with Rodger, 1888) ; determined the vapour density of hydrofluoric acid (with Hambly, 1889) ; obtained P_2O_4 (1886) and P_4O_6 (1890-1) (with Tutton), investigated critical temperatures (with Rücker, 1884) and the viscosity of liquids (with Rodger, 1894), and was responsible for some accurate atomic weight determinations. He edited a *Dictionary of Applied Chemistry*, and wrote on the history of chemistry.

Edward Divers (1837-1912), professor in Japan, worked on ammonium carbonate and carbamate (1870), discovered hyponitrites (1871) and investigated the sulphonic acids of hydroxylamine (with Haga, 1900).

Dmitri Ivanovich Mendelejeff (1834-1907), professor at St. Petersburg (now Leningrad) in Russia, made extensive researches on the thermal expansion of liquids (1861), in which he arrived at the idea of critical temperature, more completely investigated independently by Thomas Andrews (1869); his most famous contribution to chemistry was the periodic law (1869), foreshadowed by J. A. R. Newlands (1864) and put forward independently but less fully by Julius Lothar Meyer (1830-95), professor at Karlsruhe and Tübingen, in 1870. Mendelejeff later worked on the properties of solutions (which he explained by a "chemical theory") and on the critical data and compressibilities of gases (1870-85).

Bohuslav Brauner (1855-1935), professor in Prague, was a pupil of Roscoe and carried out important investigations on the chemistry of tellurium (1889) and the rare earths, being especially interested in the periodic law.

Sir William Ramsay (1852-1916), professor in University College, London, began research on organic chemistry and synthesized pyridine (1877). He then determined (from 1885) the vapour densities of substances at different temperatures and pressures (with Sidney Young, professor in Dublin), applied Eötvös' surface tension method to the determination of molecular weights of liquids (1893), discovered the inactive gases (in collaboration with Lord Rayleigh and with Professor Travers) (1894-1908), worked on radioactivity and discovered the radium emanation and its spontaneous transformation into helium (with Frederick Soddy) (1904). Ramsay put forward an early electronic theory of valency (1908-9).

Henri Moissan (1852-1907), professor in Paris, studied the metal oxides (1879-83) and fluorine compounds (1883), isolated fluorine (1886), invented the electric furnace (1892) and with it prepared metal carbides and silicon carbide (1893-4), artificial diamonds (1893), studied the nitrides of metals (1895 f.), the reduction of refractory metal oxides (1896), and (in collaboration with Professor S. Smiles) the silicon hydrides (1902).

Friedrich Raschig (1863-1928), an industrial chemist, put forward a theory of the reactions in sulphuric acid chambers (1887 f.), investigated the sulphonic acids of hydroxylamine (1887), worked out a method for the technical production of hydroxylamine (1887), and investigated nitrogen iodide (1885 f.) and the thionic acids. He discovered chloramine (1907), and a simple method for the production of hydrazine from ammonia (1908), and worked out the technical conversion of hydrazine into hydrazoic acid, the lead salt of which (lead azide) was introduced as a detonator in Germany during the war of 1914-18 and is now displacing mercury fulminate generally. Raschig was an authority on the technical production of phenol. His researches (*Schwefel- und Stickstoffstudien*, 1924) are marked by great originality.

Alfred Werner (1866-1919), professor in Zurich, put forward theories of the stereochemistry of nitrogen (with Hantzsch, 1890), and of complex compounds (1893 f.; *Neuere Anschauungen auf dem Gebiete der anorganischen Chemie*, 1905), introducing the important idea of co-ordination number, and obtained optically active substances the activity of which was due to elements other than carbon (1911 f.). Werner's views may be said to underly the modern developments of inorganic chemistry, since they broke away from the inadequate theories of structure based on the study of carbon compounds, and prepared the way for the electronic theory of valency.

Theodore William Richards (1868-1928), professor in Harvard, began to work on atomic weights in 1883. He later much improved the technique of gravimetric atomic weight determinations, introducing quartz apparatus, the bottling device (1895), the nephelometer (1904), etc. About 1903 it had become evident from physicochemical measurements that some of Stas's values were less accurate than had been supposed, and since Stas's fundamental element was really silver (not oxygen), Richards and his pupils from 1904 revised some of Stas's figures, lowering the value for silver from 107·93 to 107·88. Richards also carried out investigations on the isotopes of lead, on thermochemistry (heats of neutralization, etc., 1905 f.) and electrochemistry (1897 f.) and on the atomic volumes and compressibilities of the elements (1901-4).

Alfred Stock (1876-1946), professor in Berlin, studied the decomposition of antimony hydride (1904), sulphides of phosphorus (1906 f.) and especially the hydrides of boron and silicon (1912 f.).

Otto Ruff (1871-1939) investigated nitrogen sulphide (1903 f.) and especially many fluorides of metals and non-metals, and prepared artificial diamonds (1917).

CHAPTER XVI

THE STRUCTURE OF THE ATOM

Radioactivity

THE modern theory of atomic structure is based on the discovery of the electron and the facts of radioactivity. The cathode rays produced by electrical discharges in gases at low pressure were discovered by Plücker in 1859 and investigated by Hittorf (1869), Crookes (1876, etc.) and J. J. Thomson. The latter showed in 1897 that they are negatively charged particles much smaller in mass than the hydrogen atom (the mass now accepted is about 1/1850 that of the H atom). He supposed that these particles were constituents of all atoms and called them " corpuscles ". The name electron proposed by Johnston Stoney (who had calculated an approximate value of its charge in 1874) is now adopted.

The radioactivity of uranium was discovered by Becquerel in 1896. In 1898 Pierre and Mme. Curie and Bémont obtained from pitchblende the intensely radioactive elements polonium and radium, and the radioactivity of thorium was discovered independently by Schmidt and by Mme. Curie in 1898.* In 1899 Mme. Curie suggested that radioactive atoms are unstable and disintegrate with emission of energy, a hypothesis extended by Rutherford and Soddy in 1903. Giesel, Becquerel, P. and Mme. Curie, and Rutherford, in 1899-1900 identified the α-, β-, and γ-rays of radioactive substances. In 1901 Dorn discovered radium emanation, and in 1904 Ramsay and Soddy showed that it changes spontaneously into helium. Regener, and Rutherford, afterwards proved that the α-rays consist of positively charged helium atoms.

Isotopes

The atoms of radium and of thorium undergo a succession of changes, each characterized by a definite rate of disintegration, some

* Mme. Curie, *Thesis*, Paris, 1903 ; *Chem. News*, 1903, lxxxviii (15 parts) ; Eve Curie, *Mme. Curie*, 1938 ; Partington, *Science Progress*, 1938, 141.

of the intermediate stages having only a very short life. The final product from both radium and thorium is lead. Radium is one of the products of the disintegration of uranium. From the number of α-particles, each of mass 4, expelled, it appeared that the atomic

FIG. 120.—MARIE SKLODOWSKA CURIE, 1867-1934, BORN IN WARSAW.

weights of the two kinds of lead should be different: uranium lead = 206, thorium lead = 208, ordinary lead being 207. This was confirmed by Soddy and by T. W. Richards independently in 1913, and Soddy called the different varieties isotopes of lead.

In 1911 Soddy pointed out that an α-ray change gave a product falling into a group of the periodic table two places lower than that of the parent substance, and in 1913 A. S. Russell, and K. Fajans, generalized this to include β-ray changes, when the product goes into a group one place higher than that of the parent substance.* The so-called displacement law governs the positions in the periodic table of the radioactive elements and their products of disintegration. Since they all have the same atomic number, all the isotopes of an element occupy the same place in the periodic table.

Russell, London.

FIG. 121.—LORD RUTHERFORD, 1871-1937.

The existence of isotopes of ordinary elements was discovered by J. J. Thomson in 1913 in the case of neon, which was found (by a method depending on the sorting out by means of magnetic and electric fields of the so-called positive rays produced by electric discharges in gases) to consist of a mixture of two kinds of atoms of masses 20 and 22, now denoted by ^{20}Ne and ^{22}Ne. A modifica-

* Soddy, *Chemistry of the Radio-Elements*, 1911, 29; Russell, *Chemical News*, 1913, cvii, 49 (31st January); Fajans, *Ber.*, 1913, xlvi, 422 (received 17 January); *Physikal. Z.*, 1913, xiv, 131 (15 February), 136, 257.

tion of the same method by Aston showed that most of the elements are mixtures of isotopes, e.g. chlorine of ^{35}Cl and ^{37}Cl, the atomic weight 35·46 being a mean value and no atom of this mass being present. Isotopes of oxygen, nitrogen, carbon, etc., were afterwards detected by spectroscopic methods, the most striking discovery being that of a heavier isotope of hydrogen of mass 2, called deuterium, by H. C. Urey of Columbia University in 1932.

Structure of the Atom

In 1911 Rutherford found that the very large deflections sometimes suffered by α-particles in passing through matter can be explained if every atom is assumed to consist of a very small positively charged nucleus surrounded by outer electrons at relatively large distances from the nucleus. The α-particle would then consist of the nucleus of helium, and although this can normally pass through the outer parts of another atom, it is deflected when it approaches the small positive nucleus. The nucleus of the hydrogen atom was called the proton, having a charge equal but opposite to that of the electron and a mass approximately the same as that of the hydrogen atom. The nuclei of other atoms are now supposed to consist of protons and neutrons, a neutron being a particle of approximately the same mass as the proton but without charge. The mass of the atom is approximately equal to the sum of the masses of the protons and neutrons ; the nuclear charge is equal to the number of protons.

The relation between the atomic number (p. 345) and the nuclear charge was elucidated in 1913-14 by Moseley *, working in Manchester and Oxford, who found that each element when bombarded by cathode rays emits X-rays of characteristic frequency, ν, determined by the atomic number, N, and he assumed that N was equal to the positive charge on the nucleus. This varied by one unit in passing from one element to the next in the periodic table. Some gaps were found in the series of atomic numbers, since filled by the discovery of missing elements such as hafnium by Coster and von Hevesy in 1923 and rhenium by Noddack and collaborators in 1925. Moseley's method permitted a complete survey of the elements and it showed that, with a few exceptions, the periodic table was complete.

* *Phil. Mag.*, 1913, xxvi, 102 ; 1914, xxvii, 703.

The theory of atomic structure was greatly extended in 1913 by Niels Bohr,* professor in Copenhagen, who showed that the Rutherford atomic model, by an application of Planck's quantum theory that light or radiation consists of quanta of energy, $\epsilon = h\nu$, where ν = frequency and h is a universal constant, can explain the spectra

Photo, by W. H. Hayles.

FIG. 122.—HENRY GWYN JEFFREYS MOSELEY, BORN 1887, KILLED IN ACTION IN GALLIPOLI, 1915.

of elements and also their position in the periodic table. A simple model of the atom was proposed in 1916 by G. N. Lewis,† professor in Berkeley, California, who supposed that the outer layer of electrons is 8 in the atoms of inert gases, and tends to become 8 in other atoms either (i) by gain or loss of outer electrons, forming negative or positive ions (as suggested by Kossel ‡ in 1916), or (ii) by sharing electrons in pairs with other atoms, forming ordinary

* *Phil. Mag.*, 1913, xxvi, 476, 857.
† *J. Amer. Chem. Soc.*, 1916, xxxviii, 762.
‡ *Ann. Physik*, 1916, xlix, 229.

valency bonds. The theory was extended to the heavier atoms by Irving Langmuir in 1919.*

The Theory of the Periodic Table

In Bohr's theory of the atom the electrons were regarded as point-charges revolving around the central positive nucleus of the atom in circular orbits. The energy state of the electron was denoted by

Century Co., N.Y.

FIG. 123.—GILBERT NEWTON LEWIS, 1875-1946.

a quantum number n, each orbit having a particular value of n. By analogy with the planetary orbits (where the inverse-square law of attraction also holds) the orbits would be expected to be ellipses and for a given value of n ellipses of varying eccentricity are pos-

* *J. Amer. Chem. Soc.*, 1919, xli, 868; 1920, xlii, 274.

sible. Sommerfeld (1915) introduced another quantum number l to take account of this and later theory showed that l like n has integral values, the highest of which is one less than n, e.g. if $n=4$, the possible values of l are 3, 2, 1, 0. If the plane of the ellipse is defined by a weak magnetic field, it can assume a number of discrete positions which are defined by another quantum number m, which

Century Co., N.Y.

FIG. 124.—I. LANGMUIR, BORN IN BROOKLYN, 1881.

has values equal to the positive and negative values of l. A fourth quantum number s defines the spin of the electron about its own axis and has two possible values $\pm\frac{1}{2}$ corresponding with two opposite directions of spin. A so-called " exclusion principle " stated by Pauli (1925) asserts that in any one atom there is never more than one electron having a given set of quantum numbers n, m, l and s.

By taking the values 1, 2, 3 and 4 for n and making use of the above relations, it is found that the maximum numbers of electrons with these values of n are 2, 8, 18 and 32. These are the numbers of the elements in the periods of the periodic table (p. 349):

$n=1$, hence $l=0$ and $m=0$; $s=\pm\frac{1}{2}$ corresponding with 2 electrons.

$n=2$, hence l is 1 or 0. For $l=0$, $m=0$; for $l=1$, m is -1, 0, or 1. There are four values of m and for each s is $\pm\frac{1}{2}$, making a total of 8.

$n=3$, hence l is 2, 1, or 0; for these three cases m is

$$(2, 1, 0, -1, -2), (1, 0, -1), \text{ and } 0;$$

there are 9 cases, for each of which $s=\pm\frac{1}{2}$, making 18 in all.

$n=4$, hence l is 3, 2, 1, or 0, giving 16 values of m, making 32 in all.

The electronic structures of the atoms of all the elements are known with some certainty from spectroscopic data. Although Bohr's atom model, with electrons revolving in orbits, is no longer accepted, the significance of the quantum numbers, with rather different interpretations, remains valid. It is sufficient to regard the electrons as distributed in " shells " corresponding with special values of n, the maximum numbers of electrons in these shells, starting with the one next to the nucleus, being 2, 8, 18 and 32. The end of a period is closed by an inert gas, and the number of electrons in the outer shell of an inert gas is 2 for helium and 8 for all the others.

The atomic nucleus is supposed to be built up from protons and neutrons, the number of protons giving the positive charge of the nucleus or the atomic number. The electrons outside the nucleus are added so as to make the atom neutral, i.e. as many as correspond with the positive charge of the nucleus. These electrons have principal quantum numbers n of 1, 2, 3, etc. For $n=1$ Pauli's principle shows that there are only two electrons possible; the hydrogen atom is composed of a proton and one electron, and the helium atom of a nucleus of two protons and two neutrons (atomic number 2) and two electrons. This completes the first period of two elements. The next electrons added have $n=2$ and Pauli's principle shows that there can be eight of them, corresponding with the eight elements in the second period, from lithium to neon.

For $n=3$ it might be expected that there would be 18 elements in the period, but what happens is that, of the 18 electrons possible, only eight are added in all, an outer shell of these being completed with argon, which ends a period of 8 elements. With potassium, the next element to argon, a new shell begins with an electron with $n=4$ and the shell of electrons with $n=3$ begins to fill up later. The $n=4$ shell contains 8 electrons when krypton is reached, and the period contains 18 elements.

In the next period a shell with $n=5$ begins to be formed from rubidium and contains 8 electrons when xenon is reached, when the $n=4$ shell has also filled up to 18 electrons. The next period beginning with caesium begins an outer shell with $n=6$; from lanthanum the $n=4$ shell begins to fill up to its maximum of 32 electrons with lutetium, the outer shells of $1(n=5)+2(n=6)$ valency electrons remaining unchanged. Thus the 15 rare-earth elements from lanthanum to lutetium all have very similar properties. From hafnium to radon the $n=5$ shell fills up to 18, and the $n=6$ shell to 8 electrons. This period, from caesium to radon, contains 32 elements.

The next period, beginning with the artificial element francium (see p. 378), contains electrons with $n=7$ in the outer shell. It is very incomplete, the last natural element being uranium (atomic number 92), and all the elements in it are radioactive. Some higher elements in the period have been obtained artificially (p. 379).

	$n=1$	2	3	4		$n=4$	5	6	7
Li	2	1	–	–	Rb	8	1	–	–
Ne	2	8	–	–	Xe	18	8	–	–
Na	2	8	1	–	Cs	18	8	1	–
A	2	8	8	–	La	18	8+1	2	–
K	2	8	8	1	Lu	32	8+1	2	–
Kr	2	8	18	8	Rn	32	18	8	–
					Fr	32	18	8	1
					U	32	18	8+4	2

The Electronic Theory of Valency

In the old theory of valency no distinction was made between ionized compounds such as sodium chloride and non-ionized substances such as carbon dioxide. The theories proposed by Kossel and Lewis recognized that they are essentially different. If the outer electrons in an atom are shown by dots, the formation of sodium chloride and of the chlorine molecule may be represented by:

$$\mathrm{Na}\cdot + .\underset{\cdot\cdot}{\overset{\cdot\cdot}{\mathrm{Cl}}}: = [\mathrm{Na}]^{+} + [:\underset{\cdot\cdot}{\overset{\cdot\cdot}{\mathrm{Cl}}};]^{-} \qquad :\underset{\cdot\cdot}{\overset{\cdot\cdot}{\mathrm{Cl}}}\,. + \cdot\underset{\cdot\cdot}{\overset{\cdot\cdot}{\mathrm{Cl}}}: = :\underset{\cdot\cdot}{\overset{\cdot\cdot}{\mathrm{Cl}}}:\underset{\cdot\cdot}{\overset{\cdot\cdot}{\mathrm{Cl}}}:$$

The first contains charged ions (already present in the crystal) held together by electrovalency with no real valency bond; the second is a neutral molecule in which the parts are held together by a valency bond or covalency, consisting of a shared pair of electrons with opposite spins. A double bond is formed by two shared pairs of electrons and a triple bond by three shared pairs:

$$:\overset{\cdot\cdot}{\mathrm{O}}:+:\mathrm{C}:+:\overset{\cdot\cdot}{\mathrm{O}}:=:\overset{\cdot\cdot}{\mathrm{O}}::\mathrm{C}::\overset{\cdot\cdot}{\mathrm{O}}: \qquad :\mathrm{N}:.+\cdot:\mathrm{N}:=:\mathrm{N}:::\mathrm{N}:$$

corresponding with the usual formulations:

$$\mathrm{O{=}C{=}O} \quad \text{and} \quad \mathrm{N{\equiv}N}.$$

In some cases a molecule contains an odd electron, such as nitric oxide $:\overset{\cdot\cdot}{\mathrm{O}}::\mathrm{N}:.$ and chlorine dioxide $:\overset{\cdot\cdot}{\mathrm{O}}::\underset{\cdot\cdot}{\overset{\cdot}{\mathrm{Cl}}}::\overset{\cdot\cdot}{\mathrm{O}}:$, such molecules being paramagnetic.

The outer shells of electrons on some atoms, which can function as valency electrons, are shown below:

$$:\mathrm{C}: \qquad :\mathrm{N}:. \qquad :\overset{\cdot\cdot}{\mathrm{O}}: \qquad :\mathrm{P}:. \qquad :\overset{\cdot\cdot}{\mathrm{S}}:$$

In formulating the structures of the ions of oxyacids Langmuir (1919) used another type of bond, which was emphasized by G. A. Perkins,* who called it a "borrowing union". This is a covalent bond in which both the electrons of the pair come from one atom instead of one from each of two atoms. Langmuir represented the sulphate ion SO_4'' as formed by the sulphur atom with six outer electrons acquiring two more and becoming an ion S'', the four

* *Philippine J. Sci.*, 1921, xix, 1.

pairs of electrons of which form bonds with oxygens, all the atoms
having octets of electrons in the structure (I), whilst the conventional valency structure has two single and two double bonds as in
sulphuric acid (II) :

(I)
$$
\begin{array}{c}
:\overset{\cdot\cdot}{O}: \\[2pt]
:\overset{\cdot\cdot}{\underset{\cdot\cdot}{O}}:\overset{\cdot\cdot}{\underset{\cdot\cdot}{S}}:\overset{\cdot\cdot}{\underset{\cdot\cdot}{O}}: \\[2pt]
:\underset{\cdot\cdot}{O}:
\end{array}
$$

(II)
$$
\begin{array}{c}
H-O \\
\diagdown \\
\end{array}
S
\begin{array}{c}
O \\
\diagup \\
\end{array}
$$

H—O⟍ ⟋O
       S
H—O⟋ ⟍O

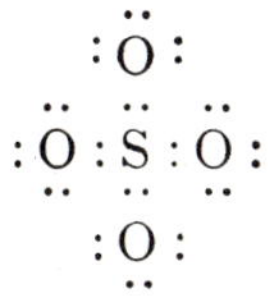

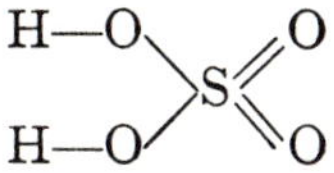

The case of nitrogen is interesting. Although the outer shell of
the atom has five electrons (see above) these do not form five bonds
and nitrogen has never a valency of five, as was formerly assumed
to be the case in ammonium compounds such as NH_4Cl and nitric
acid $HO.NO_2$. The ammonium compounds all contain the
ammonium ion $NH_4\cdot$ which is formed by the addition of a positive
proton $H\cdot$ or hydrogen nucleus to the unshared pair of electrons
on the nitrogen in the ammonia molecule, the maximum valency
of the nitrogen being four :

$$
\begin{array}{ccc}
H & & H \\
H:\overset{\cdot\cdot}{\underset{\cdot\cdot}{N}}:+H\cdot & = & H:\overset{\cdot\cdot}{\underset{\cdot\cdot}{N}}:H^+ \\
H & & H
\end{array}
$$

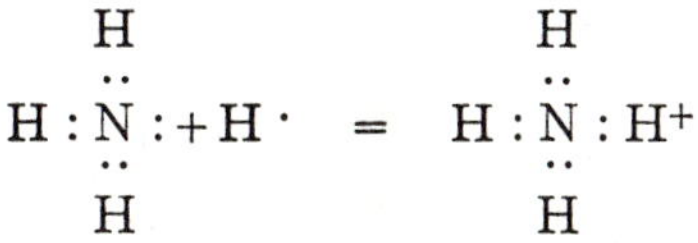

In the nitrate ion an extra electron is added, bringing up the number associated with the nitrogen atom to six and giving the ion a
negative charge ; three bonds are formed by adding oxygen atoms
to the three electron pairs in the formulation due to Langmuir :

$$
\begin{array}{c}
:\overset{\cdot\cdot}{O}: \\[2pt]
:\overset{\cdot\cdot}{\underset{\cdot\cdot}{O}}:\overset{\cdot\cdot}{N} \\[2pt]
:\underset{\cdot\cdot}{O}:
\end{array}
$$

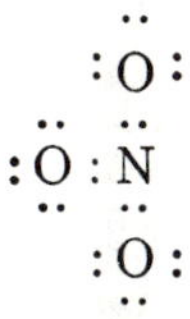

Nitric acid, however, behaves in many reactions as though it
has the structure $HO.NO_2$, e.g. in the nitration of benzene :

$$C_6H_6 + HO.NO_2 = C_6H_5NO_2 + H_2O.$$

The nitro-group NO_2 was originally formulated as containing
5-valent nitrogen, and nitric acid as (III, p. 368). Nitric acid can be
represented as containing 4-valent nitrogen (as in the ammonium

ion) by replacing one double bond between nitrogen and oxygen by a single bond formed by the electron pair on the nitrogen being donated to an oxygen atom, as shown in (IV) :

(III) 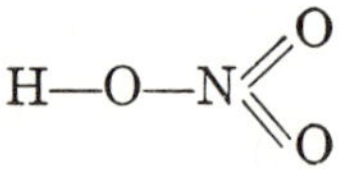 (IV) 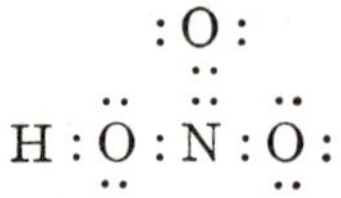

Lowry * and Sidgwick † regarded such bonds, formed by donation of a pair of electrons from one atom instead of by sharing an electron from each of two atoms, as different from ordinary single bonds.

Since the donation of an electron pair by one atom will give it a positive charge and its reception by another atom will give this a negative charge, Lowry regarded the bond as a combination of an electrovalency and a covalency, and hence formally as a double bond, which he called a " mixed double bond " or a " dative bond ", represented by $A^{+}—B^{-}$. Sidgwick, who recognized that such bonds are present in co-ordination compounds (see below), called the bond a " coordinate link " and represented it by an arrow pointing to the atom which has received a negative charge, $A \rightarrow B$. Sidgwick also formulated a rule that the maximum covalencies in the second period of the periodic table (p. 349) is 4, so that nitrogen in nitric acid (V) has four bonds, and sulphuric acid (VI) could be formulated in a similar way, with two coordinate links to the sulphur :

(V) 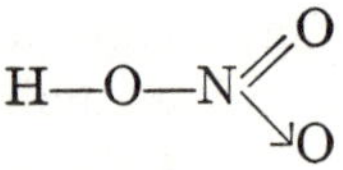 (VI) 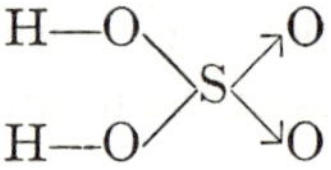 (VII) 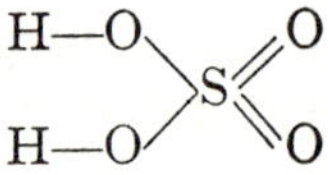

Since, however, the existence of a stable hexafluoride of sulphur, SF_6, shows that sulphur may have a maximum valency of six, there is no objection to the old structural formula (VII) of sulphuric acid, and this is now often preferred to (VI).

There are many reasons for supposing that the bonds from nitrogen to oxygen in nitric acid are all the same, whereas formula

* *Trans. Faraday Soc.*, 1923, xviii, 285 ; *J. Chem. Soc.*, 1923, cxxiii, 822 ; *J. Soc. Chem. Ind.*, 1923, xlii, 1004R, 1048R.

† *J. Chem. Soc.*, 1923, cxxiii, 725 ; *The Electronic Theory of Valency*, Oxford, 1927 ; W. G. Palmer, *Valency, Classical and Modern*, Cambridge, 1944.

(V) implies that three different kinds of bonds are involved. Pauling * suggested that the distribution of the electrons is smoothed out by rapid shifting of electron pairs into the various configurations possible in the molecule, and he called the process " resonance ". This makes all the bonds identical. In the case of molecules containing an odd electron (p. 366) this is not regarded as located on one particular atom but to be shared by resonance among the various atoms in the molecule.

It is not supposed that a compound exhibiting resonance is a mixture of molecules having the different configurations, but that all the molecules are identical and contain contributions from the various possible resonance structures. The theory has also been applied to benzene, which Kekulé's formula (p. 291, III) represents as containing alternate single and double bonds in the ring. Owing to resonance, all the bonds are the same and are intermediate in character between single and double bonds.

Atoms heavier than helium have more than one electron shell and Bohr attempted to find the arrangement of the electrons in successive shells from the atomic spectra. His scheme was modified by Bury † (1921) and Main Smith ‡ (1924). In some cases the electrons of a shell below the outer one may function as valency electrons and the varying valencies of some elements, particularly the so-called transitional elements having incomplete inner shells, can be explained. In the case of the rare-earth elements, electrons added to the atom as the nuclear charge increases go into a deep shell so that the valency does not change as it does with most other elements in such a process.

Coordination Compounds

In the early period of the theory of valency much difficulty was felt in explaining what were often called " molecular compounds ", which are formed by the addition of molecules in which all the

* *J. Amer. Chem. Soc.*, 1931, liii, 1367, 3225; 1932, liv, 988, 3570; *The Nature of the Chemical Bond*, New York, 1939. The phenomenon was recognized earlier in some theoretical discussions by others, but its application in chemistry is largely due to Pauling.

† *J. Amer. Chem. Soc.*, 1921, xliii, 1602.

‡ *J. Soc. Chem. Ind.*, 1924, xliii, 323.

valencies are saturated. Many of these are very stable. Platinic chloride $PtCl_4$ readily combines with hydrochloric acid to form H_2PtCl_6, and it was recognized that this contains a radical $PtCl_6$ which also exists in many salts such as K_2PtCl_6. The valency of platinum was taken as two in the platinous compounds such as $PtCl_2$ and four in platinic compounds such as $PtCl_4$. In H_2PtCl_6 it is apparently six. Many salts combine with ammonia to form compounds called ammines (different from amines, which contain the radical NH_2) and some of these are very stable, particularly compounds containing tervalent cobalt, the simple compounds of which are not stable. Although $CoCl_3$ is not known as such, many compounds containing it, particularly in combination with ammonia, such as $CoCl_3$, $6NH_3$, are known. Attempts were made to explain these by assuming 5-valent nitrogen, e.g. by Blomstand (1869), but the first successful theory was proposed by Alfred Werner (1866-1919) in 1893.*

Werner postulated that such compounds contain a central atom such as platinum or cobalt to which are attached by what he called " subsidiary valencies " a number of atoms or groups. These were said to be " coordinated " with the central atom (hence the name " coordination compounds " applied to such substances) and to form with it a stable complex called the " nucleus " of the compound. These atoms or groups are not ionizable. The normal or " principal valency " of the central atom is regarded as electropositive. If the metal is coordinated with neutral molecules such as ammonia its principal valency remains unchanged ; but if negative ions such as chlorine enter the nucleus the positive charge of the central atom is decreased by the amount of the entering negative charge. The nucleus may, therefore, have a positive charge, or be neutral, or (when the negative charge introduced exceeds the principal valency of the central atom) become negative and can combine with an appropriate number of positive ions outside the nucleus.

The symbol of the nucleus is enclosed in square brackets and those of ions outside it are written outside the brackets. A positive and a negative nucleus may also combine to form a neutral com-

* *Z. anorg. Chem.*, 1893, iii, 267 ; *Ber.*, 1905, xl, 15 ; *New Ideas on Inorganic Chemistry*, transl. Hedley, 1911.

pound. If the principal valencies of cobalt and platinum in the cobaltammines and platinic compounds are taken as 3 and 4, respectively, and the valencies of ferrous and ferric iron as 2 and 3, the following formulae are deduced from Werner's theory:

$$[Co(NH_3)_6]Cl_3 \text{ forming } [Co(NH_3)_6]^{\cdots} + 3Cl'$$

$$[CoCl(NH_3)_5]Cl_2 \quad [CoCl_2(NH_3)_4]Cl \quad [Co(NO_2)_3(NH_3)_3]$$
$$K_3[Co(NO_2)_6]$$

$$[Pt(NH_3)_6]Cl_4 \quad [Pt(NH_3)_5Cl]Cl_3 \quad [Pt(NH_3)_4Cl_2]Cl_2$$

$$[Pt(NH_3)_3Cl_3]Cl \quad [Pt(NH_3)_2Cl_4] \quad [Pt(NH_3)Cl_5]K \quad [PtCl_6]K_2$$

$$K_4[Fe^{II}(CN)_6] \quad K_3[Fe^{III}(CN)_6]$$

In all the above compounds the number of atoms or groups coordinated with the central atom is six, and this is called the "coordination number" of the central atom. In some cases, as with bivalent platinum, the coordination number is four:

$$[Pt(NH_3)_4]Cl_2 \quad [Pt(NH_3)_3Cl]Cl \quad [Pt(NH_3)_2Cl_2]$$
$$[Pt(NH_3)Cl_3]K \quad [PtCl_4]K_2$$

In a few cases the coordination number is eight, as in $K_4[W(CN)_8]$, rarely it is seven, five, three or two. The commonest numbers are six and four.

In all the above examples, each atom or group in the nucleus corresponds with a coordination number of one, or occupies one "coordination position"; but in some cases, as with the oxalate ion and diamines such as ethylenediamine ($NH_2.CH_2.CH_2.NH_2$), a group occupies two coordinate positions, and cases of groups occupying three, four, and six positions are known.

Werner was able to support his theory by explaining known cases of isomerism and by predicting such cases and afterwards showing that they exist. Cases of isomeric compounds are:

$$[Co(SO_4)(NH_3)_5]Br \text{ and } [CoBr(NH_3)_5]SO_4$$

$$[Cr(NH_3)_6][Cr(SCN)_6] \text{ and } [Cr(NH_3)_4(SCN)_2]\,Cr(NH_3)_2(SCN)_4]$$

$$[Cr(OH_2)_6]Cl_3 \quad [CrCl(OH_2)_5]Cl_2 + H_2O \quad [CrCl_2(OH_2)_4]Cl + 2H_2O.$$

Very interesting cases of isomerism arise on account of the different arrangement of the atoms or groups in space around the central atom, i.e. geometrical isomerism. When the coordination number is four, the arrangement may be tetrahedral or planar. In the planar case the compound $MeABX_2$ (Me = central atom) may exist in two forms, a *cis*-isomer when the two X groups are adjacent and a *trans*-isomer when they are opposite :

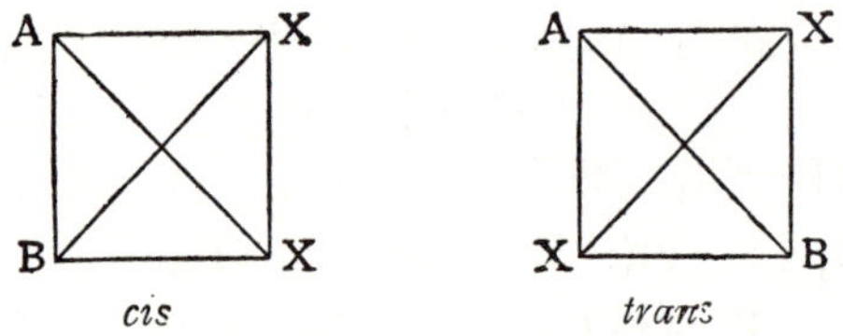

When the coordination number is six, the groups are arranged at the corners of an octahedron and two isomers of the nucleus MeR_4X_2 are :

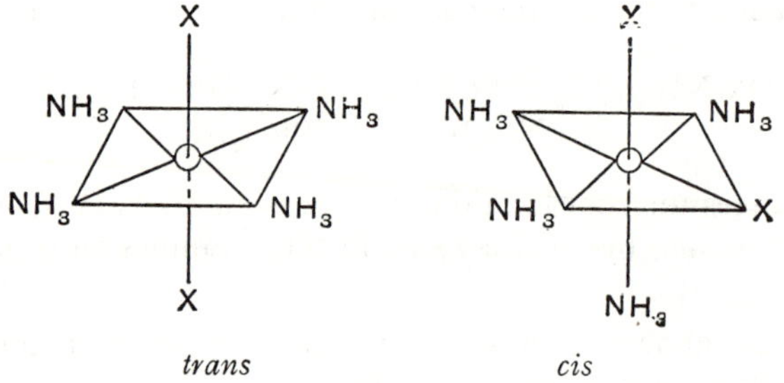

FIG. 125.—*Cis-trans* ISOMERISM IN OCTAHEDRAL CONFIGURATION.

If the molecule contains groups occupying two coordination positions, such as ethylene diamine, two *cis*-isomers and a *trans*-isomer are possible :

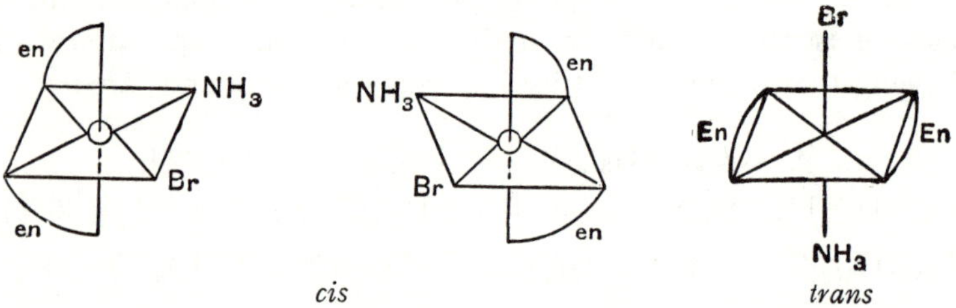

FIG. 126.—OPTICAL ACTIVITY DUE TO ENANTIOMORPHISM IN OCTAHEDRAL *cis*-CONFIGURATION.

and since the two cis-forms are mirror-images they are optically active (p. 305). The optical activity is explained very elegantly by Werner's theory, but it was suggested that it might be due to the carbon atoms in the ethylenediamine in some unexplained way. Werner (1914) disproved this by preparing a compound with high optical activity quite free from carbon:

$$[Co\{Co(OH)_2(NH_3)_4\}_3]Br_6$$

in which three groups containing cobalt are coordinated with cobalt. The arrangement around the central atom is that shown below:

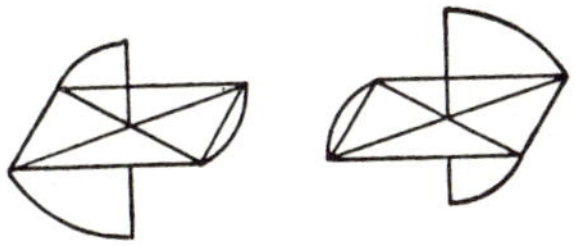

which is also known when the group occupying two coordination positions is the oxalate radical C_2O_4''. It will be noticed that the group coordinated with cobalt is positive (+3 for Co, −2 for $(OH)_2$, giving +1) and with three groups the valency of the nucleus becomes 3+3=6; this is rare, the valency of the nucleus usually becoming more positive by expulsion of negative ions from it by neutral molecules, as is illustrated in the examples on p. 371.

The explanation of coordination in terms of the electronic theory of valency was given by Sidgwick (1923, see p. 368). Groups which coordinate with metals contain lone pairs of electrons on oxygen or nitrogen, e.g. ammonia $H_3N:$ or ethylenediamine. These lone pairs are donated to the central atom to form coordinate links. Since no charge is added to or removed from the central atom by such an addition of a neutral group, its principal valency remains unaltered. But if the donated electron pair is on a negative ion such as the chlorine ion $:\ddot{\underset{..}{Cl}}:$, the positive charge of the nucleus is decreased, and if such ions are replaced by neutral molecules their expulsion takes away negative charge and the positive valency of the central atom is increased. All the formulae given above are thus easily explained.

Since the bonds formed are electron pairs, they are normal covalent bonds and Werner's name " subsidiary valencies " is not appropriate and is no longer used.

The Theory of Directed Valencies

De Broglie (1924) assumed that electrons in motion are associated with some kind of waves the lengths of which are given by Planck's constant $\boldsymbol{h}$ divided by the momentum $mv = p$ of the electron:

$$\lambda = \boldsymbol{h}/mv = \boldsymbol{h}/p$$

This was confirmed by the diffraction of beams of electrons in the same way as X-rays. Schrödinger (1926) developed a "wave-equation", which represents mathematically the distribution of charge of an electron regarded as distributed through space with a certain density of charge at a particular point. Born (1926) showed that the same results follow if the electron is regarded as a point-charge, but a probability of finding it in a particular place is calculated.

Schrödinger's wave equation is a differential equation containing a wave function ψ in the form of second differential coefficients with respect to the coordinates x, y and z. There is also a term which in the ordinary equation for wave motion contains the wave-length λ. In this case λ is taken as the length defined by de Broglie's equation above. It is assumed that ψ is finite, continuous and single valued in general for all values of the coordinates, and that the square of the wave function, ψ^2, gives the density of charge at each point. The last is an important result, since if the electron is in an atom a knowledge of the charge distribution gives an indication of the directions in which bond formation may occur.

Schrödinger's wave theory led to the conclusion that the charge density of an electron in an atom may be either spherically symmetrical, as it is in the hydrogen atom, or more prominent in certain directions, and if the formation of a valency bond is interpreted as the pairing of an electron in one atom with another electron having an opposite spin in another atom, the existence of directed valency bonds follows. That valency bonds have definite directions in space had been assumed by chemists on the basis of the facts of stereochemistry, e.g. the explanation of optical isomers requires the valencies of a carbon atom to be directed towards the corners of a tetrahedron (p. 305).

Electrons having values of the quantum number l (p. 363) equal to 0, 1, 2, 3, etc., are called s, p, d, f, etc., electrons. Pauli's principle is assumed, also the "rule of maximum multiplicity" due to Hund (1925), which states that when electrons are added successively to an atom as many levels or "orbits" are singly occupied as possible before any pairing of electrons with opposite spins occurs.

Oxygen has four *p*-electrons ($l=1$) and there are three *p*-orbits which the wave theory shows are directed along three axes *x*, *y* and *z* at right angles, the wave functions being p_x, p_y and p_z. These functions are such that, when the orbits are occupied by electrons, the charge densities are distributed in the way shown by the hour-glass shapes in Fig. 127*a*. Of the four *p*-electrons, two go singly

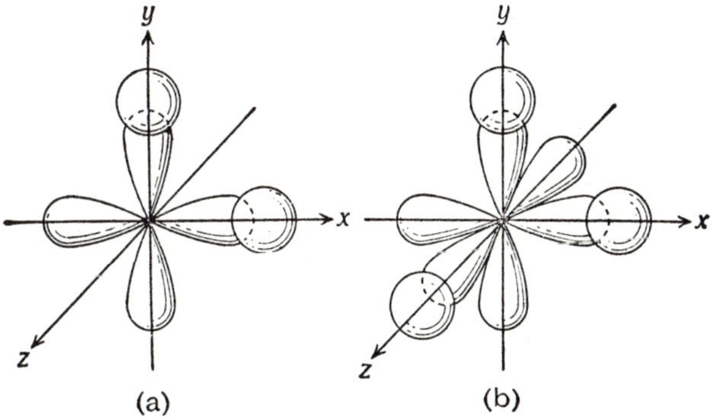

FIG. 127.—FORMATION OF DIRECTED BONDS.

into orbits according to Hund's rule, but since only one orbit is left, the other two electrons must go into it together, with opposite spins. The two unpaired *p*-electrons can combine with *s*-electrons of hydrogen atoms, which are spherical, and the maximum overlap of functions occurs when the hydrogen atoms approach along the *x*- and *y*-axes and their electrons overlap the *p*-orbits as shown. The water molecule should therefore have two O—H bonds at right angles. Owing to various secondary effects, the angle is somewhat larger than 90°, but with H_2S it is practically 90°.

Nitrogen has three *p*-electrons and these can occupy singly the p_x, p_y and p_z orbits and combine with three *s*-electrons of hydrogen atoms to form ammonia, the three N—H bonds being at right angles (Fig. 127*b*).

The normal carbon atom has two *p*-electrons and it might be expected that a compound CH_2 would be formed. The normal valency of carbon, however, is 4 and the valency bonds are arranged tetrahedrally. To produce this state the following process is supposed to occur. One of the $2s$ electrons ($n=2$, $l=0$) is given energy or excited to become a $2p$-electron ($n=2$, $l=1$), when the atom has four unpaired spins, i.e. those for the one $2s$, the original two $2p$, and the new $2p$ formed from a $2s$. The $2s$ wave function is undirected (spherical symmetry) and the three $2p$ functions would be like those for nitrogen (Fig. 127*b*). When, however, the bond energy greatly exceeds the excitation energy for raising an electron from an s to a p level, the s and p functions combine to produce four equivalent wave functions directed tetrahedrally. This process is called " hybridization ", since the wave functions are a cross between s and p functions. Exactly the same process occurs with the nitrogen atom in the ammonium ion NH_4 and this has the same tetrahedral configuration as methane CH_4.

In the case of coordination compounds, d functions ($l=2$) are also involved, and octahedral and planar configurations are formed, in agreement with the stereochemical results. This prediction of directed bonds is one of the most interesting results of the wave theory of the electron, and the relative strengths of the bonds may also be calculated.

From the earliest days of the electronic theory of valency attempts were made to explain the structure and reactions of organic compounds in terms of the behaviour of the electrons in their molecules. Information about these can be obtained from absorption spectra. The most interesting results have been found by applications of wave mechanics, i.e. considerations based on Schrödinger's wave equation. Every electron in a molecule is represented by a wave function ψ called a "molecular orbital", which is defined by quantum numbers (including a spin quantum number) which determine its energy and shape. In place of the quantum number l for an electron in an atom there is a quantum number λ, the values of which are denoted by σ, π, δ, ϕ instead of s, p, d, f in the atomic case, both sets of quantum numbers corresponding with values of 0, 1, 2, 3. A σ-electron has $\lambda=0$, a π-electron has $\lambda=1$. A single bond between two carbon atoms in a molecule is formed by a pair of electrons and is called a σ-bond. Its direction is that joining the two carbon nuclei. A double bond has four electrons and is composed of a σ-bond between the two carbon nuclei and a π-bond at right angles to this direction. In ethylene, $H_2C{=}CH_2$, there is a σ-bond between the carbons and two overlapping π-bonds in a plane at right angles to the plane containing the carbon and hydrogen atoms. In the benzene molecule there are six σ-bonds in the plane of the ring and six π-bonds with hour-glass shaped wave functions at right angles to the plane of the ring.

This representation alters somewhat the picture of double-bond formation assumed by van't Hoff (p. 306), since the bond is not now regarded as formed by two tetrahedra joined by an edge, but as a single σ-bond joining the carbons and an overlap of π-bonds at right angles to the plane of the σ-bonds. The overlapping π-bonds restrain the rotation of the two parts of the molecule, so that the final result is the same as if the tetrahedra were locked together by an edge. The type of hybridization is also different from the tetrahedral in the normal carbon atom (p. 376).

In ethylene, $H_2C{=}CH_2$, there are formed on each carbon atom three σ-bonds at angles of 120°, so-called " trigonal hybridization ", and a π-bond at right angles. One of the trigonal bonds of each atom contributes to a single bond between the two carbon atoms, the other two σ-bonds on each carbon atom form links with hydrogen atoms, the two carbon atoms and the four hydrogen atoms being in a plane.

In acetylene, $HC{\equiv}CH$, there is " digonal hybridization ", a σ-bond being formed between the two carbon atoms and two π-bonds from each in planes through the axis of the molecule and at right angles to one another.

Although it was formerly thought that two parts of a molecule could rotate freely when joined by a single bond, as in the tartaric acid molecule (p. 305), some cases of isomerism indicate that restricted rotation may appear even in such a case. This is normally present when the two parts contain large groups which cannot pass one another, and is shown in derivatives of diphenyl, which contains two benzene rings linked at one corner by a single bond.

The Transmutation of Elements

The great energy of the alpha particles emitted from radioactive elements led to the hope that by bombarding other atoms with them these might be broken up and other elements produced. It was thought that copper had been changed into lithium in this way, but this and similar changes were later shown to be illusory. The first real transmutation of one element into another was effected by Rutherford (1919), who showed that when nitrogen atoms are bombarded with alpha particles, hydrogen nuclei or protons are emitted from them. Blackett (1922) then showed that the change occurs by an alpha particle entering the nucleus of the nitrogen atom, and since the nuclear charge is raised from 7 to 9 an isotope of fluorine is formed. This emits a proton, leaving a nucleus of charge 8, an isotope of oxygen. In other nuclear changes, brought about by collisions with alpha particles, hydrogen or deuterium nuclei, or neutrons, there may be emission of hydrogen nuclei (protons), neutrons, or alpha particles.

A new type of change was first observed by Joliot and Irene Curie in 1934. In this the unstable nucleus formed by absorbing a bombarding particle emits a positive electron, a particle with the same mass as the negative electron but a positive charge. Natural radioactive elements emit beta rays or negative electrons but never positive electrons. In this type of change many otherwise unknown isotopes of elements were produced, e.g. radioactive phosphorus by bombarding aluminium with alpha particles, when the aluminium atom with a nuclear charge of 13 absorbs an alpha particle of charge 2 and the resulting particle emits a neutron and leaves a particle of nuclear charge 15, an isotope of phosphorus. This emits a positive electron and leaves a radioactive isotope of silicon with the nuclear charge 14. The chemical nature of the atom is, as was mentioned on p. 359, determined by its nuclear charge ; but it was also confirmed by chemical tests, the phosphorus isotope being precipitated along with the common element as zirconium phosphate, when the precipitate showed radioactivity with the same rate of decay as the original material.

The number of artificial elements has swelled to very large proportions. Some elements not known in nature have been obtained to fill gaps in the periodic table with atomic numbers 43 (technetium), 85 (astatine), and 87 (francium) ; these places had been left empty until the artificial elements were discovered. The most spectacular discovery was that of elements beyond uranium (which are not shown in the table on p. 349), the so-called " transuranium elements ", which are all radioactive. Natural uranium contains the isotopes of mass 235 and 238, mostly 238. When uranium 238 is bombarded with neutrons it forms an element of atomic number 93 by the combined nucleus emitting a beta ray (charge -1) and this element is neptunium. By emission of a beta ray from neptunium the element plutonium of atomic number 94 is formed. Bombardment of 238 uranium and plutonium with helium nuclei artificially accelerated (instead of natural alpha rays) gives two elements of atomic numbers 95 (americium) and 96 (curium), and by continuing the process, elements of atomic numbers 97 (berkelium), 98 (californium), and up to 100 or so have been obtained.*

The uranium isotope 235 when bombarded with neutrons undergoes fission. In this process more neutrons are liberated and the change becomes a self-supporting chain reaction with a very large liberation of energy. This is the principle of the atomic bomb and the production of nuclear energy. The fusion of protons and neutrons to form helium nuclei at the enormous temperature of an atomic-bomb explosion is the principle of the hydrogen bomb.

The dream of the alchemists was realized by the artificial production of gold from mercury by Sherr, Bainbridge and Anderson (1941); but the reverse production of mercury from gold is of more practical interest, since this mercury contains only one kind of atom and is more suitable for the production of standard spectrum lines, because the different isotopes in natural mercury give satellite lines in the spectrum.

Summary

Atomic Structure and the Electronic Theory of Valency

The cathode rays, discovered by Plücker in 1859, were shown by J. J. Thomson (1897) to consist of negative electrons. The discovery of radioactivity by Becquerel (1896) led to the discovery of radium by P. and Mme. Curie (1898). Rutherford and Soddy (1903) explained radioactivity by the hypothesis of atomic disintegration.

* Seaborg, Katz and Manning, *The Transuranium Elements*, New York, 1949.

Soddy, and independently T. W. Richards, discovered the isotopes of lead from uranium and thorium in 1913, and in the same year J. J. Thomson showed by the method of positive rays that neon is a mixture of two isotopes. The method was applied to the other non-radioactive elements by Aston, and most of these elements were found to be mixtures of isotopes.

In 1911 Rutherford suggested the " nuclear atom ", consisting of a small positive nucleus and outer electrons. Moseley in 1913 worked out a method of determining the nuclear charge from the frequency of X-rays emitted by an element. This nuclear charge is equal to the atomic number, giving the position of an element in the periodic table. An element is defined by its atomic number, and different isotopes of an element have the same atomic number.

Atomic models were devised by Bohr (1913), G. N. Lewis (1916) and I. Langmuir (1919). Lewis distinguished between ionic and non-ionic (covalent) linkages. The first are formed by transfer of electrons and production of separate charged ions (as explained by Kossel, 1916), the second, according to Lewis, by sharing of electrons in pairs, a single bond consisting of one shared pair, a double bond of two, and a triple bond of three.

A type of bond produced when both the electrons of a pair come from the same atom was used by Langmuir (1919). Special attention was drawn to it by G. A. Perkins (1921). Lowry (1923) called it a " mixed double bond " or " dative bond ", and Sidgwick (1923) a " coordinate link ", since such bonds are present in the so-called coordination compounds. Pauling (1931) suggested that different types of bonds in molecules can become equalized by rapid shifting of electrons, a process he called " resonance ". The resulting " resonance hybrid " contains contributions from the different possible electronic configurations.

Coordination compounds, formerly called " molecular compounds ", are formed by the addition of molecules in which the valencies of the atoms are apparently satisfied. They were investigated by Werner from 1893, who showed that the number of atoms or groups associated with a central atom is often 4 or 6, the " coordination number " (other coordination numbers, up to a maximum of 8, occur, but less often). Various types of isomerism, including optical isomerism, are explained by Werner's theory. Sidgwick (1923) pointed out that groups which coordinate with metals do so by contributing electron pairs to the valency shell of the metal, hence the name " coordinate link " given to this type of bond.

The structure of the periodic table was explained by Bohr and others on the basis of the four quantum numbers of the electrons and Pauli's principle (1925), that in any one atom there is never more than one electron having a given set of the four quantum numbers.

Wave mechanics, developed by Schrödinger (1926) from the idea of the " wave nature " of the electron postulated by de Broglie (1924), leads to a theory of directed valency bonds, since the charge density associated with an electron may have decided directions in space, corresponding with the directions of the valency bonds. In some cases the wave functions which replace the electrons in the theory are " hybrids ", since they are formed from different types of functions. The tetrahedral carbon valencies are produced by hydridization.

The application of the modern theory of valency in Organic Chemistry has led to new formulations of many compounds, such as ethylene and benzene, the valency bonds in which are now regarded as examples of bonds produced by hybridization instead of the conventional double bonds.

The transmutation of elements was first achieved by Rutherford (1919), who expelled hydrogen nuclei (protons) from nitrogen atoms by bombarding them with alpha particles, and since then many artificial elements have been formed by bombardment with various types of fast particles. In some cases positive electrons are emitted in the processes. In this way elements not known in nature have been discovered, filling previously vacant spaces in the periodic table, and elements heavier than uranium (the " trans-uranium elements ") have been made artificially from uranium.

A SHORT BIBLIOGRAPHY

THE following bibliography contains particulars of some works which may be consulted for further details, and also books on the History of Chemistry (of varying value). A detailed statement of the particular uses to which the various items can be put would occupy far more space than is available, but it may be said that papers published after 1800 are most easily traced in the *Royal Society Catalogue of Scientific Papers* and in the various volumes of the *Jahresbericht über die Fortschritte der Chemie* (from 1847); the tracing of literature before 1800 is an art which can be learnt only by experience. The place of publication of English books is London, of French books Paris, unless otherwise stated. General works of reference are not included. Monographs and special works are quoted in footnotes in the text.

Alembic Club Reprints, several vols., Edinburgh, var. dates.
Armitage, F. P., *A History of Chemistry*, 1906 ; 4th impression, 1928.
Bergman, T., *Physical and Chemical Essays*, transl., vol. iii, Edinburgh, 1791.
Berry, A. J., *Modern Chemistry. Some Sketches of its Historical Development*, Cambridge, 1946.
Berthelot, M. (1) *Les origines de l'alchimie*, 1885 ; (2) *Collection des anciens alchimistes grecs*, 3 vols., 1887-8 ; (3) *Introduction à l'étude de la chemie des anciens et du moyen âge*, 1889 ; (4) *La révolution chimique : Lavoisier*, 1890 ; (5) *La chimie au moyen âge*, 3 vols., 1893 ; (6) *Archéologie et histoire des sciences*, 1906.
Boerhaave, H., *A New Method of Chemistry*, transl. P. Shaw, vol. i, 1741.
Bolton, H. C., *Bibliography of Chemistry*, Washington, 1893 ; *First Supplement*, 1899 ; *Sect. viii, Academic Dissertations*, 1901 ; *Second Supplement*, 1904.
Brande, W. T., *A Manual of Chemistry*, 6th edit., vol. i, 1848.
Brown, A. C., *A History of Chemistry*, 1913 ; 2nd edit., 1920.
Catalogue des manuscrits alchimiques grecs, 8 vols., Brussels, 1924-32.
Chemical Society (London) *Memorial Lectures*, 4 vols., 1901-14-33-51; *ibid.*, *Faraday Lectures*, 1928 ; *ibid.*, *British Chemists*, 1947.
Classiques de la Science, Libraire A. Colin, various vols.
Daubeny, C., *An Introduction to the Atomic Theory*, 2nd edit., Oxford, 1850.
Delacre, M., *Histoire de la Chimie*, 1920.
Deventer, C. M. van, *Grepen uit de Historie de Chemie*, Haarlem, 1924.
Dumas, J. B. A., *Leçons sur la philosophie chimique*, 1837 ; reprinted, 1878, 1937.
Ephraim, F., *Chemische Valenz- und Bindungslehre*, Leipzig, 1928.
Feldhaus, F. M., *Die Technik der Vorzeit*, Leipzig and Berlin, 1914.
Ferguson, J., *Bibliotheca Chemica*, 2 vols., Glasgow, 1906.
Fester, G., *Die Entwicklung der chemischen Technik*, Berlin, 1922.
Figuier, L., *L'alchimie et les alchimistes*, 3rd edit., 1860.
Findlay, A., *A Hundred Years of Chemistry*, 1948.
Forbes, R. J., *Metallurgy in Antiquity*, Leiden, 1950.
Freund, Ida, *The Study of Chemical Composition*, Cambridge, 1904.
Gmelin, J. F., *Geschichte der Chemie*, 3 vols., Göttingen, 1797-9.
Graebe, C., *Geschichte der organischen Chemie*, vol. i, Berlin, 1920; vol. ii by P. Walden, Berlin, 1941.
Hilditch, T. P., *A Concise History of Chemistry*, 1911 ; 2nd edit., 1922.
Hjelt, E., *Geschichte der organischen Chemie*, Brunswick, 1916.
Hoefer, F., *Histoire de la chimie*, 2 vols., 1842-3 ; 2nd edit. (enlarged), 1866-9.

Hofmann, A. W. von, *Zur Erinnerung an vorangegangene Freunde*, 3 vols., Brunswick, 1888.

Holmyard, E. J. (1) *Chemistry to the Time of Dalton*, 1925 ; (2) *The Great Chemists*, 1928 ; (3) *Makers of Chemistry*, Oxford, 1931.

Hopkins, A. P., *Alchemy. Child of Greek Philosophy*, New York, 1934.

Index Catalogue of the Surgeon General's Office, Washington (U.S.A.), 16 vols., 1880-95 ; 2nd series, 21 vols., 1896-1916.

Jaeger, F. M., *Historische Studien*, Groningen, 1919.

Jagnaux, R., *Historie de la chimie*, 2 vols., 1891.

Jöcher, C. G., *Allgemeines Gelehrten-Lexicon*, 4 vols., Leipzig, 1750-1.

Kahlbaum, G. W. A., *Monographien aus der Geschichte der Chemie*, 8 vols., Leipzig, 1897-1904.

Kopp, H. (1) *Geschichte der Chemie*, 4 vols., Brunswick, 1843-7 (reprinted, Leipzig, 1931) ; (2) *Beiträge zur Geschichte der Chemie*, 3 pts., Brunswick, 1869-75 ; (3) *Die Entwickelung der Chemie in der neueren Zeit*, Munich, 1873 ; (4) *Die Alchemie in älterer und neuerer Zeit*, 2 pts., Heidelberg, 1886.

Ladenburg, A., *Lectures on the History of the Development of Chemistry since the Time of Lavoisier*, English transl., 1900 ; 2nd edit., 1906.

Leicester, H. M., and Klickstein, H. S., *A Source Book in Chemistry*, New York, 1952.

[Lenglet du Fresnoy, the Abbé] *Histoire de la philosophie hermétique*, 3 vols., 1742.

Lippmann, E. O. von, (1) *Abhandlungen und Vorträge zur Geschichte der Naturwissenschaften*, 2 vols., Leipzig, 1906-13 ; (2) *Entstehung und Ausbreitung der Alchemie*, 3 vols., 1919-31-54 ; (3) *Zeittafeln zur Geschichte der organischen Chemie*, Berlin, 1921 ; (4) *Beiträge zur Geschichte der Naturwissenschaften und der Technik*, 2 vols., 1923-53.

Lowry, T. M., *Historical Introduction to Chemistry*, 1915 ; 3rd edit., 1936.

Masson, I., *Three Centuries of Chemistry*, 1925.

Mercer, J. E., *Alchemy, its science and romance*, 1921.

Metzger, Hélène, (1) *Les doctrines chimiques en France*, 1923 ; (2) *Newton, Stahl, Boerhaave et la doctrine chimique*, 1930.

Meyer, E. von, *History of Chemistry*, English transl., 1891 ; 3rd edit., 1906.

Meyer, R., *Vorlesungen über die Geschichte der Chemie*, Leipzig, 1922.

Miall, S., *History of the British Chemical Industry*, 1931.

Moore, F. J., *A History of Chemistry*, 3rd edition, New York, 1939.

Muir, M. M. P., (1) *Heroes of Science ; Chemists*, 1883 ; (2) *The Alchemical Essence and the Chemical Element*, 1894 ; (3) *The Story of the Chemical Elements*, 1897 ; (4) *The Story of Alchemy and the Beginnings of Chemistry*, 1902; (5) *History of Chemical Theories and Laws*, New York, 1907.

Ostwald's *Klassiker der exakten Wissenschaften*, sev. vols., Leipzig, var. dates; Ostwald, W., *Lehrbuch der allgemeinen Chemie*, 3 vols., Leipzig, 1910-1; *ibid.*, *Elektrochemie. Geschichte und Lehre*, Leipzig, 1896.

Partington, J. R., (1) *The Composition of Water*, 1928; (2) *Origins and Development of Applied Chemistry*, 1935.

Picton, H. W., *The Story of Chemistry* [1889].

Poggendorff, J. C., *Biographisch-literarisches Handwörterbuch*, 6 vols., Leipzig, and Berlin, 1863-1940 (in progress).

Ramsay, W., (1) *Gases of the Atmosphere*, 1896, 4th edit., 1915; (2) *Introduction to the Study of Physical Chemistry*, 1904; (3) *Essays Biographical and Chemical*, 1908, 2nd impr., 1909.

Read, J., *Prelude to Chemistry*, 1936, 2nd edit. 1939; *ibid.*, *Humour and Humanism in Chemistry*, 1947; *ibid.*, *The Alchemist in Life, Literature and Art*, 1947.

Redgrove, H. S., *Alchemy Ancient and Modern*, 1911; 2nd edit., 1922.

Rodwell, G. F., *The Birth of Chemistry*, 1874.

Roscoe, H. E., and Schorlemmer, C., *Treatise on Chemistry*, vol. i, 5th edit., 1920 (also historical notices throughout vols. i-iii).

Royal Society (London) *Catalogue of Scientific Papers*, 19 vols., 1867-1925 (1800-1900; includes foreign papers).

Sarton, G., *Introduction to the History of Science*, 3 vols., Baltimore, 1927-31-47; *ibid.*, *A Guide to the History of Science*, Waltham, Mass. (U.S.A.), 1952.

Schmieder, K. C., *Geschichte der Alchemie*, Halle, 1832 (reissued Munich-Planegg, pref. dated 1927).

Schorlemmer, C., *The Rise and Development of Organic Chemistry*, Manchester, 1879; 2nd enlarged edit., London, 1894.

Singer, Mrs. D. W., *Catalogue of Latin and Vernacular Alchemical Manuscripts in Great Britain and Ireland*, 3 pts., Brussels, 1928-31.

Spielmann, J. R., *Instituts de Chymie*, transl. by Cadet, Paris, 1770, vol. ii (bibliography).

Stillman, J. M., *The Story of Early Chemistry*, New York, 1924.

Taylor, F. S., *The Alchemists*, 1951.

Thompson, R. C., *A Dictionary of Assyrian Chemistry and Geology*, Oxford, 1936; *ibid.*, *A Dictionary of Assyrian Botany*, London, 1949.

Thomson, T., (1) *A System of Chemistry*, 3rd edit., 5 vols., Edinburgh, 1807; (2) *History of the Royal Society*, 1812; (3) *History of Chemistry*, 2 vols., 1830-1, reprint, *n.d.*

Thorndike, L., *History of Magic and Experimental Science*, vols.

i-ii, London, 1923; vols. iii-iv, New York, 1934; vols. v-vi, 1941.

Thorpe, T. E., (1) *Essays in Historical Chemistry*, 1894, 2nd edit., 1902; (2) *History of Chemistry*, 2 vols., 1910-14.

Tilden, W. A., (1) *The Progress of Scientific Chemistry*, 1899, enlarged 2nd edit., 1913; (2) *Famous Chemists, the Men and their Work*, 1921, reprinted 1930.

Waite, A. E., *Lives of Alchemystical Philosophers*, 1888.

Whewell, W., *History of the Inductive Sciences*, 3rd edit., 3 vols., 1857.

Wiegleb, J. C., *Geschichte des Wachstums und der Erfindung in der Chemie*, 3 vols., Berlin, Stettin and Stuttgart, 1790-2.

Wolf, A., *A History of Science, Technology and Philosophy in the 16th and 17th Centuries*, 1936, 2nd edit., 1950; *ibid., in the Eighteenth Century*, 1938, 2nd edit., 1952.

Wurtz, A., (1) *A History of Chemical Theory from the Age of Lavoisier to the Present Time*, 1869; (2) *The Atomic Theory*, 1880 and later edits.

Z[eitlinger], H., and S[otheran], H. C., *Bibliotheca Chemico-Mathematica*, Sotheran and Co., London, 2 vols., 1921; *First Supplement*, 1932; *Second Supplement*, 1937.

INDEX OF NAMES

(*Pages containing biographical notices are in heavy type*)

INDEX OF SUBJECTS

B

C

I

M

P

Q

ORDINARY DIFFERENTIAL EQUATIONS, Morris Tenenbaum and Harry Pollard. Exhaustive survey of ordinary differential equations for undergraduates in mathematics, engineering, science. Thorough analysis of theorems. Diagrams. Bibliography. Index. 818pp. 5⅜ × 8½. 64940-7 Pa. $16.95

STATISTICAL MECHANICS: Principles and Applications, Terrell L. Hill. Standard text covers fundamentals of statistical mechanics, applications to fluctuation theory, imperfect gases, distribution functions, more. 448pp. 5⅜ × 8½. 65390-0 Pa. $9.95

ORDINARY DIFFERENTIAL EQUATIONS AND STABILITY THEORY: An Introduction, David A. Sánchez. Brief, modern treatment. Linear equation, stability theory for autonomous and nonautonomous systems, etc. 164pp. 5⅜ × 8¼. 63828-6 Pa. $5.95

THIRTY YEARS THAT SHOOK PHYSICS: The Story of Quantum Theory, George Gamow. Lucid, accessible introduction to influential theory of energy and matter. Careful explanations of Dirac's anti-particles, Bohr's model of the atom, much more. 12 plates. Numerous drawings. 240pp. 5⅜ × 8½. 24895-X Pa. $6.95

THEORY OF MATRICES, Sam Perlis. Outstanding text covering rank, non-singularity and inverses in connection with the development of canonical matrices under the relation of equivalence, and without the intervention of determinants. Includes exercises. 237pp. 5⅜ × 8½. 66810-X Pa. $7.95

GREAT EXPERIMENTS IN PHYSICS: Firsthand Accounts from Galileo to Einstein, edited by Morris H. Shamos. 25 crucial discoveries: Newton's laws of motion, Chadwick's study of the neutron, Hertz on electromagnetic waves, more. Original accounts clearly annotated. 370pp. 5⅜ × 8½. 25346-5 Pa. $10.95

INTRODUCTION TO PARTIAL DIFFERENTIAL EQUATIONS WITH APPLICATIONS, E.C. Zachmanoglou and Dale W. Thoe. Essentials of partial differential equations applied to common problems in engineering and the physical sciences. Problems and answers. 416pp. 5⅜ × 8½. 65251-3 Pa. $10.95

BURNHAM'S CELESTIAL HANDBOOK, Robert Burnham, Jr. Thorough guide to the stars beyond our solar system. Exhaustive treatment. Alphabetical by constellation: Andromeda to Cetus in Vol. 1; Chamaeleon to Orion in Vol. 2; and Pavo to Vulpecula in Vol. 3. Hundreds of illustrations. Index in Vol. 3. 2,000pp. 6⅛ × 9¼. 23567-X, 23568-8, 23673-0 Pa., Three-vol. set $41.85

CHEMICAL MAGIC, Leonard A. Ford. Second Edition, Revised by E. Winston Grundmeier. Over 100 unusual stunts demonstrating cold fire, dust explosions, much more. Text explains scientific principles and stresses safety precautions. 128pp. 5⅜ × 8½. 67628-5 Pa. $5.95

AMATEUR ASTRONOMER'S HANDBOOK, J.B. Sidgwick. Timeless, comprehensive coverage of telescopes, mirrors, lenses, mountings, telescope drives, micrometers, spectroscopes, more. 189 illustrations. 576pp. 5⅜ × 8¼. (Available in U.S. only) 24034-7 Pa. $9.95

CATALOG OF DOVER BOOKS

ROTARY-WING AERODYNAMICS, W.Z. Stepniewski. Clear, concise text covers aerodynamic phenomena of the rotor and offers guidelines for helicopter performance evaluation. Originally prepared for NASA. 537 figures. 640pp. 6⅛ × 9¼.
64647-5 Pa. $15.95

DIFFERENTIAL GEOMETRY, Heinrich W. Guggenheimer. Local differential geometry as an application of advanced calculus and linear algebra. Curvature, transformation groups, surfaces, more. Exercises. 62 figures. 378pp. 5⅜ × 8½.
63433-7 Pa. $8.95

INTRODUCTION TO SPACE DYNAMICS, William Tyrrell Thomson. Comprehensive, classic introduction to space-flight engineering for advanced undergraduate and graduate students. Includes vector algebra, kinematics, transformation of coordinates. Bibliography. Index. 352pp. 5⅜ × 8½. 65113-4 Pa. $8.95

A SURVEY OF MINIMAL SURFACES, Robert Osserman. Up-to-date, in-depth discussion of the field for advanced students. Corrected and enlarged edition covers new developments. Includes numerous problems. 192pp. 5⅜ × 8½.
64998-9 Pa. $8.95

ANALYTICAL MECHANICS OF GEARS, Earle Buckingham. Indispensable reference for modern gear manufacture covers conjugate gear-tooth action, gear-tooth profiles of various gears, many other topics. 263 figures. 102 tables. 546pp. 5⅜ × 8½. 65712-4 Pa. $14.95

SET THEORY AND LOGIC, Robert R. Stoll. Lucid introduction to unified theory of mathematical concepts. Set theory and logic seen as tools for conceptual understanding of real number system. 496pp. 5⅜ × 8¼. 63829-4 Pa. $10.95

A HISTORY OF MECHANICS, René Dugas. Monumental study of mechanical principles from antiquity to quantum mechanics. Contributions of ancient Greeks, Galileo, Leonardo, Kepler, Lagrange, many others. 671pp. 5⅜ × 8½.
65632-2 Pa. $14.95

FAMOUS PROBLEMS OF GEOMETRY AND HOW TO SOLVE THEM, Benjamin Bold. Squaring the circle, trisecting the angle, duplicating the cube: learn their history, why they are impossible to solve, then solve them yourself. 128pp. 5⅜ × 8½. 24297-8 Pa. $4.95

MECHANICAL VIBRATIONS, J.P. Den Hartog. Classic textbook offers lucid explanations and illustrative models, applying theories of vibrations to a variety of practical industrial engineering problems. Numerous figures. 233 problems, solutions. Appendix. Index. Preface. 436pp. 5⅜ × 8½. 64785-4 Pa. $10.95

CURVATURE AND HOMOLOGY, Samuel I. Goldberg. Thorough treatment of specialized branch of differential geometry. Covers Riemannian manifolds, topology of differentiable manifolds, compact Lie groups, other topics. Exercises. 315pp. 5⅜ × 8½. 64314-X Pa. $8.95

HISTORY OF STRENGTH OF MATERIALS, Stephen P. Timoshenko. Excellent historical survey of the strength of materials with many references to the theories of elasticity and structure. 245 figures. 452pp. 5⅜ × 8½. 61187-6 Pa. $11.95

TENSOR CALCULUS, J.L. Synge and A. Schild. Widely used introductory text covers spaces and tensors, basic operations in Riemannian space, non-Riemannian spaces, etc. 324pp. 5⅜ × 8¼. 63612-7 Pa. $8.95

A CONCISE HISTORY OF MATHEMATICS, Dirk J. Struik. The best brief history of mathematics. Stresses origins and covers every major figure from ancient Near East to 19th century. 41 illustrations. 195pp. 5⅜ × 8½. 60255-9 Pa. $7.95

A SHORT ACCOUNT OF THE HISTORY OF MATHEMATICS, W.W. Rouse Ball. One of clearest, most authoritative surveys from the Egyptians and Phoenicians through 19th-century figures such as Grassman, Galois, Riemann. Fourth edition. 522pp. 5⅜ × 8½. 20630-0 Pa. $10.95

HISTORY OF MATHEMATICS, David E. Smith. Nontechnical survey from ancient Greece and Orient to late 19th century; evolution of arithmetic, geometry, trigonometry, calculating devices, algebra, the calculus. 362 illustrations. 1,355pp. 5⅜ × 8½. 20429-4, 20430-8 Pa., Two-vol. set $23.90

THE GEOMETRY OF RENÉ DESCARTES, René Descartes. The great work founded analytical geometry. Original French text, Descartes' own diagrams, together with definitive Smith-Latham translation. 244pp. 5⅜ × 8½. 60068-8 Pa. $6.95

THE ORIGINS OF THE INFINITESIMAL CALCULUS, Margaret E. Baron. Only fully detailed and documented account of crucial discipline: origins; development by Galileo, Kepler, Cavalieri; contributions of Newton, Leibniz, more. 304pp. 5⅜ × 8½. (Available in U.S. and Canada only) 65371-4 Pa. $9.95

THE HISTORY OF THE CALCULUS AND ITS CONCEPTUAL DEVELOPMENT, Carl B. Boyer. Origins in antiquity, medieval contributions, work of Newton, Leibniz, rigorous formulation. Treatment is verbal. 346pp. 5⅜ × 8½. 60509-4 Pa. $8.95

THE THIRTEEN BOOKS OF EUCLID'S ELEMENTS, translated with introduction and commentary by Sir Thomas L. Heath. Definitive edition. Textual and linguistic notes, mathematical analysis. 2,500 years of critical commentary. Not abridged. 1,414pp. 5⅜ × 8½. 60088-2, 60089-0, 60090-4 Pa., Three-vol. set $29.85

GAMES AND DECISIONS: Introduction and Critical Survey, R. Duncan Luce and Howard Raiffa. Superb nontechnical introduction to game theory, primarily applied to social sciences. Utility theory, zero-sum games, n-person games, decision-making, much more. Bibliography. 509pp. 5⅜ × 8½. 65943-7 Pa. $12.95

THE HISTORICAL ROOTS OF ELEMENTARY MATHEMATICS, Lucas N.H. Bunt, Phillip S. Jones, and Jack D. Bedient. Fundamental underpinnings of modern arithmetic, algebra, geometry and number systems derived from ancient civilizations. 320pp. 5⅜ × 8½. 25563-8 Pa. $8.95

CALCULUS REFRESHER FOR TECHNICAL PEOPLE, A. Albert Klaf. Covers important aspects of integral and differential calculus via 756 questions. 566 problems, most answered. 431pp. 5⅜ × 8½. 20370-0 Pa. $8.95

CATALOG OF DOVER BOOKS

CHALLENGING MATHEMATICAL PROBLEMS WITH ELEMENTARY SOLUTIONS, A.M. Yaglom and I.M. Yaglom. Over 170 challenging problems on probability theory, combinatorial analysis, points and lines, topology, convex polygons, many other topics. Solutions. Total of 445pp. 5⅜ × 8½. Two-vol. set.
Vol. I 65536-9 Pa. $7.95
Vol. II 65537-7 Pa. $6.95

FIFTY CHALLENGING PROBLEMS IN PROBABILITY WITH SOLUTIONS, Frederick Mosteller. Remarkable puzzlers, graded in difficulty, illustrate elementary and advanced aspects of probability. Detailed solutions. 88pp. 5⅜ × 8½.
65355-2 Pa. $4.95

EXPERIMENTS IN TOPOLOGY, Stephen Barr. Classic, lively explanation of one of the byways of mathematics. Klein bottles, Moebius strips, projective planes, map coloring, problem of the Koenigsberg bridges, much more, described with clarity and wit. 43 figures. 210pp. 5⅜ × 8½. 25933-1 Pa. $5.95

RELATIVITY IN ILLUSTRATIONS, Jacob T. Schwartz. Clear nontechnical treatment makes relativity more accessible than ever before. Over 60 drawings illustrate concepts more clearly than text alone. Only high school geometry needed. Bibliography. 128pp. 6⅛ × 9¼. 25965-X Pa. $6.95

AN INTRODUCTION TO ORDINARY DIFFERENTIAL EQUATIONS, Earl A. Coddington. A thorough and systematic first course in elementary differential equations for undergraduates in mathematics and science, with many exercises and problems (with answers). Index. 304pp. 5⅜ × 8½. 65942-9 Pa. $8.95

FOURIER SERIES AND ORTHOGONAL FUNCTIONS, Harry F. Davis. An incisive text combining theory and practical example to introduce Fourier series, orthogonal functions and applications of the Fourier method to boundary-value problems. 570 exercises. Answers and notes. 416pp. 5⅜ × 8½. 65973-9 Pa. $9.95

THE THEORY OF BRANCHING PROCESSES, Theodore E. Harris. First systematic, comprehensive treatment of branching (i.e. multiplicative) processes and their applications. Galton-Watson model, Markov branching processes, electron-photon cascade, many other topics. Rigorous proofs. Bibliography. 240pp. 5⅜ × 8½. 65952-6 Pa. $6.95

AN INTRODUCTION TO ALGEBRAIC STRUCTURES, Joseph Landin. Superb self-contained text covers "abstract algebra": sets and numbers, theory of groups, theory of rings, much more. Numerous well-chosen examples, exercises. 247pp. 5⅜ × 8½. 65940-2 Pa. $7.95

Prices subject to change without notice.

Available at your book dealer or write for free Mathematics and Science Catalog to Dept. GI, Dover Publications, Inc., 31 East 2nd St., Mineola, N.Y. 11501. Dover publishes more than 175 books each year on science, elementary and advanced mathematics, biology, music, art, literature, history, social sciences and other areas.